Heavy Metals in the Marine Environment

Editors

Robert W. Furness, Ph.D.
Senior Lecturer in Zoology
University of Glasgow
Glasgow, Scotland

and

Philip S. Rainbow, Ph.D.
Reader in Marine Biology
Queen Mary College
University of London
London, England

CRC

CRC Press, Inc.
Boca Raton, Florida

Library of Congress Cataloging-in-Publication Data

Heavy metals in the marine environment / editors, Robert W. Furness and Philip S. Rainbow.

p. cm.

Bibliography: p.

Includes index.

ISBN 0-8493-5506-0

1. Heavy metals--environmental aspects. 2. Marine pollution-Environmental aspects. I. Furness, R. W. II. Rainbow, P. S.

QH545.H42H42 1990

363.73′9463--dc20

89-34792

CIP

This book represents information obtained from authentic and highly regarded sources. Reprinted material is quoted with permission, and sources are indicated. A wide variety of references are listed. Every reasonable effort has been made to give reliable data and information, but the author and the publisher cannot assume responsibility for the validity of all materials or for the consequences of their use.

Direct all inquiries to CRC Press, Inc., 2000 Corporate Blvd., N.W., Boca Raton, Florida 33431.

International Standard Book Number 0-8493-5506-0

Library of Congress Card Number 89-34792

Printed in the United States

THE EDITORS

Robert W. Furness, B.Sc., Ph.D., is a Senior Lecturer in Zoology in the Applied Ornithology Unit of the School of Biological Sciences at the University of Glasgow, Scotland.

Dr. Furness graduated in 1974 from the University of Durham, England, with a B.Sc. degree in zoology and botany, and obtained his Ph.D. degree in 1977 for studies of the population ecology of seabirds. After spending 1 year in the departments of genetics and zoology at the University of Aberdeen as a Science Research Council postdoctoral fellow, he was appointed a Lecturer in Zoology at Glasgow University, with particular responsibility for teaching Human Biology. He became a Senior Lecturer in 1988.

Dr. Furness is a member of the British Ecological Society, British Ornithologists' Union, Association for the Study of Animal Behaviour, British Trust for Ornithology, African Seabird Group, Treasurer of the British Seabird Group, Editor of *Ringing & Migration,* and member of the editorial panels of *Marine Ecology Progress Series, Cormorant,* and *Colonial Waterbirds.* He received a Visiting Scholarship to the Newfoundland Institute for Cold Ocean Science in 1986 and is convening a symposium on seabirds as monitors of marine environments at the 20th International Ornithological Congress in 1990.

Dr. Furness has been the recipient of research grants from the Natural Environment Research Council, the Nature Conservancy Council, The Royal Society, the North Atlantic Treaty Organisation, the Carnegie Trust, the Nuffield Foundation, the U.S. Center for Field Research, and private industry. He is the author or co-author of three books and of more than 100 research papers. His current major research interests include the dynamics of mercury in birds, seabirds as indicators of pollution, and the interactions between seabirds, fisheries, and fish stocks.

Philip S. Rainbow, M.A. Cantab., Ph.D. is a Reader in Marine Biology in the School of Biological Sciences and Deputy Director of the Centre for Research in Aquatic Biology (CRAB) at Queen Mary College, University of London.

Dr. Rainbow received his B.A. degree from the University of Cambridge in 1972, and M.A. Cantab. in 1976, after reading Natural Sciences (Part 2 Zoology). He obtained his Ph.D. from the University of Wales in 1975, after research in the Marine Science Laboratories, University College of North Wales (UCNW), Bangor. He has served as Lecturer and Reader at Queen Mary College since 1975.

Dr. Rainbow is a member of the Marine Biological Association of the United Kingdom and of the Estuarine and Coastal Sciences Association, a fellow of the Linnean Society of London (Council member), and a scientific fellow of the Zoological Society of London. He is also a university representative on the Executive Committee of the Field Studies Council.

Dr. Rainbow has published more than 60 research papers and has been co-editor of two books. His research has been funded by NERC (Natural Environment Research Council, U.K.), the Royal Society, NATO Scientific Affairs Division, and private industry. His current major research interests are in the significance of trace metal concentrations in marine and estuarine invertebrates.

CONTRIBUTORS

James D. Burton, Ph.D., F.R.S.C.
Professor of Oceanography
Department of Oceanography
University of Southampton
Southampton, U.K.

Roy Chester, Ph.D.
Professor of Oceanography
Oceanography Laboratories
Department of Earth Sciences
University of Liverpool
Liverpool, U.K.

Robert W. Furness, Ph.D.
Senior Lecturer in Zoology
Department of Zoology
University of Glasgow
Glasgow, Scotland, U.K.

Stephen G. George, Ph.D.
NERC Unit of Aquatic Biochemistry
Department of Biological Sciences
University of Stirling
Stirling, Scotland, U.K.

Jens C. Hansen, Dr. Med.
Associate Professor
Institute of Environmental and
Occupational Medicine
University of Aarhus
Aarhus, Denmark

W. J. Langston, Ph.D.
Zoologist
Plymouth Marine Laboratory
Citadel Hill, Plymouth, U.K.

Samuel N. Luoma, Ph.D.
Project Chief/Biologist
U.S. Geological Survey
Menlo Park, California

K. J. T. Murphy, Ph.D
Doctor
Environmental Resources, Ltd.
London, U.K.

David J. H. Phillips, Ph.D.
Mott MacDonald
Environmental Services
London, U.K.

Philip S. Rainbow, Ph.D.
Reader in Marine Biology
Queen Mary College
University of London
London, U.K.

Peter J. Statham, Ph.D.
Lecturer in Chemical Oceanography
Department of Oceanography
University of Southampton
Southampton, U.K.

David R. Thompson, Ph.D.
Department of Zoology
University of Glasgow
Glasgow, Scotland, U.K.

P. M. Walsh, M.Sc.
Department of Zoology
University of Glasgow
Glasgow, Scotland, U.K.

Eric W. Wolff, M.A.
Ice and Climate Division
British Antarctic Survey
Cambridge, U.K.

TABLE OF CONTENTS

Chapter 1

HEAVY METALS IN THE MARINE ENVIRONMENT

P. S. Rainbow and R. W. Furness

TABLE OF CONTENTS

A recent inventory of man-induced mobilization of heavy metals into the biosphere[1] concluded that "mankind has become the most important element in the global biogeochemical cycling of the trace metals." In that review article, Nriagu and Pacyna also stated that "the greatly increased circulation of toxic metals through the soil, water, and air and their inevitable transfer to the human food chain remains an important environmental issue which entails some unknown risks for future generations."

I. AIM OF THIS VOLUME

The aim of this volume is to draw together state-of-the-art reviews of knowledge on levels of heavy metals in marine environments (particularly in marine animals), the dynamic processes in these systems, toxic effects, and threats presented by heavy metals in foods of marine origin.

All heavy metals, whether biologically essential or not, have the potential to be toxic to organisms at a threshold bioavailability. Such threshold concentrations vary between metals, between species and with the physicochemical characteristics of the medium, some like copper being particularly toxic even though essential in trace amounts. Responses of animals to metals in their medium or food depend to a large extent on the ability of species to regulate levels attained in their tissues. Higher animals have the capacity to regulate levels of many metals, while marine invertebrates can regulate some within certain limits. Where animals cannot regulate physiological levels of metals, an alternative strategy is to detoxify and store metals in relatively harmless forms. Knowledge of the manner in which animals deal with potentially toxic concentrations of heavy metals is of fundamental importance in the assessment of metal pollution by analysis of metal levels in biological samples. The interaction of heavy metals with biological materials is a key theme running through this volume. Toxic effects may be reflected at the individual, population, or ecosystem level, affecting species composition and production levels, or may be of direct dietary significance to man. The global cycling of metals through the marine environment is crucially affected by biological processes.

II. DEFINITION OF HEAVY METALS

The term *heavy metals* is often used without strict definition, as discussed by Nieboer and Richardson[2] who call for the use of a chemically significant classification of metal ions based on their classification into Lewis acids and bases. Nevertheless, the authors in this volume have been given the freedom to interpret the term *heavy metals* as widely as each deems relevant. This is a pragmatic approach based, in part, on the fact that published data on elemental analysis of biological samples provide large amounts of information for certain elements, such as cadmium, copper, iron, mercury, selenium, and zinc, but very little or none for many other elements.

Heavy metals are often referred to as trace metals, occurring as they do in low concentrations in organisms, although the term *trace metal* might imply the presence of an essential requirement by organisms for that metal. Many metals are essential to life.[3,4] Of these, the major metals sodium, potassium, calcium, and magnesium are generally not considered to be heavy metals by any definition. Other metals described by at least some authors as essential to at least some organisms, usually in trace amounts, include aluminum, arsenic, chromium, cobalt, copper, iron, manganese, molybdenum, nickel, selenium, tin, vanadium, and zinc.[3,4] Aluminum falls outside most definitions of heavy metals; arsenic and selenium have variable designations, but the remaining essential trace metals are normally listed among heavy metals. Lists of nonessential heavy metals usually include cadmium, gold, lead, mercury, and silver, as well as rare, more obscure, metals (including radionuclides) of higher atomic weight.

REFERENCES

1. **Nriagu, J. O. and Pacyna, J. M.,** Quantitative assessment of worldwide contamination of air, water and soils by trace metals, *Nature,* 333, 134, 1988.
2. **Nieboer, E. and Richardson, D. H. S.,** The replacement of the nondescript term 'heavy metals' by a biologically and chemically significant classification of metal ions, *Environ. Pollut.,* B1, 3, 1980.
3. **Bowen, H. J. M.,** *Environmental Chemistry of the Elements,* Academic Press, London, 1979, chap. 8.
4. **Harrison, P. M. and Hoare, R. J.,** *Metals in Biochemistry,* Chapman and Hall, London, 1980, chap. 1.

Chapter 2

TRACE METALS IN SEAWATER

J. D. Burton and P. J. Statham

TABLE OF CONTENTS

I. INTRODUCTION

The major advances in knowledge of the concentrations and distributions of trace metals in the seas and oceans have occurred since the mid 1970s. This progress reflects major breakthroughs in procedures for essentially contamination-free (clean) sampling, the adoption of clean methodologies for handling and analysis of samples, and improvements in the analytical methods themselves. The extent of progress by the early 1980s was demonstrated by the major collection of papers edited by Wong et al.[1] Overviews of the subject have been given by Burton and Statham[2] and Bruland.[3] With the improved knowledge of the true variations in the concentrations of metals within the ocean, many systematic features in their distributions have become apparent. It has been possible to classify their marine geochemical behaviors and to interpret these in terms of the interactions of the dissolved chemical species with particulate material, which reflect the fundamental bonding properties of the elements.[4]

For the purposes of this chapter, the trace metals have been taken to include those elements with metallic or metalloid characteristics with dissolved concentrations in seawater below 1 mg/kg. This concentration is generally adopted as the lower limit for a constituent to be classed as major.[5] Attention is limited essentially to the stable elements. Concentrations are expressed in subunits of mol/kg or mol/l throughout. The emphasis is on dissolved metals, since in most environments the contributions of particulate metal fractions are low, although there are some important exceptions to this for certain metals in environments where the concentrations of total suspended particulate material are high, such as in coastal water, in bottom waters containing resuspended particles, and in waters immediately influenced by hydrothermal inputs. The dissolved fractions are conventionally defined operationally as those passing through filters of about 0.4 to 0.5-μm average pore diameter. In some studies, however, measurements have been made on unfiltered samples, sometimes with storage after acidification. This has led to other operationally defined terms being used to describe the analyzed fractions, such as total dissolvable metal. For the situations generally considered here, such fractions can be considered equivalent to the dissolved metal and are referred to as such.

As discussed by Whitfield and Turner,[4] it is now clear that the behaviors of dissolved trace metals in the oceans fall into three principal categories reflecting their chemical and geochemical characteristics. First, there is a group of metals that behave conservatively or show only minor divergences from such behavior. This means that, except in marginal sectors of the ocean influenced by local sources (e.g., rivers, glacial inputs), the concentrations of these elements alter significantly only as a result of the mixing of water bodies which have acquired different concentrations in boundary regions, as a result of such inputs or the addition or removal of water by precipitation and evaporation at the ocean surface. This characteristic reflects their low geochemical reactivity, which is generally expressed in terms of a high mean oceanic residence time ($>10^5$ year) relative to the time-scales of the water mixing processes in the oceans, and it results in close relationships of the concentrations to salinity for the metals concerned. These metals behave essentially like the major constituents of seawater but do not attain such high concentrations because they are less abundant in source materials, such as crustal rocks.

The second group of trace metals consists of elements that show some degree of positive correlation with the micronutrients, phosphate, nitrate, and dissolved silicon, which are characterized over much of the open ocean by depletion in surface waters relative to deep waters. The correlations of trace metals with these biologically essential constituents reflect the involvement of the metals with the flux of particulate material of biological origin. This flux, driven initially by primary productivity, is transmitted from the euphotic zone to deeper waters, where subsequent release to solution occurs during the microbial decomposition of organic material, or the dissolution of mineral phases, such as biogenous silica (opal) and calcium carbonate, originally produced as skeletal material by organisms in the upper ocean. The metals showing this type of behavior include not only those for which there are specific metabolic requirements but also others that are associated with the organic material through their affinity for ligands (e.g., those containing sulfur or nitrogen) occurring in organic structural molecules, or with the biogenous mineral phases. Elements showing this kind of behavior have intermediate mean oceanic residence times and have been classified as recycled.[4] The terms *bio-limiting* (for strongly surface-depleted elements) and *bio-intermediate* (for detectably but less strongly surface-depleted elements) have also been applied,[6] but this nomenclature may carry implications of biological requirements which in many cases appear unjustified. The present account concentrates upon this group of metals which are markedly affected, directly or indirectly, by the biological processes in the ocean.

The third major category contains elements that are markedly nonconservative and have short mean oceanic residence times ($<10^3$ year), so that their concentrations decrease markedly with distance from sources, which may be at the boundaries of the ocean with the atmosphere, the sea-bed, or the continental lithosphere, or within the ocean itself. They are of particular potential interest as tracers for transport and mixing in the oceans.[7] The short residence times of these metals reflect efficient removal to particles by oxidative or hydrolytic scavenging.[4]

The processes which lead to down-column transport of metals in association with particles are not yet well understood and some metals are not readily categorized or show distributions which reflect a combination of processes. The influence of biological activity in the surface ocean is not limited to the recycled metals. Thus, the removal of ^{234}Th, an isotope of the scavenged metal thorium, from the mixed layer is positively correlated with primary productivity and the flux of particulate organic carbon from the euphotic zone.[8] Biological activity may lead not only to the association of metals with biogenous particles but also to a more efficient down-column transport of all particles as a result of packaging into fecal pellets of grazing organisms.

In this account, open oceanic environments are emphasized because they have been more fully investigated and are, in general, better understood than shelf and coastal environments, as regards the factors influencing the distributions of the trace metals. The account also concentrates on the behavior of the metals in waters containing free oxygen, although some reference is made to differences in behavior of metals between these waters and suboxic or anoxic environments. No attempt has been made to discuss all the metals for which reliable data are now available. Rather, the account focuses on metals which illustrate the range of behaviors that have become recognized through recent findings.

II. METALS SHOWING ESSENTIALLY CONSERVATIVE BEHAVIOR

The trace alkali metals, lithium, rubidium, and cesium, like their more abundant counterparts, sodium and potassium, behave, according to the evidence so far available, nearly conservatively in the ocean. Data summarized by Bruland[3] indicate average concentrations at salinity 35×10^{-3} as follows: Li, 25 μmol/kg; Rb, 1.4 μmol/kg; Cs, 2.2 nmol/kg. Conservative behavior can be defined only within the limits imposed by the precision and accuracy of analytical methods, and the greater variation in reported concentrations for cesium than for lithium and rubidium may reflect poorer precision and accuracy associated with a lower concentration. A further factor that must be considered is that all elements undergo some recycling within the water column and that the particular characteristic of the conservative elements is that the recycling flux is low relative to the concentration. Since cesium is considerably lower in concentration than lithium and rubidium, even a low recycling flux may lead to detectable variability and significant changes in the ratio of concentration of cesium to salinity with depth have been reported.[9] Thallium, as Tl(+I), resembles the alkali metals in its chemical properties and it is probably the presence of the metal in this oxidation state that explains the apparently conservative behavior of the element in seawater. A mean concentration of 69 pmol/kg is reported[10] for waters from the Atlantic and the North and South Pacific Oceans, with variations which are considerable (59 to 78 pmol/kg) but comparable with the analytical precision.

Metals forming stable oxy-anions in aqueous solution also tend to have low geochemical reactivities and one such element, molybdenum, has so far been found to be conservative, although most of the data have been insufficiently precise for small variations to be detectable. Data summarized[2] for the North Atlantic Ocean and adjacent seas indicate a range of 90 to 135 nmol/l. Values for the northeast Pacific Ocean[11] average 107 nmol/l with a standard deviation of 2.5 nmol/l and an overall mean value of about 110 nmol/l is indicated. The mean oceanic residence time of molybdenum, relative to the river input, is estimated[11] to be 7×10^5 year, emphasizing the low geochemical reactivity of the molybdate anion.

Stabilization through the reaction of the uranyl ion to form a carbonato complex appears to account for the apparently conservative behavior of uranium in the oceans, with an average concentration,[12,13] normalized to $S = 35 \times 10^{-3}$, of about 13.5 nmol/l. Anderson[14] concluded from a study of particulate fluxes that the rate of cycling of uranium in the water column was insufficient to generate a detectable vertical gradient in the concentration of the dissolved metal. The measurements of Chen et al.[13] show a very narrow range of concentrations, but nevertheless suggest that those in the Pacific Ocean are about 1% higher than those in the Atlantic Ocean. These workers conclude, on the basis that the concentrations are more variable than those predicted by a mixing model, that the extent of the variations observed reflects slightly nonconservative behavior.

A highly unusual instance of conservative behavior is provided by methylated species of germanium. Lewis et al.[15] measured monomethylgermanium and dimethylgermanium in four widely separated oceanic profiles and found concentrations of, respectively, 326 ± 14 pmol/l and 97 ± 9 pmol/l; an example is given in Figure 1. The origin of these methylated species is not known. Their remarkable stability contrasts markedly with other methylated forms of metals in the ocean, which are observed only as transient products of biological activity.

III. RECYCLED METALS

The recycled metals range from those showing closely nutrient-like behavior, with very pronounced changes in concentration in vertical oceanic profiles, to those with vertical gradients which may be difficult to detect analytically. In the extreme, the distinction between those latter elements and the conservative elements may be dependent on the precision of analytical data.

The first trace metal which was shown to have a very closely nutrient-like distribution in the open ocean was cadmium,[16,17] an element with no known biological function but with an evident affinity for biogenous particulate material, and this metal serves to illustrate the main features that result from this resemblance. Typical profiles for dissolved cadmium in the basins of the North Pacific[18] and North Atlantic[19] Oceans are shown in Figure 2 together with corresponding data for phosphate. They show marked depletion in surface waters, where concentrations below 0.05 nmol/kg are commonly observed, and a rapid increase in concentration to a maximum in the lower main thermocline region, and a slight decrease to relatively constant values in deep water. This distribution is characteristic for phosphate and nitrate which are cycled in the formation and decomposition of soft organic tissues and a close correlation of cadmium and phosphate is evident from these data. Bruland[18] found that for the North Pacific, the relationship was expressed by

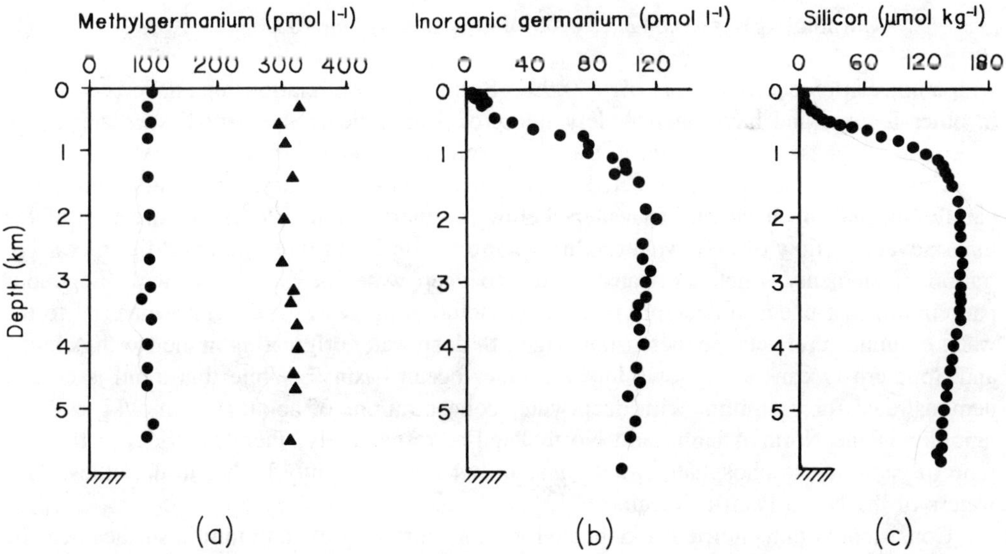

FIGURE 1. Vertical profiles of (a) monomethylgermanium (▲) and dimethylgermanium (●) in the northwestern Pacific Ocean (25°00′N, 170°05′E), (b) inorganic germanium at the same station. In (c), the vertical profile of dissolved silicon is shown for comparison. (From Lewis B. L., Froelich, P. N., and Andreae, M. O., *Nature,* 313, 303, 1985; and Froelich, P. N., Hambrick, G. A., Andreae, M. O., Matlock, R. A., and Edmond, J. M., J. Geophys. Res., 90, 1133, 1985. With permission.)

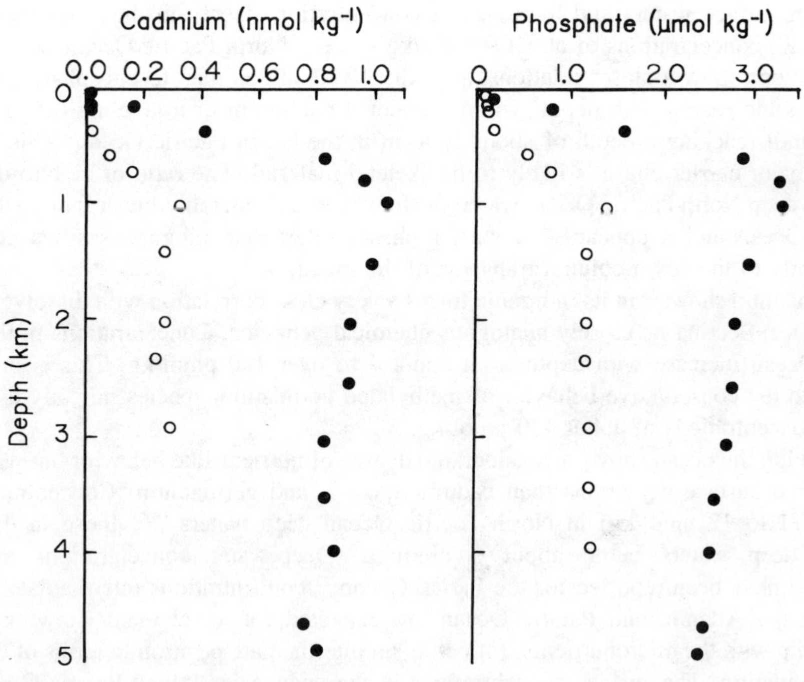

FIGURE 2. Vertical profiles of dissolved cadmium and of phosphate in the North Atlantic (○) and North Pacific (●) Oceans. (From Bruland, K. W., *Chemical Oceanography,* Vol. 8, Riley, J. P. and Chester, R. Eds., Academic Press, London, 1983, 157. With permission.)

$$Cd(nmol/kg) = (0.347 \pm 0.007)PO_4(\mu mol/kg) - (0.068 \pm 0.017) \quad (1)$$

with a correlation coefficient, *r*, of + 0.992. Similarly close relationships have been found in other regions and have been widely interpreted to indicate that soft tissues act as the major carrier phase for cadmium in transport from the euphotic zone.

This close relationship is reflected in an increase in the concentration of cadmium, again paralleling that of phosphate, in waters below the thermocline, along the direction of the main advective flow of deep water. This feature is consistent with the build-up in concentration of elements which are transported into deep water in association with biogenous particulates that undergo decomposition or dissolution processes that return material to the water column. It reflects the increasing "age" of deep water originating in the North Atlantic and Southern Oceans as it flows into the other ocean basins.[6] While this trend is clearly demonstrated for cadmium, with deep-water concentrations of about 0.3 nmol/kg and 0.9 nmol/kg in the North Atlantic and North Pacific, respectively, there is evidence that the ratio of cadmium to phosphate is not constant but is significantly higher in the oldest deep waters of the North Pacific Ocean.[19]

Consistent with its nutrient-like behavior, concentrations of cadmium in surface oceanic waters show a significant increase in areas of upwelling. Thus in the North Pacific Ocean, surface concentrations[18] increase from an average of 0.0014 nmol/kg in the central ocean areas within the gyre circulation, to 0.056 nmol/kg in waters of the California Current region and 0.16 nmol/kg in a coastal upwelling regime.

It has proved much more difficult to obtain accurate data for zinc in oceanic waters, because of greater problems of control of contamination, and information on this metal therefore remains sparse. It has been shown, however, that zinc is also highly depleted in mid-ocean surface waters and increases in concentration steeply in the main thermocline region, with concentrations of about 9 nmol/kg in deep North Pacific Ocean waters.[18] The concentrations show a closer relationship to dissolved silicon than to phosphate, with concentrations increasing with depth, without either a maximum or a leveling off of concentration, until reaching a depth of about 2000 m in the North Pacific Ocean. This suggests that the major carrier phase is likely to be skeletal material. The ratio of zinc to dissolved silicon in deep North Pacific Ocean waters is, however, considerably lower than in the North Atlantic Ocean and it appears that carrier phases other than biogenous silica contribute significantly to the down-column transport of the metal.

Germanium shows[20] in its inorganic forms a very close correlation with dissolved silicon (Figure 1), reflecting its closely analogous chemical behavior. Concentrations in the North Pacific Ocean increase with depth from about 7 to over 100 pmol/kg. This is in marked contrast to the conservative behavior of methylated germanium species, already discussed, at total concentrations of about 420 pmol/kg.

Nickel in the ocean shows a considerable degree of nutrient-like behavior but with much less marked surface depletions than cadmium, zinc, and germanium. Concentrations are highest (11 to 12 nmol/kg) in North Pacific Ocean deep waters,[18,21] those in the North Atlantic deep waters being about 6 nmol/kg. Deep-water concentrations averaging 9.4 nmol/l have been reported for the Indian Ocean;[22] concentrations intermediate between those for the Atlantic and Pacific Ocean are expected for an element showing a close relationship with the micronutrients, reflecting an intermediate position in terms of the deep-water circulation. The surface concentrations in the open Atlantic and Pacific Oceans are, however, similar at about 2 nmol/kg; in the North Pacific surface concentrations increase in the California Current[18] to 3.7 nmol/kg.

The range of factors significantly influencing the distribution of copper in oceanic waters is greater than that for the metals already considered. Nutrient-like behavior is reflected in surface water concentrations that are generally depleted relative to those in deeper waters and which are lowest in the mid-ocean regions within gyre circulations. Thus surface concentrations in the North Pacific Ocean decrease from 1.2 nmol/kg in the California Current to 0.4 to 0.5 nmol/kg in the central oceanic region[18] while deep water concentrations[18,23] are typically about 3 nmol/kg. For the North Atlantic Ocean, concentrations in the upper water column of 1.1 to 1.7 nmol/l are reported[24] with increases to about 2 to 4 nmol/l in deep waters. The application of a vertical advective-diffusive model[25] to profiles in the North Pacific Ocean has demonstrated, however, that there is a significant removal of copper from the deep waters by scavenging on particles.[18,23] As is shown later, such a process generally leads to a pattern of interocean variability which is the inverse of that seen for the recycled elements, i.e., for scavenged metals, the older deep waters of the Pacific Ocean show lower concentrations than recently formed deep waters in the Atlantic Ocean. That this is not the case with copper is probably attributable to inputs of the metal at the sediment-water interface. In the Atlantic Ocean, bottom water concentrations, particularly above red clay deposits, have been shown to be significantly elevated,[24] reaching about 10 nmol/l, and marked increases to similar concentrations occur in the bottom waters in some Pacific Ocean profiles.[18,23]

Each of the distributions considered so far is governed essentially by the behavior of the element in a single oxidation state, although it has been shown[26] that Cu(+I) plays a significant role in the marine chemistry of copper in the euphotic zone, in addition to Cu(+II) which is the thermodynamically stable oxidation state. With some other elements, however, changes in oxidation state play an important part throughout the ocean in the recycling of the element.

Selenium is an important example of this latter group of elements. It occurs in solution in seawater in several oxidation states. Selenium (+VI) and selenium (+IV) occur as the oxy-anions, selenate and selenite, respectively. A further fraction of selenium has been defined operationally which in oceanic waters can be assumed to be organic selenide, Se(−II), probably largely as seleno-amino acids. Extensive measurements by Measures and collaborators[27-29] established the main features in the distributions of Se(+IV) and total dissolved Se. It was found that Se(+IV) was highly depleted in surface waters and increased in concentration with depth. In a typical Pacific Ocean vertical profile, concentrations ranged from about 50 pmol/l in the surface waters to about 800 pmol/l in deep waters; total dissolved selenium concentrations increased with depth from about 550 pmol/l near the surface to 2200 pmol/l at depth. This recycled behavior was reflected in similar profiles but with lower concentrations (by about 30 to 40%) in the Atlantic Ocean.

In these studies, the difference between the concentrations of Se(+IV) and total dissolved Se was attributed to Se(+VI), although it was recognized that the operational analytical definition of the difference fraction did not rule out the presence of organic forms of selenium [Se(−II)]. Subsequent work by Cutter and Bruland[30] has demonstrated that organic selenide is in fact an important fraction of the total dissolved element in the upper oceanic water column. This is shown by data for a station in the eastern tropical North Pacific Ocean (Figure 3). These show that in the deep water, Se(+IV) and Se(+VI), the only significant species, are present at rather similar concentrations (about 0.9 and 1.2 nmol/kg, respectively); these species, particularly Se(+IV), are highly depleted in the mixed layer where organic selenide is the dominant form, with an average concentration of 0.9 nmol/kg in the upper 60 m. Surface transects reported by these authors show Se(+VI) concentrations to be highly depleted in the oligotrophic waters within the oceanic gyres of the North and South Pacific Oceans and higher in the upwelling regimes of the equatorial current region and the California Current. Organic selenide has a more variable concentration but, except for the California Current region, is at relatively high concentrations, accounting typically for some 80% of the dissolved element in surface waters. Although the speciation in the upper water column established in this work differs from that suggested by Measures and collaborators,[27-29] the

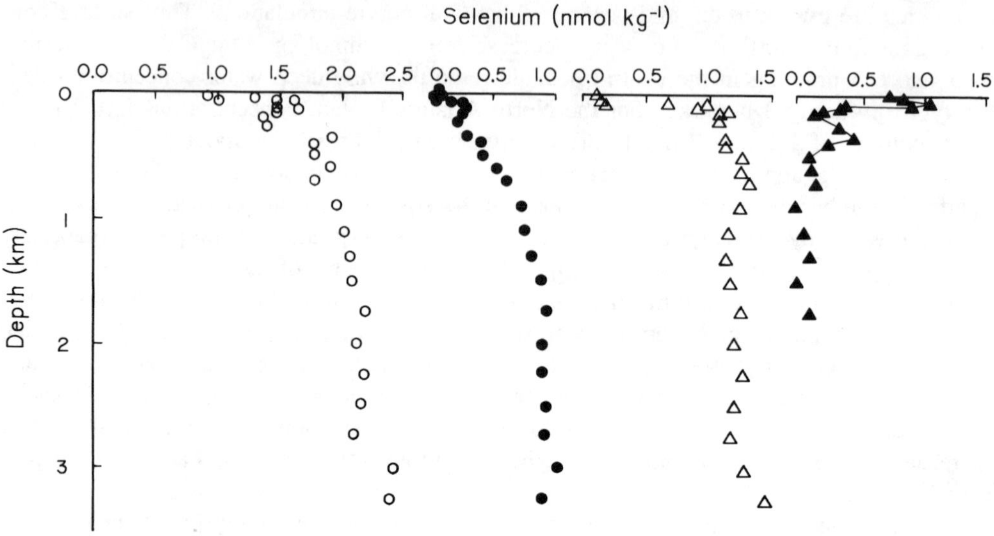

FIGURE 3. Vertical profiles of dissolved selenium in the eastern tropical North Pacific Ocean (18°N, 108°W): total selenium species (○), Se(+IV) (●), Se(+VI) (△), organic selenide [Se(−II)] (▲). Substantial denitrification occurs in the water column between 150 and 600 m. (From Cutter, G. A. and Bruland, K. W., *Limnol. Oceanogr.*, 29, 1179, 1984. With permission.)

data for total dissolved selenium and Se(+IV), and those for Se(+VI) in deep waters, are analytically coherent.

The data in Figure 3 show there to be a secondary maximum in organic selenide, centered at 350 m, within the suboxic oxygen minimum zone which occurs at this station. Coincidence of this maximum with a zone of high concentrations of amino acids, together with the findings[31] of associations of selenide with total dissolved amino acids under reducing conditions in Saanich Inlet, led to the suggestion that the selenide may occur as seleno-amino acids. There are indications that total dissolved selenium and selenate show minima at about the same depth range as the zone of denitrification which occurs in this profile as a result of the suboxic conditions in the oxygen minimum. Similar minima were noted in some profiles by Measures et al.[29] at depths that suggested that these features could arise by advection from regions with intense oxygen minima and the possibility that they reflect a process of deselenification was discussed.

Whereas the distributions of some of the recycled metals reflect dominantly the scavenging of the elements by particulate material, including biogenous phases, leaving the euphotic zone, rather than the involvement of the element in the metabolism of organisms, the presence of a high proportion of organic selenide in surface waters argues for metabolic biological cycling within organisms in the case of selenium. There is evidence[32] that Se(+IV) is preferentially taken up by plankton relative to Se(+VI). The role of subsequent metabolic reductive pathways in the cycling of selenium has been discussed by Burton et al.[33] and Cutter and Bruland.[30] The cycling model of the latter workers suggests that in addition to the release of organic selenide in the upper water column, this form constitutes the major fraction of the particulate selenium transported into the deep ocean, where it undergoes oxidation to Se(+IV) and Se(+VI). The presence of Se(+IV) as a major fraction of total selenium in the deep sea is unexpected on thermodynamic grounds and must reflect a slow rate of oxidation. From a kinetic model, Cutter and Bruland[30] estimated a value of 1150 years for the mean life of Se(+IV) in the deep ocean, relative to conversion to Se(+VI).

Arsenic can also occur in seawater as oxyanions in two different oxidation states, namely thermodynamically stable As(+V) (arsenate), and As(+III) (arsenite); methylated forms

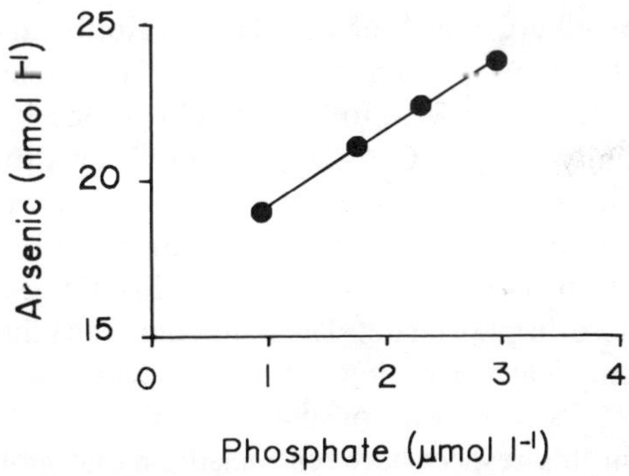

FIGURE 4. Relationship between the concentrations of phosphate and dissolved arsenic in deep waters of different oceanic regions. Points represent, in order of increasing concentration, the North Atlantic, South Atlantic, Northeast Indian, and Pacific Oceans. (From Middelburg, J. J., Hoede, D., van der Sloot, H. A., van der Weijden, C. H., and Wijkstra, J., *Geochim. Cosmochim. Acta,* 52, 2871, 1988. With permission.)

are also significant in some surface waters.[34,35] The distribution of arsenic shows only a relatively minor effect of cycling by transport from the euphotic zone as a result of association with particles and subsequent release. Andreae[35] showed that, while in the North Pacific Ocean concentrations of arsenate increase significantly with depth below the euphotic zone to about 24 nmol/l in deep waters, the increase in total dissolved arsenic is less marked. This is because, whereas arsenate is the only significant species in deep ocean water, arsenite is a significant although variable species in the euphotic zone, with concentrations of 0.2 to 4 nmol/l. In these waters, monomethylarsonic and dimethylarsinic acids were also detected, with concentrations of methylated arsenic of less than 2 nmol/l. Concentrations at stations in the South Atlantic Ocean[36] showed a small but significant difference in concentration of total dissolved arsenic, with mean values of 19.9 nmol/l for waters above 110 m and 21.1 nmol/l for those below. For the North Atlantic Ocean, Middelburg et al.[37] found similar evidence of a significant, although relatively minor, effect of transport from the euphotic zone in association with particles. Concentrations of arsenate in surface waters (14 to 18 nmol/l) were lower than those in deep waters (18 to 20 nmol/l). Variable concentrations of arsenite (0.05 to 4 nmol/l) were found in surface waters; these concentrations include any methylated forms present.

Middelburg et al.[37] have used their data and those from other investigations to examine the interocean variability of arsenic in deep waters. The average concentrations (nmol/l) in waters below 200 m increase systematically: North Atlantic, 19; South Atlantic, 21; Northeast Indian, 22.5; Pacific, 24. They show a consistent relationship to the concentrations of phosphate (Figure 4), with a molar ratio $\Delta As/\Delta P$ of 2.5×10^{-3}, which is closely similar to the ratio of 2.9×10^{-3} in which the elements occur in marine algae.[35]

The vertical distribution of oxidation states for arsenic is similar to that found in oxygenated water columns for a number of elements, with thermodynamically unstable reduced forms occurring in the upper water column primarily as a result of reductive biological processes, but not persisting in deep waters much below the euphotic zone. The high proportion of Se(+IV) throughout deep waters, as described above, is an unusual and important exception. In contrast, the major consequence of the reduction and methylation of arsenic by phytoplankton is evidently release back to the upper water column and relatively efficient reconversion to As(+V). Microbial demethylation of dimethylarsenic has been shown to occur in coastal waters[38] and the rate of bacterial oxidation of As(+III) is relatively rapid.[39] The major consequence of uptake of arsenic by microorganisms in the euphotic zone appears to be interconversion of the different chemical species in the surface waters, with only a minor fraction becoming associated in forms that are transported with the biogenous particles into deeper water.

Measurements by Middelburg et al.[37] of vanadium in North Atlantic Ocean waters show no clear evidence of systematic variations with depth, in agreement with most previous findings; concentrations varied from 31.8 to 33.9 nmol/l. In the Pacific Ocean,[40] however, values normalized to salinity 35×10^{-3} show a significant difference between surface waters (32.7 nmol/l) and deep waters (36.4 nmol/l). This evidence of recycling is confirmed by the interocean differences in deep-water concentrations (nmol/l) discussed by Middelburg et al.:[37] North Atlantic, 32.6; Northeast Indian, 35.6; Pacific, 36.4. The differences are closely correlated with those in phosphate. Thus although vanadium, as expected from its occurrence as an oxy-anion, clearly has a low flux in association with particles, from surface to deep waters, relative to its reservoir concentration, the consequences of recycling are detectable. The contrast in this respect between vanadium and molybdenum is reflected in the mean oceanic residence time, relative to the river input, for the former metal of about 5×10^4 year, an order of magnitude lower than that for molybdenum. It is possible that more precise analyses for molybdenum could provide evidence for its recycling.

Of those metals which show distributions considerably influenced by recycling and which have not been discussed here, there have been particularly important studies of barium (see summaries in References 2 and 3), stimulated partly by its analogy with ^{226}Ra which is of interest as a tracer, and the rare earth elements. The latter form a group with generally similar chemical behaviors but with systematic differences in physicochemical properties; cerium with its different oxidation-reduction chemistry is exceptional in not behaving as a recycled element. The valuable insights into oceanic chemistry provided by these elements have been discussed by Elderfield.[41]

IV. SCAVENGED METALS

The general characteristics of the scavenged metals are well illustrated by the behavior of aluminum. This metal is strongly hydrolyzed and highly particle-reactive. Concentrations in deep North Atlantic Ocean waters, below about 2000 m, are typically[42-44] in the range of 25 to 30 nmol/l. Variations with depth are complex, showing regional differences and, in the upper water column, temporal variability, but some general features are apparent. For example, relatively high concentrations (29 to 43 nmol/l) occur in the mixed layer in the Sargasso Sea,[44] followed by an overall decrease (although with a secondary maximum centered on about 300 m) to a broad minimum, with concentrations as low as 10 nmol/l, centered on about 1400 m. Below this depth, concentrations increase with depth, typically to about 25 nmol/l. Surface transects in the northwestern Atlantic Ocean show a marked decrease in concentrations in the boundary current region[43,44] and it appears that atmospherically transported particles may be an important source of aluminum in surface waters, as suggested by Hydes[45] for the northeastern Atlantic Ocean.

Measures et al.[44] consider that the major changes they observed in concentrations with depth are attributable to a large extent to differences in the original concentrations of the metal in source waters. They suggest that high concentrations in the Lower North Atlantic Deep Water reflect high concentrations in the upper waters north of Iceland which form the source, by convection, for this water mass. The northwestern European shelf region is suggested as the initial major source for this aluminum. The lower concentrations in the Upper North Atlantic Deep Water were considered to reflect its large component of Labrador Sea Water which has sources in Arctic surface waters that are known to be low in aluminium. In the Sargasso Sea, the signal due to Mediterranean Outflow Water, which also contributes to the Upper North Atlantic Deep Water, is weak. At its source, however, the Mediterranean Outflow Water is highly enriched in aluminum,[46] probably because of dissolution from particles entering from the atmosphere; a concentration as high as 111 nmol/l was found for the Western Mediterranean Deep Water and in the Gulf of Cadiz the Outflow water contained 72 nmol/l.

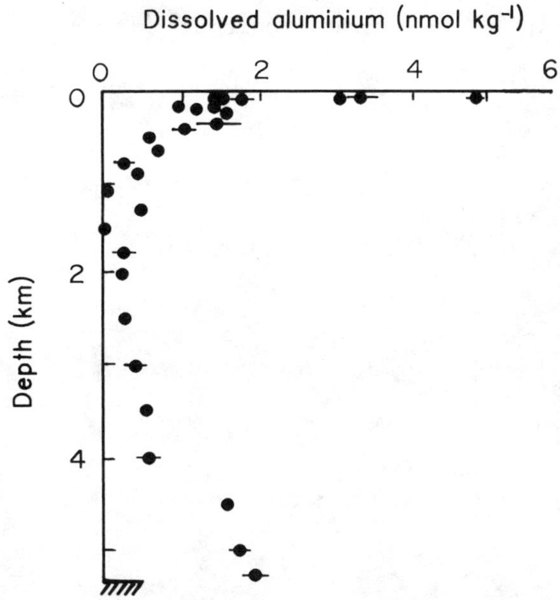

FIGURE 5. Vertical profile of concentration of dissolved aluminium in the central North Pacific gyre (28°15′N, 155°07′W). (From Orians, K. J. and Bruland, K. W., *Earth Planet. Sci. Lett.*, 78, 397, 1986. With permission.)

Vertical profiles for the Pacific Ocean[47] show pronounced depletion relative to Atlantic Ocean waters, reflecting the short residence time of aluminum in oceanic waters. In the deep waters of the central gyre regions, concentrations are in the range 0.5 to 2.0 nmol/kg, about two orders of magnitude lower than in Mediterranean deep waters. Because of the low concentrations and the less complex pattern of deep currents compared with the Atlantic Ocean, the effects of boundary inputs are more clearly defined. The profile given in Figure 5, shows enhanced surface concentrations, attributable to atmospheric inputs, and the effects of a bottom source, either from dissolution at the sediment interface or diffusion from the pore waters of bottom sediments.

The vertical distribution of gallium, an element which shows a generally high degree of geochemical coherence with aluminum, has been examined by Orians and Bruland,[48] in the first investigation to use modern approaches. A profile for this metal in the central North Pacific gyre shows a general similarity to that of aluminum. Concentrations increase from surface values of 12 pmol/l to a subsurface maximum of 18 pmol/l, below which a broad minimum (7 to 16 pmol/l) occurs from 500 to 1000 m, followed by an increase with depth to 30 pmol/l in bottom waters. Concentrations in the upper waters of the Gulf Stream were higher, with a surface maximum of about 56 pmol/l. These findings indicate a considerable enrichment of gallium over aluminum in the ocean relative to the crustal abundances of the two metals. This is partly attributable to a higher residence time, relative to scavenging processes, for gallium.

The distribution of manganese in oceanic waters, like that of aluminum, basically reflects the high particle reactivity, and consequently short oceanic residence time, of the metal. Its chemistry is, however, more complex since the element undergoes a change in oxidation state between oxidizing and reducing environments and there is a marked difference between these oxidation states in terms of geochemical mobility. Under the conditions prevailing when free oxygen is present at the concentrations characteristic of most oceanic waters, thermodynamic considerations indicate that manganese should occur as Mn(+IV), in which oxidation state the solubility of the element is very low. Under reducing conditions, however, Mn(+II) becomes the thermodynamically stable oxidation state and the element is then

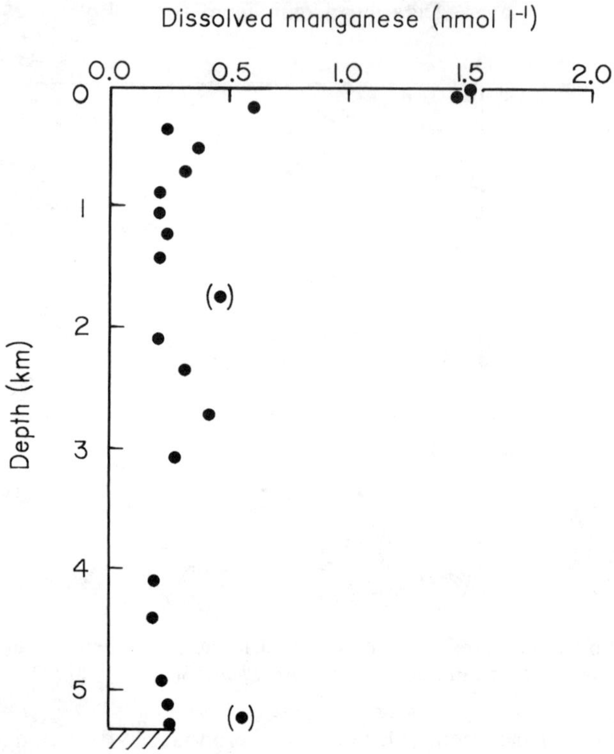

FIGURE 6. Vertical profile of concentration of dissolved manganese in the eastern North Atlantic Ocean (19°22′N, 12°51′W). Points in parentheses are probably affected by sampling or analytical errors. (From Statham, P. J., Burton, J. D., and Hydes, D. J., *Nature,* 313, 565, 1985. With permission.)

much more soluble. The oxidation of Mn(+II) to Mn(+IV) is mediated mainly at particle surfaces and, particularly in oceanic regimes, the scavenging of Mn(+II), involving processes of both adsorption and oxidation, is sufficiently slow for enhanced concentrations of manganese to be observed in waters adjacent to regions of input or release of Mn(+II) and these signatures on water bodies can be persistent over considerable times and hence spatial scales.

Over much of the ocean, where there are no significant localized inputs to the intermediate and deep waters, the general features in the distribution of manganese are similar to those shown in Figure 6. Characteristic features are a maximum or several maxima in the mixed layer, with highest concentrations typically of about 1 to 3 nmol/l, a marked decrease in concentration within the upper few hundred meters, with sometimes a minimum in the upper thermocline, and low, rather uniform, values in the deep water. Such profiles have been reported for many parts of the Atlantic and Pacific Oceans as summarized in Reference 7. In isolated bottom waters where anoxic conditions have developed, as in the Black Sea and Cariaco Trench, concentrations exceed those in oxygenated waters by several orders of magnitude.[49]

The enhanced concentrations in surface waters reflect the advection of manganese from sources at the ocean margin (river inputs and release from reducing sediments) in some areas,[50,51] but there is substantial evidence for eolian inputs in more remote areas.[52-54] Deep-water concentrations of manganese do not show such pronounced changes with the age of these waters as were noted for aluminum, partly because scavenging is evidently sufficiently rapid for even recently formed deep waters to be already markedly depleted, partly because trends are obscured by the influence of local sources. Concentrations of manganese in the northwestern Atlantic Ocean have been reported[55,56] in the range of about 0.3 to 0.6 nmol/l whereas values for the northeastern Atlantic Ocean and Mediterranean deep waters have been reported[54,57] to average about 0.25 nmol/l. Concentrations as low as 0.1 nmol/l have been found in some deep Pacific Ocean waters,[50,58] although overall values for deep waters, excluding those clearly influenced by hydrothermal inputs, range from these low values to concentrations as high as 0.5 nmol/kg in deep waters of the Californian Current region.

Pronounced anomalies in deep water concentrations can be produced by inputs of hydrothermal fluids, highly enriched in Mn(II), at ocean-ridge spreading centers, and lateral spreading from the source areas. Close to the ridge axis in the region of the East Pacific Rise, concentrations in the depth range of the hydrothermal plume injections, can exceed 35 nmol/kg, and slight anomalies attributable to such injections can be detected in the Pacific Ocean at distances of the order of 10^3 km from source areas.[59] Similar inputs occur on the Mid-Atlantic Ridge,[60] but dispersal appears to be more constrained than in the Pacific Ocean because of the topographic characteristics of the median rift valley.

Concentrations of manganese are markedly enhanced in intermediate waters in parts of the eastern North Pacific Ocean which are characterized by suboxic conditions in the oxygen minimum. For example, Martin et al.[61] found concentrations of 0.5 to 1.4 nmol/l, decreasing in the offshore direction, in the core of a manganese maximum centered on 600 to 700 m in a section between 29°N and 36°N, and higher concentrations up to 4.7 nmol/kg have been recorded[51] close to the continental slope at about 47°N. The enhanced concentrations reflect both transport from the continental slope region and release from particles sinking from surface waters into the zone where conditions become reducing. Smaller anomalies of dissolved manganese can be observed[52,54] in association with less pronounced oxygen minima, where concentrations are below about 100 μmol/l.

Information on cobalt in the ocean is too inadequate to enable interpretation of its geochemical behavior. Concentrations are generally below 100 pmol/l in the open ocean. Some reported features suggest a resemblance to manganese, with relatively high concentrations at or near the surface and a decrease with depth. Thus Bruland[3] cites data for the California Current region showing a shallow maximum (about 100 pmol/kg) and deep water concentrations as low as about 10 pmol/kg, while Jickells and Burton[62] found surface water concentrations in the Sargasso Sea in the range of 18 to 44 pmol/kg and a rapid decrease to less than 15 to 20 pmol/l at 200 to 600 m. In the latter area, there was, however, a slow increase in concentration with depth in the deep waters. An increase in concentration with depth in offshore waters in the northeastern Pacific Ocean is reported,[63] contrasting with the manganese-like profile found in near-shore waters;[64] deep-water concentrations were about 20 to 30 pmol/kg in both environments.

The behavior of iron in natural waters parallels that of manganese to some extent, but whereas Mn(+II) persists in oxidizing oceanic environments sufficiently long to play a major role in the transport of the element, Fe(+II) is more rapidly oxidized. The high reactivity of dissolved iron towards particles is shown by estimates of the residence time with respect to removal in the upper 500 m of the Pacific Ocean, ranging from 1 to 20 years in different regimes.[58] Iron(+II) occurs at high concentrations in hydrothermal fluids, occurs also in suboxic and anoxic environments, and may be formed by reduction in the upper ocean. It is reported[65] to account for a major part of the dissolved iron in waters above reducing sediments off Peru, but generally these sources do not produce the distinctive and persistent features for iron in oxygenated waters that are seen for manganese.

The low concentrations of iron in oceanic waters, and the ubiquity of the element as a contaminant in sampling analysis, have made the accurate measurement of iron particularly difficult and there are very few reliable data for the element, despite its geochemical importance. Concentrations reported for dissolved iron in estuarine and near-shore waters often include a considerable contribution of colloidal iron which is not removed by filtration (see, for example, Reference 66). For oceanic waters, an unknown, but in some circumstances probably significant, fraction of the iron reported as dissolved, by operational definition, is colloidal.

Surface concentrations of dissolved iron in the Pacific Ocean along a section from the California Current to the mid-ocean region[58] decrease offshore from values as high as 0.75 to 1.25 nmol/kg to about 0.25 to 0.5 nmol/kg. Vertical profiles are characterized by increases at depths where highly depleted oxygen minima occur. In these waters, concentrations are generally greater than 1 nmol/kg and reach 6 nmol/kg. Decreased concentrations occur in deeper waters, the range being similar to that in the upper water column. Thus, on the basis of present data, iron does not show the distribution of a typical scavenged metal, even though its behavior is predicted to be of that type.

It has been shown[67] experimentally for nutrient-rich surface waters of the sub-Arctic northeastern Pacific Ocean that addition of iron in amounts of nmol/l leads to increased utilization of nitrate suggesting that the availability of iron may limit the growth of phytoplankton in these waters.

Considerable efforts were made in the early to mid-1970s to establish the concentration of mercury in ocean waters, largely because of concern about pollution by the metal in certain coastal waters and elevated, although natural, concentrations in the tissues of tuna and swordfish. The results obtained were, however, inaccurate because of contamination problems and it is only recently that some of the major features in the oceanic geochemistry of the metal have begun to be established. Analytical data for mercury generally refer to measurements on unfiltered samples and to the fraction released by reduction to the metal under acidic conditions. This "reactive" fraction is discussed here simply as the concentration of mercury.

The mean oceanic residence time of mercury has been estimated as 350 years.[68] Input to the ocean is dominated by atmospheric pathways which are estimated to transport more than ten times the riverine flux.[69] Concentrations recently measured[68,70,71] in oceanic waters are generally below 10 pmol/l.

Profiles for the Sargasso Sea and the slope region north of the Gulf Stream[68] showed similar features with surface concentrations of about 4 pmol/l and similar or lower values below the thermocline, with maximum values of about 10 pmol/l in the thermocline. It was suggested that the maximum in the slope region is attributable to advection of mercury following release from marginal sediments as a consequence of postdepositional processes. The maximum in the Sargasso Sea profile is attributed to transport along isopycnals, a mechanism which has also been considered to account for increased concentrations of lead at similar depths in this region.[72] Comparisons[68] of deep water concentrations of mercury in the northwest Atlantic and central North Pacific Oceans suggest a decrease in the older waters by a factor of about two to three, consistent with the estimated residence time and the behavior of mercury as a scavenged metal.

The acquisition of accurate data on lead at the low concentrations that occur in deep Pacific Ocean waters represent one of the most striking achievements in the application of the clean methodologies developed from the mid-1970s onwards. Schaule and Patterson[73,74] showed that concentrations of dissolved lead in Pacific surface waters increased from a region some 200 km offshore of California to the central areas within the oceanic gyre, from about 25 to 75 pmol/kg, reflecting the interaction of a dominantly atmospheric source, advection, and regional differences in scavenging efficiency. In the Central Pacific Ocean, a marked latitudinal variation is apparent.[75] The maximum concentrations of dissolved lead, which occur in subsurface waters at about 400 m, decrease from 68 pmol/kg at 33°N to 12 pmol/kg at 20°S. Below the maxima, concentrations decrease to about 2500 to 3000 m depth below which they show little variation around a value of about 5 pmol/kg.

Surface water concentrations in the North Atlantic Ocean[72,76-78] are higher than those in

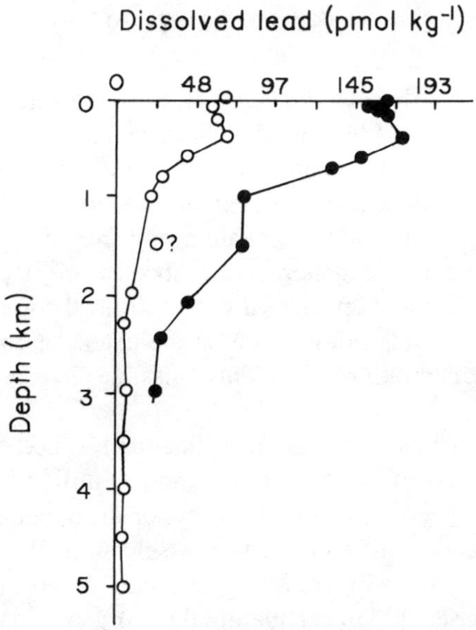

FIGURE 7. Vertical profiles of dissolved lead in the central northeastern Pacific Ocean (32°41′N, 145°00′W) (○) and in the Sargasso Sea (34°15′N, 66°17′W) (●). (From Schaule, B. K. and Patterson, C. C., *Trace Metals in Sea Water,* Wong, C. S., Boyle, E., Bruland, K. W., Burton, J. D., and Goldberg, E. D., Eds., Plenum Press, New York, 1983, 487. With permission.)

the North Pacific Ocean, but the trends in vertical profiles are similar. For example, values reported by Brugman et al.[78] for surface waters in the northeastern Atlantic Ocean average 130 pmol/l; profiles in the western and eastern basins are similar, with concentrations decreasing to about 20 to 25 pmol/l in deep water. Profiles for the Atlantic and Pacific Oceans are compared in Figure 7.

These features are consistent with a dominant influence on the upper ocean of anthropogenic lead transported to the ocean through the atmosphere and the penetration of this material into subsurface waters. Flegal et al.[79] showed that the isotopic composition of dissolved lead in surface waters of the central Pacific Ocean, between 19°N and 15°S, was consistent with an anthropogenic source. In the northeastern Pacific Ocean, similar measurements[80] showed that the lead in surface waters was dominantly from Asian sources. That at 400 m originated, however, from North American sources, a feature which can be explained by lateral advection as proposed by Lee et al.[81] to account for vertical profiles of lead, bismuth, and plutonium in the North Pacific Ocean. Flegal and Patterson[75] showed that the magnitude of the upper water maxima in various ocean regions are correlated with the magnitudes of the input rates of lead from the atmosphere. It was estimated that in the North Pacific and Atlantic Oceans, surface water concentrations were at least 8 to 20 times above the natural values, while a 2-fold increase, at least, had occurred in South Pacific waters.

Tellurium, like selenium, occurs in oceanic waters in the +IV and +VI oxidation states, but its behavior as a scavenged metal contrasts with that element. At three stations, in the North and South Atlantic Oceans and in the Panama Basin,[82] concentrations of total dissolved tellurium decrease with depth from surface concentrations of about 1.3 to 1.6 pmol/l. At the latter two stations, measurements of Te(+IV) show that the overall decrease is contributed to by a fall in the concentration of the metal in both oxidation states, although the scavenging of Te(+VI) is somewhat more marked.

V. SPECIATION

Analytical procedures for dissolved trace metals in seawater are generally intended to measure the total concentration of the various chemical species present in filtered samples, with, in some cases, distinction between different oxidation states of metals. In relatively few cases are metals present dominantly as simple cations or anions; generally they occur as a variety of complexes formed with inorganic, and in some cases organic, ligands. As has been seen for some metals, organometallic species are also important. Most of the information available on chemical speciation comes from theoretical models and from analytical approaches which provide information on the extent of complexation rather than on the occurrence of specific complexes. It is only possible here to provide a brief summary of some recent findings.

The complexation of metals with inorganic ligands has been evaluated theoretically in numerous studies on the basis of ion-association models utilizing the stability constants for the complexes concerned. There has often been major disagreement between the results of such studies, largely reflecting differences in the selection of complexes included in the models and discrepancies in the values taken for the stability constants. A critical evaluation of data was made by Turner et al.[83] in calculating the inorganic speciation for 58 trace metals in model seawater at pH 8.2, 25°C, and 1 atm. These calculations indicate that Al^{3+}, Fe^{3+}, and Ga^{3+} are almost wholly associated with hydroxyl ions. The free cation is a major species for Co^{2+} (58% of total species), Fe^{2+} (66%), Mn^{2+} (58%), Ni^{2+} (47%), Tl^{+} (53%), and Zn^{2+} (40%); chloro-complexes are also important for all these elements. For the trace alkali metal cations, the free forms account for more than 90% of the total concentrations, 99% in the case of Li^{+}. In contrast, Cd^{2+} is 97% associated as chloro-complexes and Cu^{2+} is 79% associated with carbonate ions, while for Pb^{2+} both chloro- and carbonato-complexes are important. Arsenic(+III) occurs mainly (87%) as $As(OH)_3^{\circ}$, and As(+V) almost entirely as $HAsO_4^{2-}$. The speciation of germanium is dominated by $H_4GeO_4^{\circ}$ (86%) whereas the MoO_4^{2-} ion is virtually completely unprotonated.

The role of organic material in complexing metals generally remains uncertain. Extensive detailed studies have been carried out only on copper. The organic complexing ligands in seawater have not been fully characterized. One category of information that is available relates to the occurrence of fractions of copper which can be defined operationally by analytical techniques and equated with organically complexed metal. For example, Hanson et al.[84] have measured the copper that is nonlabile in analysis by anodic stripping voltammetry and that which is removed from seawater on a solid phase which retains relatively hydrophobic material. Results for warm-core ring waters in the Gulf Stream and Sargasso Sea showed high proportions of total copper to be in these fractions, with values as high as 90% in waters with high biological activity.

A second category of information comes from measurements of the complexing capacity of samples of seawater for copper and the calculation from these data of conditional stability constants that can be used to predict the extent of organic association. For oligotrophic waters of the Sargasso Sea, Hanson et al.[84] estimated by this approach that 40 to 96% of the copper was organically complexed. Fractions of organically associated copper exceeding 98% in Atlantic Ocean waters have been reported by other workers.[85,86] There is also evidence for organic associations of nickel in seawater.[87]

The implications of these findings are largely uncertain in terms of the interpretation of biogeochemical cycling processes, but several considerations suggest their importance. Thus, there is evidence that the effects of a trace metal on growth of microorganisms are related not to the total concentration of the dissolved element, but to the activity of the free ion,[88] which is affected by both the activity coefficient of the free ion and the extent of complexation of the ion with specific ligands. The interactions between dissolved metals and particulate matter must be considerably influenced not only by the surface properties of particles but also the chemical speciation of the metal in solution. Lee and Edmond[82] attribute the difference in behavior between Te(+ VI) and Se(+ VI) to their contrasting speciation, Te(+ VI) being present as $Te(OH)_6^\circ$ rather than an oxy-anion as is the case with Se(+ VI). The dominant dissolved species for gallium in $Ga(OH)_4^-$, whereas for aluminium some 50% of the dissolved metal occurs as the uncharged complex, $Al(OH)_3^\circ$. It has been suggested[48] that this difference may be a factor conferring a lower particle reactivity on the heavier element.

Questions of chemical speciation are also relevant to the design of analytical procedures for trace metals and the interpretation of analytical findings, since the possibility has to be considered that methods intended to determine the total dissolved concentrations may not measure strongly complexed fractions of certain metals. While this contention has not been disproved, there is at present no compelling evidence in support of it. Bruland et al.[89] intercompared methods using different initial treatments and methods of concentration and determination for copper, cadmium, and lead in samples of seawater from various environments. The results supported the view that widely used methods employing separation using complexing ligands and determination by atomic absorption spectrophotometry provide information on the total dissolved trace metals in acidified samples.

VI. SHELF AND COASTAL WATERS

This account has concentrated on the distributions of trace metals in oceanic waters. In these environments, the relatively long residence times of the intermediate and deep waters relative to exchange with the surface waters enable the integrated effects of cycling of the metals over comparable time scales to be seen. Even a single deep ocean profile for a trace metal can give considerable insight into the behavior of the element.

Following the breakthrough in quality of analytical data for trace metals in oceanic waters, more investigations of shelf and coastal waters are being undertaken. These waters receive inputs from rivers which generally have concentrations of metals that are higher than those in seawater, and in the case of polluted rivers often much higher. One major factor determining distributions in such environments is the mixing of these river waters with seawater entering the shelf region as seen for example in the North Sea.[90] There are, however, other factors, particularly the influence of benthic and eolian inputs, and interactions between dissolved species and particulate material, which may be important in producing the ''patchiness'' in concentrations which has been noted in these waters. Larger data bases are needed to establish and interpret the variations in these dynamic environments, particularly the evaluation of sources and sinks through the comparison of observed distributions with those predicted for conservative constituents. Little is known yet in detail about the possible seasonal variations in concentrations of trace metals in response to the marked seasonal cycle of biological activity in temperate shallow seas.

A detailed summary of investigations on shelf waters lies outside the scope of this account. An example is given, however, in Figure 8 of the distribution of several metals during summer in surface waters sampled on a track covering the major regions of the northwestern European shelf seas.[91] This shows the relatively low concentrations associated with Atlantic Ocean waters entering the North Sea and English Channel and the rather localized effects of river inputs to the coastal areas.

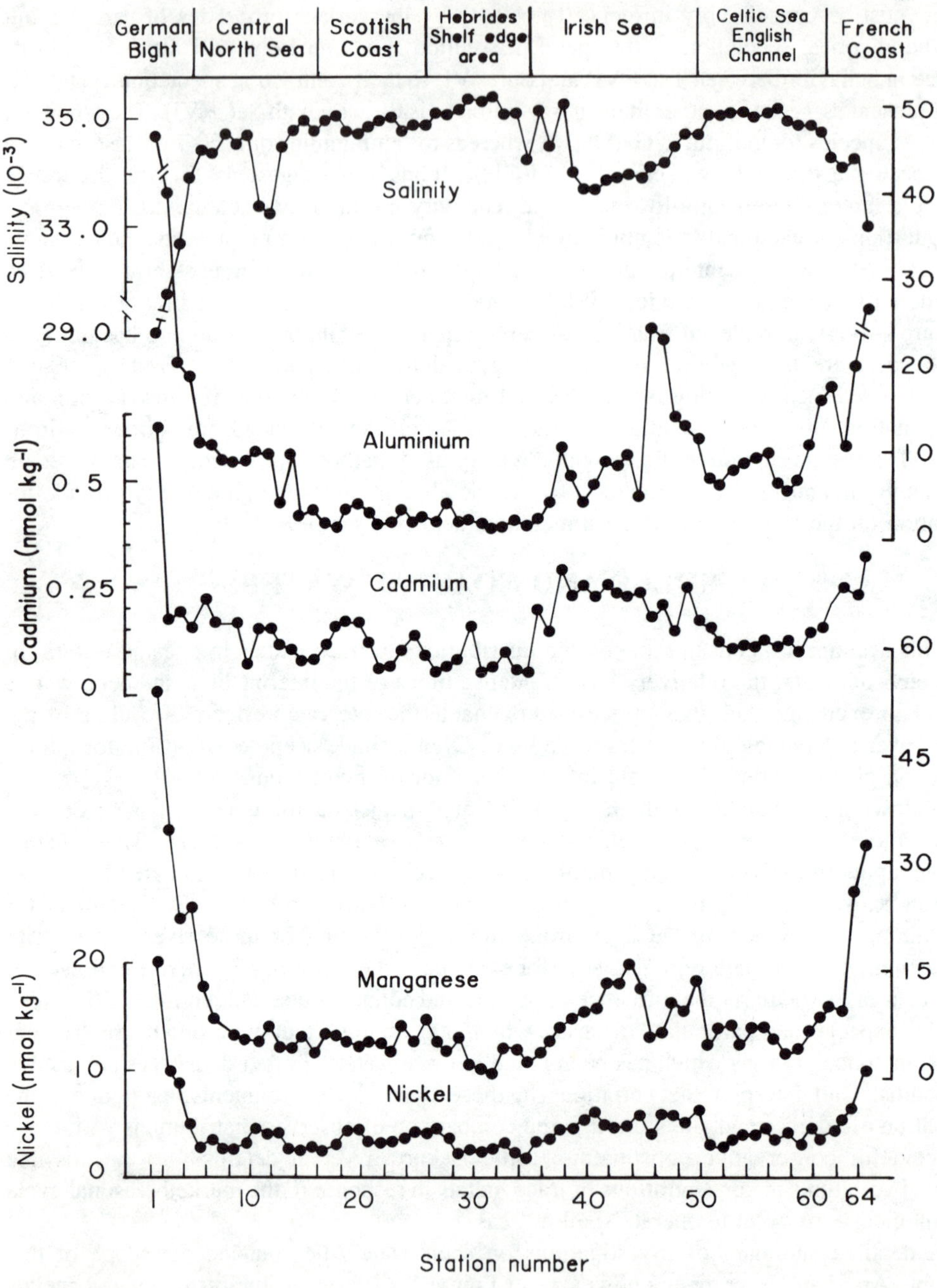

FIGURE 8. Variations in concentrations of aluminum, cadmium, manganese and nickel in surface northwestern European shelf waters. Station numbers refer to a section from off Hamburg (station 1) to off Le Havre (station 64), through the North Sea, the Hebridean shelf edge area, the Irish Sea, and the Western Approaches. (From Kremling, K. and Hydes, D., *Cont. Shelf Res.*, 8, 89, 1988. With permission.)

REFERENCES

1. **Wong, C. S., Boyle, E., Bruland, K. W., Burton, J. D., and Goldberg, E. D., Eds.,** *Trace Metals in Sea Water,* Plenum Press, New York, 1983.
2. **Burton, J. D. and Statham, P. J.,** Occurrence, distribution, and chemical speciation of some minor dissolved constituents in ocean waters, in *Environmental Chemistry,* Vol. 2, Specialist Periodical Report, Royal Society of Chemistry, London, 1982, 234.
3. **Bruland, K. W.,** Trace elements in sea-water, in *Chemical Oceanography,* Vol. 8, Riley, J. P. and Chester, R., Eds., Academic Press, London, 1983, 157.
4. **Whitfield, M. and Turner, D. R.,** The role of particles in regulating the composition of seawater, in *Aquatic Surface Chemistry,* Stumm, W., Ed., John Wiley & Sons, New York, 1987, 457.
5. **Wilson, T. R. S.,** Salinity and the major elements of seawater, in *Chemical Oceanography,* Vol. 1, 2nd ed., Riley, J. P. and Skirrow, G., Eds., Academic Press, London, 1975, 365.
6. **Broecker, W. S.,** *Chemical Oceanography,* Harcourt-Brace Jovanovich, New York, 1974.
7. **Burton, J. D. and Statham, P. J.,** Trace metals as tracers in the ocean, *Philos. Trans. R. Soc. London Ser. A,* 325, 127, 1988.
8. **Bruland, K. W. and Coale, K. H.,** Surface water $^{234}Th/^{238}U$ disequilibria: spatial and temporal variations of scavenging rates within the Pacific Ocean, in *Dynamic Processes in the Chemistry of the Upper Ocean,* Burton, J. D., Brewer, P. G., and Chesselet, R., Eds., Plenum Press, New York, 1986, 159.
9. **Folsom, T. R.,** Gradient of caesium in the ocean, *Nature,* 248, 216, 1984.
10. **Flegal, A. R. and Patterson, C. C.,** Thallium concentrations in seawater, *Mar. Chem.,* 15, 327, 1985.
11. **Collier, R. W.,** Molybdenum in the northeast Pacific Ocean, *Limnol. Oceanogr.,* 30, 1351, 1985.
12. **Ku, T.-L., Knauss, K. G., and Mathieu, G. G.,** Uranium in open ocean: concentration and isotopic composition, *Deep-Sea Res.,* 24, 1005, 1977.
13. **Chen, J. H., Edwards, R. L., and Wasserburg, G. J.,** ^{238}U, ^{234}U and ^{232}Th in seawater, *Earth Planet. Sci. Lett.,* 80, 241, 1986.
14. **Anderson, R. F.,** Concentration, vertical flux, and remineralization of particulate uranium in seawater, *Geochim. Cosmochim. Acta,* 46, 1293, 1982.
15. **Lewis, B. L., Froelich, P. N., and Andreae, M. O.,** Methylgermanium in natural waters, *Nature,* 313, 303, 1985.
16. **Boyle, E. A., Sclater, F., and Edmond, J. M.,** On the marine geochemistry of cadmium, *Nature,* 263, 42, 1976.
17. **Martin, J. H., Bruland, K. W., and Broenkow, W. W.,** Cadmium transport in the California Current, in *Marine Pollutant Transfer,* Windom, H. L. and Duce, R. A., Eds., Heath, Lexington, 1976, 159.
18. **Bruland, K. W.,** Oceanographic distributions of cadmium, zinc, nickel, and copper in the North Pacific, *Earth Planet. Sci. Lett.,* 47, 176, 1980.
19. **Bruland, K. W. and Franks, R. P.,** Mn, Ni, Cu, Zn and Cd in the western North Atlantic, in *Trace Metals in Sea Water,* Wong, C. S., Boyle, E., Bruland, K. W., Burton, J. D., and Goldberg, E. D., Eds., Plenum Press, New York, 1983, 395.
20. **Froelich, P. N., Hambrick, G. A., Andreae, M. O., Mortlock, R. A., and Edmond, J. M.,** The geochemistry of inorganic germanium in natural waters, *J. Geophys. Res.,* 90, 1133, 1985.
21. **Sclater, R. F., Boyle, E., and Edmond, J. M.,** On the marine geochemistry of nickel, *Earth Planet. Sci. Lett.,* 31, 119, 1976.
22. **Danielsson, L.-G.,** Cadmium, cobalt, copper, iron, lead, nickel and zinc in Indian Ocean water, *Mar. Chem.,* 8, 199, 1980.
23. **Boyle, E. A., Sclater, F. R., and Edmond, J. M.,** The distribution of dissolved copper in the Pacific, *Earth Planet. Sci. Lett.,* 37, 38, 1977.
24. **Moore, R. M.,** The distribution of dissolved copper in the eastern Atlantic Ocean, *Earth Planet. Sci. Lett.,* 41, 461, 1978.
25. **Craig, H.,** A scavenging model for trace elements in the deep sea, *Earth Planet Sci. Lett.,* 23, 149, 1974.
26. **Moffett, J. W. and Zika, R. G.,** Photochemistry of copper complexes in seawater, in *Photochemistry of Environmental Aquatic Systems,* Symposium Series 327, American Chemical Society, Washington, D.C., 1987, 116.
27. **Measures, C. I. and Burton, J. D.,** The vertical distribution and oxidation states of dissolved selenium in the northeast Atlantic Ocean and their relationship to biological processes, *Earth Planet. Sci. Lett.,* 46, 385, 1980.
28. **Measures, C. I., McDuff, R. E., and Edmond, J. M.,** Selenium redox chemistry at GEOSECS I re-occupation, *Earth Planet. Sci. Lett.,* 49, 102, 1980.
29. **Measures, C. I., Grant, B. C., Magnum, B. J., and Edmond, J. M.,** The relationship of the distribution of dissolved selenium IV and VI in three oceans to physical and biological processes, in *Trace Metals in Sea Water,* Wong, C. S., Boyle, E., Bruland, K. W., Burton, J. D., and Goldberg, E. D., Eds., Plenum Press, New York, 1983, 73.

30. **Cutter, G. A. and Bruland, K. W.,** The marine biogeochemistry of selenium: a re-evaluation, *Limnol. Oceanogr.,* 29, 1179, 1984.
31. **Cutter, G. A.,** Selenium in reducing waters, *Science,* 217, 829, 1982.
32. **Wrench, J. J. and Measures, C. I.,** Temporal variations in dissolved selenium in a coastal ecosystem, *Nature,* 299, 431, 1982.
33. **Burton, J. D., Maher, W. A., Measures, C. I., and Statham, P. J.,** Aspects of the distribution and chemical form of selenium and arsenic in ocean waters and marine organisms, *Thalassia Jugosl.,* 16, 155, 1980.
34. **Andreae, M. O.,** Distribution and speciation of arsenic in natural waters and some marine algae, *Deep-Sea Res.,* 25, 391, 1978.
35. **Andreae, M. O.,** Arsenic speciation in seawater and interstitial waters: the influence of biological-chemical interactions on the chemistry of a trace element, *Limnol. Oceanogr.,* 24, 440, 1979.
36. **Statham, P. J., Burton, J. D., and Maher, W. A.,** Dissolved arsenic in waters of the Cape Basin, *Deep-Sea Res.,* 34, 1353, 1987.
37. **Middelburg, J. J., Hoede, D., van der Sloot, H. A., van der Weijden, C. H., and Wijkstra, J.,** Arsenic, antimony and vanadium in the North Atlantic Ocean, *Geochim. Cosmochim. Acta,* 52, 2871, 1988.
38. **Sanders, J. G.,** Microbial role in the demethylation and oxidation of methylated arsenicals in seawater, *Chemosphere,* 8, 135, 1979.
39. **Scudlark, J. R., and Johnson, D.L.,** Biological oxidation of arsenite in seawater, *Estuar. Coastal Shelf Sci.,* 14, 693, 1982.
40. **Collier, R. W.,** Particulate and dissolved vanadium in the North Pacific Ocean, *Nature,* 309, 441, 1984.
41. **Elderfield, H.,** The oceanic chemistry of the rare-earth elements, *Philos. Trans. R. Soc. London Ser.* A, 325, 105, 1988.
42. **Hydes, D. J.,** Aluminium in seawater: control by inorganic processes, *Science,* 205, 1260, 1979.
43. **Moore, R. M. and Millward, G. E.,** Dissolved particulate interactions of aluminium in ocean waters, *Geochim. Cosmochim. Acta,* 48, 235, 1984.
44. **Measures, C. I., Edmond, J. M., and Jickells, T. D.,** Aluminium in the northwest Atlantic, *Geochim. Cosmochim. Acta,* 50, 1423, 1986.
45. **Hydes, D. J.,** Distribution of aluminium in waters of the North East Atlantic 25°N to 35°N, *Geochim. Cosmochim. Acta,* 47, 967, 1983.
46. **Measures, C. I. and Edmond, J. M.,** Aluminium as a tracer of the deep outflow from the Mediterranean, *J. Geophys. Res.,* 93, 591, 1988.
47. **Orians, K. J. and Bruland, K. W.,** The biogeochemistry of aluminium in the Pacific Ocean, *Earth Planet. Sci. Lett.,* 78, 397, 1986.
48. **Orians, K. J. and Bruland, K. W.,** Dissolved gallium in the open ocean, *Nature,* 332, 717, 1988.
49. **Brewer, P. G.,** Minor elements in sea water, in *Chemical Oceanography,* Vol. 1, 2nd ed., Riley, J. P. and Skirrow, G., Eds., Academic Press, London, 1975, 415.
50. **Landing, W. M. and Bruland, K. W.,** Manganese in the North Pacific, *Earth Planet. Sci. Lett.,* 49, 45, 1980.
51. **Jones, C. J. and Murray, J. W.,** The geochemistry of manganese in the northeast Pacific Ocean off Washington, *Limnol. Oceanogr.,* 30, 81, 1985.
52. **Klinkhammer, G. P. and Bender, M. L.,** The distribution of manganese in the Pacific Ocean, *Earth Planet. Sci. Lett.,* 46, 361, 1980.
53. **Kremling, K.,** The distribution of cadmium, copper, nickel, manganese and aluminium in surface waters of the open Atlantic and European Shelf area, *Deep-Sea Res.,* 32, 531, 1985.
54. **Statham, P. J. and Burton, J. D.,** Dissolved manganese in the North Atlantic Ocean, 0 — 35°N, *Earth Planet. Sci. Lett.,* 79, 55, 1986.
55. **Yeats, P. A. and Bewers, J. M.,** Manganese in the western North Atlantic Ocean, *Mar. Chem.,* 17, 255, 1985.
56. **Jickells, T. D.,** Studies of Trace Elements in the Sargasso Sea and the Mid-Atlantic Bight, Ph.D. thesis, University of Southampton, Southampton, U.K., 1986.
57. **Statham, P. J., Burton, J. D., and Hydes, D. J.,** Cd and Mn in the Alboran Sea and adjacent North Atlantic: geochemical implications for the Mediterranean, *Nature,* 313, 565, 1985.
58. **Landing, W. M. and Bruland, K. W.,** The contrasting biogeochemistry of iron and manganese in the Pacific Ocean, *Geochim. Cosmochim. Acta,* 51, 29, 1987.
59. **Klinkhammer, G. and Hudson, A.,** Dispersal patterns for hydrothermal plumes in the South Pacific using manganese as a tracer, *Earth Planet. Sci. Lett.,* 79, 241, 1986.
60. **Klinkhammer, G., Elderfield, H., Greaves, M., Rona, P., and Nelsen, T.,** Manganese geochemistry near high-temperature vents in the Mid-Atlantic Ridge rift valley, *Earth Planet. Sci. Lett.,* 80, 230, 1986.
61. **Martin, J. H., Knauer, G. A., and Broenkow, W. W.,** VERTEX: the lateral transport of manganese in the northeast Pacific, *Deep-Sea Res.,* 32, 1405, 1985.

62. **Jickells, T. D. and Burton, J. D.,** Cobalt, copper, manganese and nickel in the Sargasso Sea, *Mar. Chem.,* 23, 131, 1988.
63. **Martin, J. H.,** Iron and cobalt in north-east Pacific waters, *Trans. Am. Geophys. Un.,* 66, 1291, 1985.
64. **Knauer, G. A., Martin, J. H., and Gordon, R. M.,** Cobalt in north-east Pacific waters, *Nature,* 297, 49, 1982.
65. **Hong, H. and Kester, D. R.,** Redox state of iron in the offshore waters of Peru, *Limnol. Oceanogr.,* 31, 512, 1986.
66. **Figuères, G., Martin, J.-M., and Meybeck, M.,** Iron behaviour in the Zaire estuary, *Neth. J. Sea Res.,* 12, 329, 1978.
67. **Martin, J. H. and Fitzwater, S. E.,** Iron deficiency limits phytoplankton growth in the north-east Pacific subarctic, *Nature,* 331, 341, 1988.
68. **Gill, G. A. and Fitzgerald, W. F.,** Vertical mercury distributions in the oceans, *Geochim. Cosmochim. Acta,* 52, 1719, 1988.
69. **Gill, G. A. and Fitzgerald, W. F.,** Mercury in surface waters of the open ocean, *Global Biogeochem. Cycles,* 3, 199, 1987.
70. **Gill, G. A. and Fitzgerald, W. F.,** Mercury sampling of open ocean waters at the picomolar level, *Deep-Sea Res.,* 32, 287, 1985.
71. **Dalziel, J. A. and Yeats, P. A.,** Reactive mercury in the central North Atlantic Ocean, *Mar. Chem.,* 15, 357, 1985.
72. **Boyle, E. A., Chapnick, S. D., Shen, G. T., and Bacon, M. P.,** Temporal variability of lead in the western North Atlantic, *J. Geophys. Res.,* 91, 8573, 1986.
73. **Schaule, B. K. and Patterson, C. C.,** The occurrence of lead in the northeast Pacific, and the effects of anthropogenic inputs, in *Lead in the Marine Environment,* Branica, M. and Konrad, Z., Eds., Pergamon Press, Oxford, 1980, 31.
74. **Schaule, B. K. and Patterson, C. C.,** Lead concentrations in the northeast Pacific: evidence for global anthropogenic perturbations, *Earth Planet. Sci. Lett.,* 54, 97, 1981.
75. **Flegal, A. R. and Patterson, C. C.,** Vertical concentration profiles of lead in the central Pacific at 15°N and 20°S, *Earth Planet. Sci. Lett.,* 64, 19, 1983.
76. **Schaule, B. K. and Patterson, C. C.,** Perturbations of the natural lead depth profile in the Sargasso Sea by industrial lead, in *Trace Metals in Sea Water,* Wong, C. S., Boyle, E., Bruland, K. W., Burton, J. D., and Goldberg, E. D., Eds., Plenum Press, New York, 1983, 487.
77. **Boyle, E. and Huested, S.,** Aspects of the surface distributions of copper, nickel, cadmium, and lead in the North Atlantic and North Pacific, in *Trace Metals in Sea Water,* Wong, C. S., Boyle, E., Bruland, K. W., Burton, J. D., and Goldberg, E. D., Eds., Plenum Press, New York, 1983, 379.
78. **Brugman, L., Danielsson, L.-G., Magnusson, B., and Westerlund, S.,** Lead in the North Sea and the north-east Atlantic Ocean, *Mar. Chem.,* 16, 47, 1985.
79. **Flegal, A. R., Schaule, B. K., and Patterson, C. C.,** Stable isotopic ratios of lead in surface waters of the Central Pacific, *Mar. Chem.,* 14, 281, 1984.
80. **Flegal, A. R., Itoh, K., Patterson, C. C., and Wong, C. S.,** Vertical profile of lead isotopic compositions in the north-east Pacific, *Nature,* 321, 689, 1986.
81. **Lee, D. S., Edmond, J. M., and Bruland, K. W.,** Bismuth in the Atlantic and North Pacific: a natural analogue to plutonium and lead?, *Earth Planet. Sci. Lett.,* 76, 254, 1986.
82. **Lee, D. S. and Edmond, J. M.,** Tellurium species in seawater, *Nature,* 313, 782, 1985.
83. **Turner, D. R., Whitfield, M., and Dickson, A. G.,** The equilibrium speciation of dissolved components in freshwater and seawater at 25°C and 1 atm pressure, *Geochim. Cosmochim. Acta,* 45, 855, 1981.
84. **Hanson, A. K., Jr., Kester, D. R., Sakamoto-Arnold, C. M., and Huizenga, D. L.,** Copper complexation in Sargasso Sea and Gulf Stream warm-core ring waters, *Mar. Chem.,* 23, 181, 1988.
85. **Buckley, P. J. M. and van den Berg, C. M. G.,** Copper complexation profiles in the Atlantic Ocean. A comparative study using electrochemical and ion exchange techniques, *Mar. Chem.,* 19, 281, 1986.
86. **Kramer, C. J. M.,** Apparent copper complexation capacity and conditional stability constants in North Atlantic waters, *Mar. Chem.,* 18, 335, 1986.
87. **van den Berg, C. M. G. and Nimmo, M.,** Determinations of interactions of nickel with dissolved organic material in seawater using cathodic stripping voltammetry, *Sci. Total Environ.,* 60, 185, 1987.
88. **Morel, F. M. M.,** *Principles of Aquatic Chemistry,* John Wiley & Sons, New York, 1983, 300.
89. **Bruland, K. W., Coale, K. H., and Mart, L.,** Analysis of seawater for dissolved cadmium, copper and lead. An intercomparison of voltammetric and atomic absorption methods, *Mar. Chem.,* 17, 285, 1985.
90. **Kremling, K., Wenck, A., and Pohl, C.,** Summer distribution of dissolved Cd, Co, Cu, Mn, and Ni in central North Sea waters, *Dtsch. Hydrogr. Z.,* 40, 103, 1987.
91. **Kremling, K. and Hydes, D.,** Summer distribution of dissolved Al, Cd, Co, Cu, Mn and Ni in surface waters around the British Isles, *Cont. Shelf Res.,* 8, 89, 1988.

Chapter 3

METALS IN THE MARINE ATMOSPHERE

R. Chester and K. J. T. Murphy

TABLE OF CONTENTS

I. INTRODUCTION

In recent years, it has become increasingly apparent that atmospheric transport is an important route by which trace metals, and other substances are delivered to the oceans. In the present chapter, attention is confined to particulate atmospheric components, which in this context are referred to as aerosols. However, it must be stressed that gaseous phases can undergo condensation reactions which can result in the enrichment of the particulate material with some volatile elements; i.e., gas-to-particle reactions can be an important source of metals to the aerosols. The transport of the aerosols within the troposphere takes place via the major wind systems which operate on a global scale. These are the trades, the westerlies, and the polar easterlies, which are separated by pressure zones. Aerosols are dispersed about the surface of the planet by these various zonal wind systems and by transfer between them. However, the troposphere is a reservoir in which aerosols have a relatively short residence time, usually in the order of days for those having radii in the range ~0.1 to ~10 μm, and from which they are removed at about the same rate as they enter. In order to assess the relative importance of this atmospheric pathway for the transport of metals to the sea surface, it is necessary to have data on (1) the sources of the metals to the marine atmosphere, (2) the concentrations of the metals in the marine atmosphere over various oceanic regions, (3) the mechanisms by which the metals are removed from the marine atmosphere, (4) the magnitudes of the net metal fluxes to the sea surface, and (5) the fate of the atmospherically transported metals in the biogeochemical cycles which control the distributions of metals in the ocean system. Each of these topics are addressed below in the light of recent research.

II. THE SOURCES OF PARTICULATE METALS TO THE MARINE ATMOSPHERE

The particulate and gaseous material supplied to the atmosphere can originate from either natural or anthropogenic sources, and when considering the elemental chemistry of both of these, it is useful to distinguish between low temperature and high temperature generation processes. This is important because the forms in which the elements are present in the atmosphere, i.e., their speciation, can be strongly dependent on the temperature at which they were released from their parent hosts.

The **natural** sources of metals to the atmosphere include the following: (1) the Earth's surface which supplies particulate components to the atmosphere via the low temperature mechanical mobilization of surface deposits by wind erosion; (2) the ocean surface which supplies particulate components to the marine atmosphere during the low temperature formation of sea salts by mechanical action; the sea surface can also be a source of some gaseous components; (3) volcanic activity which releases particulate material, mainly ash, together with a gaseous phase formed from high temperature volatilization processes; (4) the biosphere from which metals are released into the atmosphere via the high temperature burning of vegetation and the emission of particulate and vapor phases from plant surfaces and soils.

Anthropogenic, or pollutant, sources of metals to the atmosphere include: fossil fuel burning, waste incineration, mining, the processing of ores, the production of chemicals, and the agricultural utilization of pesticides, herbicides, etc. Many of these anthropogenic processes involve high temperature emissions.

Recently, there have been a number of attempts to assess the relative source strengths of the natural and anthropogenic processes which supply metals to the atmosphere (see e.g., Lantzy and Mackenzie;[1] Nriagu;[2] Weisel et al.[3]). Chester[4] summarized the data from these various compilations and drew the following conclusions. (1) The inputs of some elements,

TABLE 1
Geometric Mean EF_{crust} Values in the World Aerosol

EF_{crust} range	Element
~0 — 10^1	Si, Nb, Al, Yb, Tm, Hf, Sc, Y, Ti, Zr, Er, Sr, Ho, Th, Lu, Tb, Fe, F, K, Sm, Dy, Ga, Mg, Ce, Pu, Ca, La, U, Gd, Be, Nd, Rb, Co, Mn, Tl, Na, Pr, Ba, Li, Cr
~10^1 — 10^2	Cs, V, W, B, Ni, Ge, H, In, Cu
~10^2 — 10^3	Mo, Bi, Zn, As, I, Hg, S, Cl, Au, Ag, Sn
~10^3 — 10^4	Sb, Pb, Br, Cd, Te, Se, C, N

Note: Data from Reference 5. Elements in each class are arranged in order of increasing EF_{crust} value.

e.g., Al, Fe, Mn, Co, and Sc are dominated by the injection of crustal material into the atmosphere. (2) The inputs of other elements (e.g., Na, K, Mg, and possibly some trace metals) are dominated by the generation of sea salts. (3) Then inputs of a number of metals, including V, Sn, Ni, Cu, Cd, Zn, As, Se, Sb, Mo, Ag, Hg, and Pb are dominated by anthropogenic inputs. It must be stressed, however, that some of these inputs are much more localized than others. For example, the earth's crust and the sea surface are globally important sources, whereas anthropogenic inputs tend to be centralized in a number of heavily industrialized regions which are mainly in the Northern Hemisphere. Volcanic activity also tends to be concentrated in a number of belts.

The marine aerosol is composed of materials from the various sources described above. These materials have different chemical compositions, with the result that the overall 'make-up' of the marine aerosol varies in response to the extent to which the various components are present. Because the source strengths of the individual inputs vary locally, the concentrations of many particulate metals in the marine atmosphere also vary in the same way. This variation in metal concentrations can extend over several orders of magnitude, and to illustrate this, the concentration ranges of a number of metals in the marine atmosphere are listed in Table 5. Because of these large concentration variations, it is useful to be able to identify the sources of individual elements in the marine aerosol. This is usually achieved by means of an enrichment factor which employs an indicator element that is known to be derived predominantly from one specific source. The three most important sources of metals to the marine aerosol are the Earth's crust, the ocean surface, and anthropogenic activity.

Aluminum is the most commonly used indicator element for the **crustal** source. As with most sources, there are problems involved in selecting a composition for the crustal precursor material, but that of the average crustal rock is normally used for the calculation of EF_{crust} values, according to the equation:

$$EF_{crust} = (E/Al)_{air}/(E/Al)_{crust} \quad (1)$$

in which(E/Al) air is the ratio of the concentrations of an element E and Al in the aerosol and $(E/Al)_{crust}$ is the ratio of the concentrations in average crustal rocks.

Because of the various constraints involved in the calculation, EF_{crust} values should only be treated as order of magnitude indicators of the crustal source. Thus, values close to unity are taken as an indication that an element has a mainly crustal origin, and those >10 are considered to indicate that a substantial portion of the element has a noncrustal origin.

Rahn et al.[5] have given data on the EF_{crust} values for some 70 elements in the *World Aerosol*, and the geometric means of these are listed in Table 1. From this table it can be seen that over half the elements have EF_{crust} values which are <10, i.e., they are present

TABLE 2
Concentrations and EF_{crust} Values of Trace Metals in Aerosols from the Atlantic Westerlies and Northeast Trades

	Westerlies		Northeast trades	
Element	Concentration (ng/m^3)	EF_{crust}	Concentration (ng/m^3)	EF_{crust}
Al	48	1.0	5925	1.0
Fe	36	1.2	3865	1.0
Mn	1.7	3.9	65	1.0
Cr	0.49	13	10	1.5
V	1.0	20	15	1.6
Ni	0.51	16	6.6	1.3
Co	0.05	2.7	2.1	1.2
Cu	1.0	30	4.5	1.2
Zn	7.5	292	16	3.8
Pb	6.0	1490	6.9	9.3
Cd	0.11	1259	0.12	9.5

Note: Data from Reference 6.

in the aerosol in roughly crustal proportions. These are termed the crustal, or nonenriched elements (NEE). The other elements have EF_{crust} values which range between >10 to $\sim 5 \times 10^3$, i.e., their concentrations are not controlled by inputs from crustal sources. These elements are termed the anomalously enriched elements (AEE).

Chester[4] has pointed out that the NEE will almost always retain their character in all aerosols, However, the degree to which an AEE is actually enriched can vary considerably as the mutual proportions of the various components in an aerosol change. This can be illustrated by comparing the elemental compositions of two contrasting North Atlantic aerosol populations; one from the westerlies which receives a significant pollutant input of metals and one from the north-east trades which is dominated by the input of crustal material from the Sahara Desert. Data for the concentrations of a number of metals and their EF_{crust} values, in these two aerosol populations are given in Table 2. From this table, it can be seen that the metals can be divided into two general groups. Group 1 contains the NEE Fe, Mn, and the crustal indicator element Al. Both Fe and Mn, and by definition Al, have EF_{crust} values <10 in both populations; i.e., they are crust-dominated even in the westerlies aerosol. Group 2 contains the AEE Cu, Zn, and Pb. All three elements have EF_{crust} values >10 in the westerlies aerosol indicating that here noncrustal sources predominate. However, in the aerosol from the northeast trades, the EF_{crust} values of the elements fall to <10; i.e., in this population, they are crust-controlled.

It is apparent, therefore, that under some conditions, e.g., when there are relatively large injections of crustal material into the atmosphere, the AEE can in fact behave as NEE.

Sodium is commonly used as the indicator element for the **oceanic** source, and in the past, it has been the practice to use bulk seawater as the precursor source composition. In this way the EF_{sea} was calculated according to the equation

$$EF_{sea} = (E/Na)_{air}/(E/Na)_{seawater} \quad (2)$$

in which, $(E/Na)_{air}$ is the ratio of the concentrations of an element E and Na in the aerosol and $(E/Na)_{seawater}$ is the ratio of their concentrations in *bulk* seawater. However, it is the oceanic microlayer, a thin film at the air/sea interface, and not bulk seawater which is the real precursor material for many of the trace metals associated with sea salts. The reason

TABLE 3
BIMS EF Values Determined Over the North Atlantic Ocean

Element	BIMS EF_{sea}
K	1.0
Mg	1.0
Al	5×10^3
Co	6×10^1
Cu	8×10^2
Fe	1×10^4
Mn	1×10^3
Pb	4×10^3
V	1×10^2
Zn	2×10^4

Note: Data from Reference 3. See text for explanation of BIMS EF values.

for this is that when the sea salts are generated by bubble bursting, they skim off part of the microlayer, and since the layer itself contains enhanced concentrations of many trace metals, sea-salt generation can significantly fractionate the metals relative to bulk seawater. Recently, attempts have been made to identify the actual seawater aerosol precursor material by collecting samples of the microlayer emitted into the atmosphere via the bursting of bubbles; e.g., by using the Bubble Interfacial Microlayer Sampler (BIMS) (see e.g., Fasching et al.,[7] Piotrowicz et al.[8]) In the BIMS experiments, bubbles were artificially generated and the particles they formed as they burst clear of the water were collected on filters. The net concentration of a metal on these particles was then used, together with the bulk seawater concentration, to calculate a BIMS EF_{sea} value. For example, in a recent study, Weisel et al.[3] carried out a series of BIMS experiments for a number of trace metals by generating bubbles from various depths in the open ocean western North Atlantic. The net concentration of each metal on the bubble-generated sea salts was then used, together with its North Atlantic surface seawater concentration, to calculate a BIMS EF_{sea} value. The geometric BIMS EF_{sea} means for all bubble depths for the elements studies are given in Table 3, and on this basis, the elements can be divided into two broad groups.

Group 1 contains K and Mg. These two elements have average BIMS EF_{sea} values of unity, which are similar to those in the bulk aerosol; i.e., these elements, together with Na, retain their original bulk seawater ratios on the bubble-generated salts and do not undergo fractionation across the air/sea interface. Group 2 contains Al, Fe, Mn, Co, V, Cu, Zn, Pb, and Sc. These elements have relatively high BIMS EF_{sea} values, ranging from 10 for Sc to 20,000 for Zn, and apparently undergo considerable fractionation at the sea surface.

Despite the very real difficulties inherent in assessing the fractionation of elements at the air/sea interface, the BIMS EF_{sea} data provide at least a first approximation of the degree to which such fractionation has taken place. Further, by making the assumption that the BIMS EF_{sea} values found for the North Atlantic apply globally, Weisel et al.[3] introduced the important concept of a relative oceanic enrichment factor (REF_{sea}), which is the aerosol EF_{sea} value (calculated with respect to bulk seawater) divided by the BIMS EF_{sea} value. The REF_{sea} values, therefore, directly relate the enrichment of an element in a marine aerosol to the fractionation which occurs at the sea surface, and so should offer a more realistic assessment of the ocean as a source material. These values should approach unity if the sea surface is a significant source for an element. This concept is evaluated with respect to a number of marine aerosol populations in Table 4 in which the individual populations are listed in order of increasing remoteness from continental metal sources. In the table, the

TABLE 4
Sea Surface Sources of Elements in Marine Aerosol Populations

	North Atlantic		North Pacific			
			Hawaii		Enewetak	
Element	BIMS EF_{sea} as a % of aerosol EF_{sea}	REF_{sea}	BIMS EF_{sea} as a % of aerosol EF_{sea}	REF_{sea}	BIMS EF_{sea} as a % of aerosol EF_{sea}	REF_{sea}
Al	0.5	200	1.6	6	5	20
Co	0.75	133	6	17	6	17
Mn	0.25	400	12	8	5	20
Cu	1.3	75	11	8.75	27	3.75
V	1.1	90	25	4	50	2
Zn	1	100	3	30	20	5
Pb	0.80	125	0.80	125	10	7.5

Note: Original data from Reference 3. See text for explanation of REF_{sea}.

BIMS EF_{sea} values are expressed as a percentage of the EF_{sea} value (i.e., that obtained using bulk seawater), and the REF_{sea} values are also given. From these various data, it can be seen that for Al, Co, Mn, and Pb, the BIMS EF_{sea} values are <10% of the EF_{sea} values, and for these elements, the sea surface is a trivial source even at the remote site (Enewetak). For V, Cu, and Zn, however, the BIMS EF_{sea} values are >20% of the EF_{sea} values for the Enewetak population, indicating that at some remote sites, the sea surface becomes a non-trivial source for these elements. However, in order to fully assess the importance of the sea surface as a source of metals to the marine atmosphere, the recycling of previously deposited aerosols and gaseous emissions involved in subsequent gas-to-particle conversions must also be considered.

There are a number of sources which can supply AEE to the marine atmosphere and these include those involving high temperature processes. However, these high temperature processes do not, in general, have readily identifiable indicator elements and it has been common practice in the past to assume that elemental concentrations in excess of those assigned to crustal and sea-salt sources have a high temperature, mainly pollutant source.

In addition to enrichment factors, a number of other parameters can be used to identify the sources of metals in aerosols. One of the most widely used of these is particle size fractionation, and it is now known that elements from the main primary sources are associated with specific particle size ranges (see e.g., Duce et al.[9,10]). This particle size/source relationship can be summarized as follows. (1) Elements emitted during anthropogenic (and volcanic) high temperature processes are found mainly on particles with mass median diameters <~0.5 μm. (2) Crust-associated elements (crustal source) occur chiefly on particles in the size range ~1 to ~3 μm. (3) Sea salt-associated elements (oceanic source) are associated mainly with particles in the size range ~3 to ~7 μm.

III. GEOGRAPHICAL VARIATIONS IN THE ELEMENTAL COMPOSITION OF THE MARINE AEROSOL

It is apparent from the discussion in the previous section that a number of sources contribute material to the marine aerosol. The contribution made by an individual source to a specific aerosol will vary as a function of the source strength. In turn, the source strength will vary, among other factors, in relation to the remoteness of the aerosol from the source location. As a first step to understanding metal concentration variations in the marine atmosphere, the variations can be assessed in geographical terms, i.e., on the basis of "remoteness" from the major sources.

A considerable amount of data is now available on the concentrations of elements in marine aerosols from a wide variety of environments. Chester[4] made a compilation of some of these data in which the marine locations were arranged in an increasing order of "remoteness from the primary sources"; this data set is reproduced in Table 5. Chester[4] summarized the principal atmospheric concentration trends which emerge from this data set as follows.

1. The highest concentrations of the AEE are found over coastal seas which are relatively close to the continental anthropogenic sources.
2. The highest concentrations of the NEE are found over regions close to the continental arid land and desert sources.
3. The concentrations of both the AEE and the NEE decrease with "remoteness" from the continental sources, and for those regions for which sufficient data are available, the general rank order is coastal seas > North Atlantic > North Pacific > South Pacific.
4. The EF_{crust} values are highest in aerosols from the coastal seas, but those for the Samoa aerosol are higher than those for the less "remote" Enewetak aerosol, which suggests that the residence times of the small-sized AEE-containing anthropogenic particles are longer than those of the larger NEE-containing mineral particles.

It can be seen from these data that a "veil" of trace metals is present over the whole ocean, although the concentrations of metals in the "veil" vary considerably from one location to another. The metals will be deposited onto the sea surface from this "veil", and the magnitudes of the fluxes involved will reflect the ambient concentrations and so will also vary from one marine location to another. However, before attempting to assess the magnitudes of these fluxes, it is necessary to understand the nature of the processes which control the deposition of the aerosols to the sea surface.

A. THE DEPOSITION OF AEROSOLS TO THE SEA SURFACE

The deposition of aerosols from the atmosphere is controlled by a combination of dry fall-out and precipitation scavenging. Dry fall-out is a continuous process which is affected by factors, such as wind speed and particle size. Precipitation scavenging, or wet deposition, is a noncontinuous or random process which involves the removal of material from the atmosphere by rainfall, and the magnitude of the flux involved will be dependent on the concentration of a metal in rain water and the amount of rain that falls. Data are now becoming available for the trace metal composition of rain at a variety of marine locations and a selection of these were summarized by Chester;[4] the data set is reproduced in Table 6. From this table, it can be seen that the concentrations of the AEE in rain water are highest in samples taken close to the major continental pollution sources. Wet and dry deposition processes combine to remove material from the atmosphere, and, in general, the sea surface deposition of the AEE, which tend to be concentrated on small particles, is dominated by wet deposition, whereas that of the larger sized sea salt and mineral aerosol particles is dominated by dry deposition.

IV. ATMOSPHERIC FLUXES TO THE SEA SURFACE

The combination of wet and dry deposition delivers atmospheric metals and other components to the sea surface in the form of a gross atmospheric deposition flux. However, once they have been deposited, aerosol components can be recycled back across the air/sea interface during the generation of sea salts. Information is now becoming available on the extent to which this recycling affects individual trace metals at some locations. For example,

TABLE 5
Concentrations (1) and EF Values (2) of Trace Metals in Marine Aerosols

Concentration units (per m³ of air)	Coastal seas, close to pollution sources						Coastal seas and open-ocean, close to crust sources					
	North[11a] Sea		Keil[12] Bight		Western[13] Med.		Eastern[14] Med.		N. Atlantic[6] (northeast trades)		N. Arabian[15] Sea	
	1	2	1	2	1	2	1	2	1	2	1	2
Al (ng)	159	1.0	394	1.0	324	1.0	915	1.0	5925	1.0	1184	1.0
Fe (ng)	330	3.0	369	1.4	—	—	570	0.9	3865	1.0	908	1.1
Sc (pg)	82	1.9	—	—	—	—	—	—	—	—	—	—
Mn (ng)	21	11	15	3.3	—	—	12	1.1	65	1.0	15	1.1
Ni (ng)	10	69	4	11	—	—	—	—	6.6	1.2	3.9	3.6
Co (pg)	360	7	—	—	—	—	—	—	2100	1.2	670	1.9
Cr (ng)	4.5	23	—	—	—	—	1.6	1.4	10	1.4	4.3	3.0
V (ng)	12	46	9.7	15	—	—	—	—	15	1.5	—	—
Cu (ng)	—	—	7.7	29	—	—	4.9	8.0	4.5	1.1	3.2	4.0
Zn (ng)	153	1132	57	170	—	—	—	—	16	3.2	4.3	4.3
Cd (ng)	—	—	—	—	1.9	2413	—	—	0.12	8.3	0.05	17
Se (ng)	1.5	15466	1.6	6657	—	—	—	—	—	—	—	—
Sb (pg)	2900	7600	—	—	—	—	—	—	—	—	—	—
Pb (ng)	147	6082	53	885	40	812	7.4	53	6.9	7.7	2.9	16

Concentration units (per m³ at air)	Open-ocean											
	Tropical[16] N. Atlantic		N. Atlantic[17] (Bermuda)		N. Atlantic[6] Westerlies		S. Atlantic[6] Westerlies		N. Pacific[10] (Enewetak)		S. Pacific[18] (Samoa)	
	1	2	1	2	1	2	1	2	1	2	1	2
Al (ng)	160	1.0	140	1.0	48	1.0	2.7	1.0	21	1.0	0.72	1.0
Fe (ng)	100	0.9	94	1.0	36	1.1	2.6	1.4	17	1.2	0.21	0.4
Sc (pg)	35	0.8	20	0.5	—	—	—	—	5	0.9	0.04	0.2
Mn (ng)	2.2	1.2	1.2	0.7	1.7	3.1	0.11	3.5	0.29	1.2	0.005	0.6
Ni (ng)	0.64	4.4	0.08	0.6	0.51	12	0.02	8.1	—	—	—	—

Co (pg)	80	1.6	—	—	50	3.4	10	12	8	1.3	0.37	1.7
Cr (ng)	0.43	2.2	0.28	1.6	0.49	8.4	0.17	52	0.09	3.5	—	—
V (ng)	0.54	2.1	—	—	1.0	13	0.03	6.8	0.08	2.3	—	—
Cu (ng)	0.79	7.4	0.90	10	1.0	31	0.29	161	0.045	3.2	0.013	27
Zn (ng)	4.4	32	3 2	27	7.5	184	1.8	784	0.17	9.5	0.07	114
Cd (ng)	—	—	0.19	558	0.11	943	0.02	3048	0.004	78	—	—
Se (ng)	0.43	4405	0.13	1522	—	—	—	—	0.13	10148	0.09	204918
Sb (pg)	110	286	30	89	—	—	—	—	4.0	79	0.20	116
Pb (ng)	9.9	407	3 5	164	6.0	882	0.97	2364	0.12	38	0.016	146

[a] Numbers refer to reference source.

TABLE 6
Concentrations of Elements in Precipitation Over Marine Regions (Units, μg/l)

	North Sea[11a]	Bay of Bengal[19]	Bermuda[20]	N. Pacific[21] (Enewetak)	S. Pacific[18] (Samoa)
Na	19,000	14,000	3,400	1,100	2,500
Mg	—	1,600	490	170	270
K	—	1,400	160	39	88
Ca	—	1,700	310	50	75
Cl	32,000	25,000	6,800	2,000	4,700
Br	—	—	—	7.1	12
Al	105	—	—	2.1	16
Fe	84	30	4.8	1.0	0.42
Mn	<12	—	0.27	0.012	0.020
Sc	0.010	0.016	—	0.00023	—
Th	—	0.12	—	0.00091	—
Co	0.17	0.95	—	—	—
I	—	—	—	1.2	0.021
V	3.7	—	—	0.018	—
Zn	35	100	1.15	0.052	1.6
Cd	—	—	0.06	0.0021	—
Cu	15	—	0.66	0.013	0.021
Pb	—	—	0.77	0.035	0.014
Ag	—	—	—	0.0056	—
Se	—	—	—	0.021	0.026

[a] Numbers refer to Reference source.

Arimoto et al.[21] estimated that at Enewetak (North Pacific), the percentage of wet deposition associated with recycled sea salts can be substantial; values for individual elements included 15% (Zn), 30% (V), and 48% (Cu). It is apparent, therefore, that recycling must be taken into account when net atmospheric fluxes to the sea surface are determined, and recently a number of models have been designed for this purpose (see e.g., Arimoto et al.[18,21]). Compared to these, most of the earlier models for the atmospheric deposition of trace metals to the sea surface were rather crude (see, e.g., Bruland et al.,[22] Duce et al.,[17] Chester et al.,[23] Walsh et al.,[24] Buat-Menard and Chesselet,[16] and Chester[25]), and although they do not generally include an attempt to assess the extent of surface recycling, they do constitute the bulk of the data which are available in the literature for atmospheric metal fluxes to the oceans. A summary of the data on the atmospheric fluxes of trace metals to the sea surface for a variety of oceanic regions is given in Table 7; however, it must be pointed out that these fluxes have been derived using a variety of deposition models and that, with the exception of those for Enewetak (North Pacific) and Samoa (South Pacific), they have generally not been corrected for sea surface recycling. Nonetheless, Chester[4] pointed out that a strong geographical trend can be identified from the data in Table 7, indicating that the strengths of the air-to-sea fluxes decrease by orders of magnitude as the ''degree of remoteness'' of a site from the major aerosol sources increases.

V. THE FATE OF ATMOSPHERICALLY TRANSPORTED TRACE METALS IN THE OCEANS

In the preceding section, a number of ''best estimates'' were given for the fluxes of trace metals from the atmosphere to the ocean surface. These atmospherically transported trace metals have a different genetic history from those brought to the oceans via river run-

TABLE 7
Fluxes of Trace Metals to the Sea Surface (units, ng/cm² y)

	New York Bight[9a]	North Sea[11]	Western Med.[26]	South Atlantic Bight[27]	Bermuda[17]	North Atlantic northeast trades[28]	Tropical North Atlantic[16]	Tropical North Pacific; total net deposition[21]	South Pacific; total net deposition[18]	North Atlantic; westerlies[29]	North Pacific; westerlies[29]	South Pacific; westerlies[29]
Al	6,000	30,000	5,000	2,900	3,900	97,000	5,000	1,200	132—1,800	—	—	—
Sc	—	5	1	—	0.6	—	1.1	0.18	0.06	—	—	—
V	—	480	—	—	5	—	17	7.8	—	—	—	—
Cr	—	210	49	—	9	111	14	—	—	—	—	—
Mn	—	920	—	60	45	570	70	9.0	3.6	—	—	—
Fe	5,700	25,500	5,100	5,900	3,000	48,000	3,200	560	47—337	—	—	—
Co	—	39	3.5	—	1.2	12	2.7	—	0.25	—	—	—
Ni	—	260	—	390	3	67	20	—	—	—	—	—
Cu	—	1,300	96	220	30	48	25	8.9	4.4—7.9	—	—	—
Zn	1,400	8,950	1,080	750	75	152	130	67	2.4—5.8	—	—	—
As	—	280	54	45	3	—	—	—	—	—	—	—
Se	—	22	48	—	3	—	14	4.2	0.8	—	—	—
Ag	—	—	3	—	—	—	0.9	—	—	—	—	—
Cd	30	43	13	9	4.5	—	5	0.35	—	—	—	—
Sb	—	58	48	—	1.0	—	3.5	—	—	—	—	—
Au	—	—	0.05	—	—	—	0.1	—	—	—	—	—
Hg	—	—	5	24	—	—	2.1	—	—	—	—	—
Pb	3.900	2,650	1,050	660	100	32	310	7.0	1.4—2.8	170	50	3
Th	—	4	1.2	—	—	—	0.9	0.61	0.036	—	—	—

[a] Number refers to reference source.

off because they reach the sea surface directly without passing through the estuarine filter. The question which must therefore be addressed is what is the fate of these atmospherically transported trace metals once they reach the air/sea interface, either by a one-step process or following one or more recycling stages?

Initially, the first constraint on the behavior of the atmospherically transported trace metals will be the extent to which they are solubilized in seawater. This is important since the physical state of the metals, i.e., whether they are in the particulate or the dissolved phase, affects both their subsequent involvement in the biogeochemical cycles and their residence times in seawater; for example, Walsh and Duce[30] pointed out that the residence times of dissolved substances are generally much longer than those of particulate material. If a trace metal is solubilized from atmospherically deposited material, it will then be in a form which can be taken out of solution either (1) by processes associated with the biomass in which case it will follow *nutrient-type* profiles (i.e., surface depletion/depth enrichment), or (2) by uptake onto inorganic particles in which it will follow *scavenging-type* profiles (i.e., surface enrichment/depth depletion). The elements which are associated with both types of removal mechanisms then undergo down-column transport to eventually become incorporated into the sediment reservoir, often via several recycling processes. However, it is apparent that the critical processes which control the subsequent fates of metals deposited to the sea surface from the atmosphere occur at, or close to, the air/sea interface where the aerosols cross the microlayer.

The particulate-dissolved equilibrium reactions which affect the atmospherically deposited metals are, therefore, central to our understanding of the subsequent fates of the metals in the water column, and these reactions are dependent on the manner in which the metals are "packaged" in the aerosols, i.e., on their solid-state speciation, and on the chemical nature of the aqueous phase, i.e., seawater and rain water, with which they come into contact.

A number of studies in the literature have reported data on the solubility of metals from atmospherically deposited material. However, there are very considerable problems involved in such studies. These involve a number of parameters and can be summarized as follows:

A. THE SEAWATER MEDIUM

Seawater is a very complex medium which contains a variety of dissolved and particulate phases, many of which can potentially affect the extent to which elements are solubilized from aerosols. The three most important parameters which must be considered in relation to the seawater used in the aerosol solubility experiments are

1. The concentrations of suspended solids in the seawater
2. The concentrations of dissolved organics in the seawater
3. The time period over which the aerosols are exposed to the seawater

The amount of suspended solids is important because the particulates can rapidly "scavenge" elements released from the aerosols, and so decrease the "real" solubility values. One way of overcoming this problem is to use filtered seawater samples in the aerosol solubility experiments, although it can be argued that this does not reproduce ambient conditions. Further, some aerosol particles can themselves act as scavengers so that the amount, and kind, of aerosol material used in laboratory experiments can affect the apparent seawater solubility of an element from an aerosol. Dissolved organic matter is important in relation to the solubility of aerosols for two principal reasons. (1) It can affect the speciation of dissolved elements by forming complexes, which may affect their solubilities by, for example, competing with any ambient particulate matter thus keeping the elements in a "soluble" form. (2) It can coat the aerosols with an electronegative coating which may ''trap'' elements and make them less soluble. Problems arising from the presence of dissolved organic matter in the seawaters used for the solubility experiments can be overcome by irradiating the samples with ultraviolet light to destroy the dissolved organics, although again this does not reflect ambient conditions. The time over which the aerosol is exposed to seawater is important for assessing the relative solubilities of different elements in an absolute sense; for example, some elements may take a longer time to reach their optimum solubility than others, and this may cover periods of days or perhaps even longer. The experimental leaching time is also important from the point of view of relating the experimental solubility data to actual environmental oceanic conditions, e.g., the time a particle deposited at the sea surface actually spends in the upper water column. In this context, it has been estimated that the residence time of a particle in the upper ~100 m of the water column is ~100 h.[31]

B. THE DEPOSITION MODE

Aerosols are removed from the atmosphere by wet and dry processes (see above). Wet deposition involves precipitation scavenging, during which the aerosols are in contact with rain water which can have a pH as low as ~3 to ~4 over coastal regions, and ~5.5 over remote marine areas. Even over these remote areas, the pH of the rain is considerably lower than that of surface seawater (~7.8 to ~8.2), and this has the potential of increasing the solubilities of some elements from aerosols (see, e.g., Maring and Duce[31]). In order to take account of the lower pH in rain waters it is therefore necessary to simulate the rain itself in the aerosol solubility experiments, e.g., using artificial rain at selected pH values, if the solubility of elements from rain precipitated aerosols in seawater is to be evaluated. Thus, the direct exposure of aerosol samples to seawater can only be used to simulate the dry deposition mode to the sea surface.

C. THE AEROSOL COMPOSITION

A number of authors have investigated the partitioning of trace metals among the various components of aerosols (see, e.g., Lum et al.[32], Chester et al.[33]). The results of investigations such as these indicate that the partitioning, or solid-state speciation of the AEE is very different in pollutant-dominated and crust-dominated aerosols. The principal difference lies in the manner in which the AEE are partitioned between loosely held (surface-associated) and refractory (lattice-associated) host fractions; a much higher percentage of the total content of metals, such as Cu, Zn, and Pb being present in surface-associated hosts in the pollutant-dominated than in the crust-dominated aerosols; in contrast, Cd was mainly held in surface-associated hosts in both pollutant- and crust-dominated aerosols. Partitioning differences such as these are important because the surface-associated metals are potentially far more soluble than the lattice-associated metals. As a result, the solubility of metals from aerosols will vary as a function of the extent to which components of pollutant and crustal origins are mixed together in an individual aerosol population.

All the factors discussed above constrain aerosol/seawater reactivity and must be taken into consideration when assessing the value of the aerosol solubility experiments reported in the literature. A number of these experiments are described below. They have been selected in order to illustrate the manner in which the constraining factors operate, and an attempt is made to relate the results obtained to ''real'' environmental situations.

Three investigations will serve to illustrate how the partitioning of an element affects its solubilization from an aerosol. In one of the first of the solubility studies, Walsh and Duce[30] simulated the ''dry'' deposition of filter-collected aerosols to the sea surface. They reported that a major fraction of the anthropogenic (noncrustal) V was released from suburban coastal aerosols within <2 min of the exposure of the filter to seawater, and that during this process, the V EF_{crust} values of the particulates fell from >25 to ~10. This early work suggested, therefore, that the manner in which an element is partitioned between the crustal and noncrustal components of an aerosol exerts a major control on the extent to which it is solubilized in seawater.

Hodge et al.[34] simulated the ''dry'' deposition of filter-collected aerosols to seawater using an experimental leaching time of 3 h and produced the first extensive data set for the solubility of atmospherically transported elements in seawater. The authors investigated the solubilities of a series of elements in two contrasting aerosol populations; an anthropogenic-rich population from southern California and a continental dust-rich population from Baja, CA. A summary of their solubility data is given in Table 8 from which it can be seen that there are considerable differences in the solubilities of a number of the elements between the two populations. For example, Pb, Zn, and Cu were solubilized to a much greater extent from the polluted aerosols in which they had relatively high EF_{crust} values than from the dust-rich samples in which the EFs were lower. These trends can be related in a general way to the partitioning of the elements among the aerosol components, the greater release of the AEE Pb, Zn, and Cu from the polluted aerosols being entirely consistent with their solubilization from 'mobile' surface-associated positions on particle surfaces and from specific compounds with which they become associated during high temperature volatilization processes. In contrast to Pb, Zn, and Cu, cadmium is found mainly in loosely held forms in both the pollutant and the crustal ''end-members'' and so is relatively soluble from both types of aerosol on contact with seawater.

Chester et al.[35] gave data on the dry deposition mode solubilities of a series of NEE and AEE from filter-collected marine aerosols after a 1-h leach in filtered seawater. The study involved three types of marine aerosol, an urban ''end-member'', a crust-dominated ''end-member'', and a series of ''mixed'' samples. The findings can be summarized as follows. (1) The NEE Al and Fe were <1% seawater soluble from the crust-dominated ''end-member'' and <7% soluble from both the urban ''end-member'' and the ''mixed'' samples. (2) The AEE Cu, Zn, and Pb were all <1% seawater soluble from the crust-dominated ''end-member'', but their solubilities from the urban ''end-member'' increased dramatically to reach values of ~33% for Cu, ~70% for Zn, and ~50% for Pb. The seawater solubilities of both Zn and Pb for the mixed samples were intermediate between those of the two ''end-members'' and were related to the manner in which the elements were partitioned among the aerosol components. (3) The NEE Mn was significantly seawater soluble from all the aerosols with values ranging from ~17% for the crustal 'end-member' to ~30% for the urban ''end-member.''

The effects of varying the seawater leaching time, the pH of the leaching solution, and the concentrations of organic matter present can be illustrated with reference to the study reported by Maring and Duce.[31] These authors reported data which showed that for the dry deposition mode, the seawater leaching time affected the solubility of Al from North Pacific aerosols. In fact, the element was shown to have a bimodal solubility with ~5 to ~6% of the total Al being solubilized in ≤6 h and an additional ~3 to ~4% being released within ~60 h. The authors suggested that this bimodal solubility was related to two forms of Al in the aerosols, the ''more soluble'' fraction represented Al which had already been weathered in the soil profile, and the ''less soluble'' fraction represented Al removed from the alumino-silicate matrix within seawater. The study also showed that dissolved organic matter in seawater apparently has little effect on the Al solubility. However, in a later publication, the same authors[36] demonstrated that the solubilization of Cu from aerosols increased in the presence of dissolved organic matter presumably as a result of the Cu being complexed more strongly by the organic matter than by the aerosols. Simulation of the wet deposition mode, using artificial rain water at pH 5.5, revealed that the rate, and perhaps the quantity, of solubilized Al increased relative to that in seawater at pH 8.

A variety of seawater leaching times, and also particulate/water ratios, have been used in the aerosol seawater solubility studies reported in the literature. One feature to emerge

TABLE 8
The Solubility of Atmospheric Trace Metals in Seawater[4,34]

Atmospheric Population	Al	Fe	Mn	Ni	Co	Cr	V	Ag	Zn	Cu	Pb	Cd
Anthropogenic-rich[a]												
Mean concentration (ng/m^3 of air)	900	610	11	11	<0.4	<1.6	13	<0.05	25	16	560	0.25
Mean % soluble in seawater	0.56	1.1	47	47	25	12.5	31	80	68	28	39	84
Mean EF_{crust}	1.0	0.99	1.0	14	1.5	1.5	9.0	—	33	26	4148	16
Dust-rich[b]												
Mean concentration (ng/m^3 of air)	3380	2100	46	2.8	<0.5	4.0	<11	<0.03	18	20	150	0.20
Mean % soluble in seawater	0.09	0.19	34	28	20	10	18	33	24	14.5	13	80
Mean EF_{crust}	1.0	0.91	1.2	0.92	0.49	0.99	2.0	—	6.2	8.4	296	25

[a] Southern California.
[b] Baja, California.

TABLE 9
The Seawater Solubilities of Elements From Aerosols Over Marine Regions (% Total Element)

Element	Coastal regions; aerosol dominated by anthropogenic sources	Coastal regions; aerosol dominated by crustal sources	Open-ocean "mixed-source" aerosols
Al	5	~<5	5
Fe	7.5	~<7.5	7.5
Mn	45	<20	35
Ni	50	<25	40
Co	25	<20	22.5
Cr	12.5	<10	10
V	30	<20	25
Cu	30	<5	30
Pb	50	<5	30
Zn	70	<5	45
Cd	85	—	80

Note: Data from various sources quoted in text; values are "rounded off" first order estimates only.

from these solubility studies is that some elements can be released rapidly from the aerosols on contact with seawater. Others may be released at different times if they are present in different forms. However, there are difficulties in attempting to model long-time seawater solubilities, e.g., to approach the ~100-h particle residence time in the upper oceanic layer in laboratory experiments which are designed to be applied to the "real" oceanic environment. For example, the aerosol particles may be coated with organics; they may be taken into the gut of filter-feeding organisms where the pH can be very much lower than in seawater, or they may be aggregated with other particles. As a consequence, it is unlikely that atmospherically deposited particles will retain their original individual identities in the upper water column following their deposition to the sea surface. It is necessary to stress, therefore, that aerosol seawater solubilities obtained on the basis of laboratory experiments must only be interpreted within strictly identified "operationally defined" limits.

By taking all the available data into consideration, very crude estimates can be made of the solubility of elements from aerosols deposited to the sea surface and these are listed in Table 9. From the rather sparse data available in the literature, it would appear that the solubility of an element from a marine aerosol is dependent to a large extent on its source. Crustal and sea-salt elements will retain their solubility characteristics in all aerosol populations. However, the solubilities of the AEE will vary as the proportions of crustal and pollutant components vary and in order to relate this to their general distributions in the marine atmosphere, the solubility data in Table 9 have been listed in terms of three characteristically different oceanic environments. Thus, despite the various constraints which it is necessary to impose on aerosol seawater solubility studies, a number of overall trends can be identified from the solubility data in this table.

1. In coastal regions where anthropogenic inputs are highest, significant fractions of the AEE are soluble in seawater.
2. For those coastal and open-ocean regions in which desert-derived crustal dusts dominate the aerosol population, and in which EF_{crust} values for all elements (i.e., including the AEE) are relatively low, considerably smaller amounts of the relatively soluble elements will enter the dissolved phase on contact with seawater. In these areas, however, the deposition fluxes of the atmospheric elements can be relatively very high, so that the soluble fractions may become significant, especially in those regions which receive dust "pulses".

3. Over open-ocean regions, the solubilities of elements from aerosols will depend on the extent to which components from various sources, e.g., anthropogenic and crustal, are mixed together in the total particle population.

The estimates for the solubilities of elements from these "mixed" aerosols given in Table 9 represent an attempt to assess average values from the published data.

It must be stressed, however, that the aerosol solubility data given in Table 9 should be regarded as no more than first order approximations.

The extent to which an element is solubilized from an aerosol into seawater will control its initial fate in the oceanic mixed layer. However, as Moore et al.[37] pointed out, the determination of seawater solubility alone will not necessarily give a good indication of how the element actually enters the biological systems because particles from which elements have not been directly solubilized can enter the gut system of organisms where they will encounter solutions having a considerably lower pH than seawater. Thus, the "biological solubility" of some elements will be increased relative to their "seawater solubility". However, the seawater solubility data do provide a very general estimate of that fraction of atmospheric deposition which will enter the oceanic pool of dissolved elements and, therefore, allows a comparison to be made between the magnitude of the dissolved loads supplied to seawater from the various primary sources. In this manner, the atmospheric deposition of trace metals can be placed into a global oceanic context.

VI. RELATIVE MAGNITUDES OF THE PRIMARY FLUXES TO THE OCEANS

The principal primary sources of trace metals to the oceans are river run-off, atmospheric deposition, and hydrothermal exhalations. Of the fluxes associated with these three sources, those from river run-off and atmospheric deposition deliver material directly to the surface ocean where they enter the mixed layer and take part in the biogeochemical cycles which control the distributions of trace metals in seawater. It is of interest, therefore, to assess the strength of the atmospheric flux relative to that carried to the oceans by rivers.

A number of authors have compared the input of fluvial and atmospheric trace metals in coastal regions, and the atmospheric/fluvial flux ratios for a number of these are given in Table 10A. These ratios show that under some conditions, the supply of most of the metals from atmospheric deposition is at least of the same order as that from river run-off and that for some metals, e.g., Pb, it is considerably greater. The atmospheric fluxes are for the total aerosol population, but those for the fluvial flux are for the dissolved phase only and to make the flux comparison more viable, Chester[4] recalculated the atmospheric fluxes for the South Atlantic Bight on a seawater-soluble basis in order to give an estimate of the dissolved atmospheric trace metal fluxes. These data are given in Table 10B, and it is apparent that even after making this correction, the atmospheric fluxes of Pb, Zn, and Cd exceed their fluvial fluxes.

Clearly, atmospheric deposition can be a significant, and sometimes dominant, source of dissolved trace metals to some coastal zones. Much less information is available, however, on the relative magnitudes of the atmospheric and fluvial dissolved trace metal fluxes to the ocean surface on a global scale. For this reason, an attempt has been made here to calculate "first order" atmospheric and fluvial dissolved trace metal fluxes to the World Ocean. To make these estimates, the following data were used. (1) For the atmospheric inputs, it was assumed that the deposition fluxes given in the literature for the tropical North Atlantic, and North Pacific, and the South Pacific (Table 7) offered a representative cross-section of those affecting all marine regions; however, those for the North and South Pacific were adjusted for regional areas on the basis of the details given in Table 11. A weighted average atmos-

TABLE 10
Relative Fluvial and Atmospheric Fluxes to Some Coastal Oceanic Regions

A. Ratio of Total Atmospheric Flux to Fluvial Flux

Element	South Atlantic Bight[a]	New York Bight[a]	North Sea[a]	Western Mediterranean[b]
Fe	5.8	6.4	1.7	—
Mn	0.6	—	0.8	—
Cu	1.9	—	1.9	—
Ni	1.7	—	1.3	—
Pb	9.5	20	6.8	6.2
Zn	2.3	3.1	1.9	0.8
Cd	2.7	3.1	1.1	—
As	2.1	1.0	1.7	—
Hg	22	—	2.1	0.8

B. Estimated Fluvial and "Soluble" Atmospheric Fluxes to the South Atlantic Bight (Units, g × 10[6])

Element	Fluvial flux[a]	Total atmospheric flux[a]	Soluble atmospheric flux[c]	Ratio: soluble atmospheric flux to fluvial flux
Fe	950	5500	413	0.43
Mn	91	57	20	0.29
Cu	110	210	63	0.57
Ni	220	370	148	0.85
Pb	65	620	186	4.8
Zn	310	710	320	1.6
Cd	3	8	6.4	2.3

[a] Data from Reference 27.
[b] Data from Reference 38.
[c] Solubility data from Table 9.

pheric deposition was then obtained from the data in Table 11, and this was then extrapolated to the whole ocean surface; these atmospheric fluxes are given in Table 12. (2) To calculate the fluvial inputs, the net global river fluxes given by Chester[4] were used and it was assumed that the total global river run-off was spread in an even layer over the World Ocean; the net fluxes obtained in this way are listed in Table 12. Collier and Edmond[39] also made a general comparison between the fluvial and the atmospheric inputs of a number of metals to the ocean surface, and some of their data are also listed in Table 12.

From this "first order" comparison, it can be seen that for Al, Mn, and Cu, the fluvial dissolved fluxes to the global ocean surface exceed those from the atmosphere by an order of magnitude. For Zn and Pb, however, the atmospheric dissolved fluxes dominate the input to the surface ocean. The fluxes given in Table 12 are for the global ocean. However, the concentrations of elements and, therefore, their deposition fluxes to the sea surface vary both temporally and spatially in the marine atmosphere. These variations can range over several orders of magnitude, and in an attempt to take account of the variations, atmospheric deposition fluxes have been calculated on a regional basis. Fluvial fluxes have also been estimated in this way and the data are listed in Table 13. From this table, it can be seen that atmospheric deposition has the most influence on the surface waters of the North Atlantic and the least on those of the South Pacific. In terms of open-ocean areas, therefore, it is

TABLE 11
Atmospheric Deposition Fluxes to the Sea Surface Over a Number of Marine Regions (μg/cm²/year)

Element	North Atlantic[a]	North Pacific[b,d]	South Pacific[c,d]
Al	5.0	1.8	0.13
Fe	3.2	0.82	0.047
Mn	0.07	0.0113	0.0036
Ni	0.02	—	—
Co	0.0027	0.00016	0.000025
Cr	0.014	—	—
V	0.017	0.0069	—
Cu	0.025	0.0071	0.0044
Pb	0.31	0.0074	0.0014
Zn	0.13	0.065	0.0058
Cd	0.002	0.00029	—

[a] Data from Buat-Menard and Chesselet[16] for the tropical North Atlantic. Deposition fluxes for all elements calculated on the basis of a global total deposition (wet and dry) rate of 1 cm/s.

[b] Data from Arimoto et al.[21] for Enewetak, North Pacific.

[c] Data from Arimoto et al.[18] for Samoa, South Pacific.

[d] Wet deposition fluxes given by Arimoto et al.[18,21] were calculated using two methods: (1) those based on the concentration of elements in rain and the total rainfall at the two islands, and (2) those based on scavenging ratios. In order to avoid any bias which may be introduced by unrepresentative total rainfall over the islands compared to the open-ocean regions, wet deposition rates for both Enewetak and Samoa have been estimated from scavenging ratios only; both wet and dry deposition rates have been adjusted to take account of sea surface recycling.

the North Atlantic which receives the highest input of trace metals from the atmosphere. Under these ''extreme'' conditions, the atmospheric dissolved fluxes for Mn, Ni, Co, Cr, V, and Cu are all an order of magnitude less than their fluvial dissolved fluxes; for Al, Fe, and the Cd, the two fluxes are the same order of magnitude, but for Zn and Pb, the atmospheric fluxes are dominant.

The fluxes given in Tables 12 and 13 are very crude and although they show the dominance of the atmospheric pathway for the deposition of some metals, e.g., Pb, to open-ocean surface waters, they may considerably underestimate the importance of atmospheric deposition for other elements. For example, much of the Al which escapes the estuarine environment, i.e., the net fluvial flux in Table 12, may in fact be removed into coastal sediments with the result that in open-ocean waters, the atmospheric source becomes more important than is indicated by the crude flux comparison given above (see, e.g., Orians and Bruland[41]). Further, the near-shore inputs of some elements, such as Cu, can be prevented from reaching the open-ocean by coastal nutrient traps (see, e.g., Boyle et al.[42]). Such a trap operates off the coast of California with the result that in open-ocean waters of the North Pacific, atmospheric deposition is a more important source of dissolved Cu to the surface waters than river run-off; in this region, however, the supply of dissolved Cu to the surface waters from upwelling has a similar magnitude to that from atmospheric deposition (see, e.g., Bruland[43]). Another factor which must be considered is the solubilization of elements from crustal dust which reaches the ocean surface in the form of intermittent ''pulses''. These ''pulses'' deliver large quantities of crustal dust to the sea surface over short periods and can lead to the setting up of non-steady-state conditions. As a result of this, solubilization from crustal dusts can dominate the supply of some elements to surface

TABLE 12
Net Fluvial and Atmospheric Fluxes to the Global Sea Surface (μg/cm²/year)

	Net global fluvial dissolved flux		Net global total atmospheric input			
Element	Present work[a]	Collier and Edmund[39]	Present work[b]	Collier and Edmund[39]	Estimated average seawater solubility from aerosols[c] (%)	Net global dissolved atmospheric flux[d]
Al	0.27	0.13—0.67	1.85	0.27—5.4	5	0.088
Fe	0.085	<0.73	1.0	1.12—3.35	7.5	0.075
Mn	0.085	0.093	0.021	0.016—0.044	35	0.0074
Ni	0.0082	0.0035	<0.02	0.003	40	<0.008
Co	0.0021	—	0.00068	—	22.5	0.00015
Cr	0.0052	—	<0.014	—	10	<0.0014
V	0.010	—	<0.010	—	25	<0.0028
Cu	0.018	0.019	0.010	0.00064—0.0095	30	0.0033
Pb	0.001	—	0.074	—	30	0.022
Zn	0.0079	<0.0065	0.055	0.022—0.013	45	0.025
Cd	0.00088	0.00022	<0.00096	0.0023	80	<0.00077

[a] Calculated from the net fluvial flux data given in Chester[4] assuming a global ocean area of 352.6 × 10⁶ km² (Baumgartner and Reichel[40]); at a global river discharge of 37,400 km³/year, which is included in the net flux estimates, this would yield a layer 10.6 cm spread evenly over the whole ocean surface.

[b] Calculated from the tropical N. Atlantic (column 1), N. Pacific (column 2), and S. Pacific (column 3) atmospheric deposition fluxes given in Table 11 by scaling them to global fluxes weighted on an areal basis for each oceanic region.

[c] See text.

[d] Net total atmospheric flux adjusted for the average seawater solubility of the individual elements.

TABLE 13
Atmospheric Dissolved vs. Fluvial Dissolved Inputs to Regions of the World Ocean (μg/cm²/year)

	Oceanic Region					
	North Atlantic		North Pacific		South Pacific	
Element	Fluvial dissolved flux[a]	Atmospheric dissolved flux	Fluvial dissolved flux	Atmospheric dissolved flux	Fluvial dissolved flux	Atmospheric dissolved flux
Al	0.56	0.25	0.24	0.09	0.12	0.0066
Fe	0.18	0.24	0.076	0.062	0.038	0.0035
Mn	0.18	0.025	0.076	0.004	0.038	0.0013
Ni	0.017	0.008	0.0074	—	0.0037	—
Co	0.0044	0.00061	0.0019	0.00004	0.0009	0.000006
Cr	0.011	0.0014	0.0047	—	0.0023	—
V	0.022	0.0043	0.0094	0.0017	0.0047	—
Cu	0.037	0.0075	0.016	0.0021	0.0079	0.0013
Pb	0.0022	0.093	0.0009	0.0022	0.0005	0.0004
Zn	0.016	0.059	0.0071	0.029	0.0035	0.0026
Cd	0.0018	0.0016	0.0008	0.00023	0.0004	—

[a] No attempt has been made to adjust the North Atlantic fluvial flux for European and North American anthropogenic inputs.

seawater. For example, Kremling[44] showed that there was an increase in "total dissolvable Mn" in North Atlantic surface waters underlying the north-east trades and attributed this to solubilization from Saharan dust.

VII. CONCLUSIONS

1. There is a "veil" of aerosol-associated trace metals present in the atmosphere over the whole of the World Ocean.
2. The concentrations of trace metals in this aerosol "veil" vary from one region to another by several orders of magnitude.
3. On the basis of relationship to crustal sources, the metals in marine aerosols can be divided into nonenriched (NEE), i.e., crust-dominated, and anomalously enriched (AEE), i.e., non-crust-dominated types.
4. The aerosols are delivered to the sea surface by a combination of wet and dry deposition processes and the fluxes involved vary in magnitude in the general order North Atlantic > North Pacific > South Pacific.
5. The initial fate of an atmospherically deposited trace metal in seawater depends on the extent to which it can be solubilized from the solid phase. This, in turn, depends on the manner in which the metal is partitioned between the various aerosol phases which is a function of the source of the metal. As a result, metal solubilities vary between aerosols of different origin, the solubilities of the AEE being greatest from aerosols which have a large pollutant fraction, e.g., those deposited to coastal surface waters.
6. The dissolved concentrations of some dissolved metals, e.g., Pb, in coastal waters are dominated by dissolution from atmospherically deposited material, and even in open-ocean waters, solubilization from aerosols can play a significant role in controlling the surface water distributions of metals, such as Al, Mn, and Cu.

REFERENCES

1. **Lantzy, R. J. and Mackenzie, F. T.,** Atmospheric trace metals: global cycles and assessment of man's impact, *Geochim. Cosmochim. Acta,* 43, 511, 1979.
2. **Nriagu, J. O.,** Global inventory of natural and anthropogenic emissions of trace metals to the atmosphere, *Nature,* 279, 409, 1979.
3. **Weisel, C. P., Duce, R. A., Fasching, J. L., and Heaton, R. W.,** Estimates of the transport of trace metals from the ocean to the atmosphere, *J. Geophys. Res.,* 89, 11607, 1984.
4. **Chester, R.,** *Marine Geochemistry,* Allen and Unwin, in press, 1990.
5. **Rahn, K. A., Borys, R. D., Shaw, G. E., Shutz, L., and Jaenicke, R.,** Long range impact of desert aerosol on atmospheric chemistry: two examples, in *Saharan Dust,* Morales, Ed., John Wiley & Sons, New York, 1979, 243.
6. **Murphy, K. J. T.,** The Trace Metal Chemistry of the Atlantic Aerosol, Ph.D. thesis, University of Liverpool, Liverpool, U.K., 1985.
7. **Fasching, J. L., Courant, R. A., Duce, R. A., and Piotrowicz, S. R.,** A new surface microlayer sampler utilising the bubble microtome, *J. Rech. Atmos.,* 8, 649, 1974.
8. **Piotrowicz, S. R., Duce, R. A., Fasching, J. L., and Weisel, C. P.,** Bursting bubbles and their effect on the sea to air transport of Fe, Cu and Zn, *Mar. Chem.,* 7, 307, 1979.
9. **Duce, R. A., Ray, B. J., Hoffman, G. L., and Walsh, P. R.,** Trace metal concentration as a function of particle size in marine aerosols from Bermuda, *Geophys. Res. Lett.,* 3, 339, 1976.
10. **Duce, R. A., Arimoto, R., Ray, B. J., Unni, C. K., and Harder, P. J.,** Atmospheric trace metals at Enewetak Atoll. I. Concentrations, sources and temporal variability, *J. Geophys. Res.,* 88, 5321, 1983.
11. **Cambray, R. S., Jeffries, D. F., and Topping, G.,** An estimate of the input of atmospheric trace elements into the North Sea and Clyde Sea (1972 — 1973), AERE R773, United Kingdom Atomic Energy Authority AERE Harwell, Her Majesty's Stationery Office, London, 1975.

12. **Schneider, B.,** Source characterisation for atmospheric trace metals over Kiel Bight, *Atmos. Environ.*, 21, 1275, 1987.
13. **Dulac, F., Buat-Menard, P., Arnold, M., Ezat, U., and Martin, D.,** Atmospheric input of trace metals to the Western Mediterranean Sea. I. Factors controlling the variability of atmospheric concentrations, *J. Geophys. Res.*, 92, 8437, 1987.
14. **Saydam, A. C.,** The elemental Chemistry of Eastern Mediterranean Atmospheric Particulates, Ph.D. thesis, University of Liverpool, Liverpool, U.K., 1981.
15. **Sanders, G. S.,** Metals in Marine Atmospheric Particles, Ph.D. thesis, University of Liverpool, Liverpool, U.K., 1983.
16. **Buat-Menard, P. and Chesselet, R.,** Variable influence of the atmospheric flux of trace metal on chemistry of oceanic suspended matter, *Earth Planet. Sci. Lett.*, 42, 399, 1979.
17. **Duce, R. A., Hoffman, G. L., Ray, B. J., Fletcher, J. S., Wallace, G. T., Fasching, J. L., Piotrowicz, S. R., Walsh, P. R., Hoffman, E. J., Miller, J. M., and Heffter, J. L.,** Trace metals in the marine atmosphere: sources and fluxes, in *Marine Pollutant Transfer,* Windom, H. L. and Duce, R. A., Eds., Lexington, Toronto, 1976.
18. **Arimoto, R., Duce, R. A., Ray, B. J., Hewitt, A. D., and Williams, J. W.,** Trace elements in the atmosphere of American Samoa: concentrations and deposition to the tropical South Pacific, *J. Geophys. Res.*, 92, 8465, 1987.
19. **Mahadevan, T. N., Sadasivan, S., and Mishra, U. C.,** Chemical composition of precipitation in tropical marine atmosphere around Indian Subcontinent, *Sci. Total Environ.*, 24, 275, 1982.
20. **Jickells, T. D., Knap, A. H., and Church, T. M.,** Trace metals in Bermuda rainwater, *J. Geophys. Res.*, 89, 1423, 1984.
21. **Arimoto, R., Duce, R. A., Ray, B. J., and Unni, C. K.,** Atmospheric trace elements in Enewetak Atoll. II. Transport to the ocean by wet and dry deposition, *J. Geophys. Res.*, 90, 2391, 1985.
22. **Bruland, K. W., Bertine, K., Koide, M., and Goldberg, E. D.,** History of metal pollution in Southern California coastal zone, *Environ. Sci. Technol.*, 8, 425, 1974.
23. **Chester, R., Sharples, E. J., Murphy, K. J. T., Saydam, A. C., and Sanders, G. S.,** The atmospheric distribution of lead over a number of marine regions, *Mar. Chem.*, 13, 57, 1983.
24. **Walsh, P. R., Duce, R. A., and Fasching, J. L.,** Considerations of the enrichment, sources and fluxes of arsenic in the troposphere, *J. Geophys. Res.*, 84, 1719, 1979.
25. **Chester, R.,** Particulate aluminium fluxes in the Eastern Atlantic, *Mar. Chem.*, 11, 1, 1982.
26. **Arnold, M., Seghaier, A., Martin, D., Buat-Menard, P., and Chesselet, R.,** Geochimie de l'aerosol marin au — dessus de la Mediterranee Occidentale, *VI Journees Etud. Pollution,* Cannes CIESM, 27, 1982.
27. **Windom, H. L.,** Comparison of atmospheric and riverine transport of trace elements to the continental shelf environment, in *River Input to Ocean Systems,* Martin, J. M., Burton, J. D., and Eisma, D., Eds., UNEP and UNESCO Workshop, Rome, 1981.
28. **Chester, R., Griffiths, A. G., and Hirst, J. M.,** The influence of soil-sized atmospheric particulates on the elemental chemistry of the deep sea sediments of the North Eastern Atlantic, *Mar. Geol.*, 32, 141, 1979.
29. **Settle, D. M. and Patterson, C. C.,** Magnitudes and sources of precipitation and dry deposition fluxes of industrial and natural leads to the North Pacific at Enewetak, *J. Geophys. Res.*, 87, 8857, 1982.
30. **Walsh, P. R. and Duce, R. A.,** The solubilisation of anthropogenic, atmospheric vanadium in sea water, *Geophys. Res. Lett.*, 3, 1015, 1976.
31. **Maring, H. B. and Duce, R. A.,** The impact of atmospheric aerosols on trace metal chemistry in open ocean surface seawater. I. Aluminium, *Earth Plan. Sci. Lett.*, 84, 381, 1987.
32. **Lum, K. R., Betteridge, J. S., and Macdonald, P. R.,** The potential availability of P, Al, Cd, Co, Cr, Cu, Fe, Mn, Ni, Pb and Zn in urban particulate matter, *Environ. Technol. Lett.*, 3, 57, 1982.
33. **Chester, R., Murphy, K. J. T., Towner, J., and Thomas, A.,** The partitioning of elements in crust-dominated marine aerosols, *Chem. Geol.*, 54, 1, 1986.
34. **Hodge, V., Johnson, S. R., and Goldberg, E. D.,** Influence of atmospherically transported aerosols on surface ocean water composition, *Geochem. J.*, 12, 7, 1978.
35. **Chester, R., Lin, R. J., and Murphy, K. J. T.,** The solubility of atmospheric trace metals in seawater following dry mode deposition, in prep.
36. **Maring, H. B. and Duce, R. A.,** The impact of atmospheric aerosols on trace metal chemistry in open ocean surface seawater, II. Copper, in press.
37. **Moore, R. M., Milley, J. E., and Chatt, A.,** The potential for biological mobilisation of trace elements from aeolian dust in the ocean and its importance in the case of iron, *Oceanol. Acta,* 7, 221, 1984.
38. **Buat-Menard, P.,** Particle geochemistry in the atmosphere and oceans, in *Air-Sea Exchange of Gases and Particles,* Liss, P. S. and Slinn, W. G. N., Eds., D. Reidel, Hingham, MA, 1983, 455.
39. **Collier, R. and Edmond, J.,** The trace element geochemistry of marine biogenic particulate matter, *Prog. Oceanog.*, 13, 113, 1984.

40. **Baumgartner, A. and Reichel, E.,** *The World Water Balance: Mean Annual Global, Continental and Maritime Precipitation, Evaporation and Run-off,* Elsevier, Amsterdam, 1975.
41. **Orians, K. J. and Bruland, K. W.,** Dissolved aluminium in the central North Pacific, *Nature,* 316, 427, 1985.
42. **Boyle, E. A., Huested, S. S., and Jones, S. P.,** On the distribution of copper, nickel and cadmium in the surface waters of the North Atlantic and North Pacific Oceans, *J. Geophys. Res.*, 86, 8048, 1981.
43. **Bruland, K. W.,** Oceanographic distributions of cadmium, zinc, nickel and copper in the North Pacific, *Earth Planet. Sci. Lett.*, 47, 176, 1980.
44. **Kremling, K.,** The distribution of cadmium, copper, nickel, manganese and aluminum in surface waters of the open Atlantic and European shelf area, *Deep-Sea Res.*, 32, 531, 1985.

Chapter 4

PROCESSES AFFECTING METAL CONCENTRATIONS IN ESTUARINE AND COASTAL MARINE SEDIMENTS

Samuel N. Luoma

TABLE OF CONTENTS

I. INTRODUCTION

A fundamental property of trace metals is that most react strongly with particles. Metal reactions with specific types of particle surfaces are well understood;[1-5] and the implications of particle reactivity for the fate of metals in marine and estuarine environments have been discussed extensively.[6-9] The high concentrations of metals in sediments, resulting from particle reactivity, enhances the importance of sediments as a direct source of these potential toxins for biota in some environments.[5,10,11]

Metal contamination originating from human activities often can be identified from analyses of the concentrated reservoir of metals in sediments more readily than from analyses of the low concentrations that occur in solution.[12] However, metal concentrations in sediments may contain a substantial natural contribution from the earth's crust. Differentiating the influences of human activities from the natural background is an important challenge in studies of sediment contamination with metals. Sensitive detection of anomalous metal concentrations is important. Small increases in metal enrichment in sediments can be biologically significant in some environments.[11] Furthermore, large changes in fluid phase concentrations of metals (either in interstitial water or the water column) can result from small changes in concentrations of sediment-bound metals.[13]

This paper discusses processes that influence metal concentrations in sediment and evaluates the approaches most effective in separation of anthropogenic contamination from natural influences. The discussion is limited to estuarine and coastal marine bed sediments and emphasizes behavior that is common to the heavy metals Ag, Cd, Cu, Ni, Pb, Zn, Hg, and Cr. Many of the approaches and considerations employed to study sediment contamination with metals have been reviewed elsewhere.[9,12] This paper is meant to supplement those discussions, with specific reference to marine and estuarine environments.

II. WHOLE SEDIMENT CONCENTRATIONS OF METALS IN MARINE/ESTUARINE SEDIMENTS

Methods of determining metal concentrations in sediments include:[14] (1) total analyses that either decompose the entire sediment or analyze directly for all metal in the sediment; (2) strong decomposition techniques that release a large proportion of the metal from particles, including most of the metal associated with the lattices of resistant minerals; or (3) partial extractions designed to remove some fraction of the metal. Techniques 1 and 2 both commonly are described in analyzing "total metal". The differences between the two, in most sediments, are largest for components such as Fe or Al. Partial extractants are employed to "selectively" remove metal forms from sediments,[15] or to "nonselectively" remove an operationally defined nonresidual fraction of metal.[16] Analyses of nonresidual metal may be profitably employed to improve detection of contamination (see later discussion). Recent studies have demonstrated significant limitations in the use of the selective chemical extraction approach.[17-21]

Total concentrations reported for individual metals from whole marine or estuarine sediments range over more than three orders of magnitude. Multiple inputs of metals from human activities occur in many environments and contribute to the heterogeneity of concentrations in sediments. Natural processes also contribute to variability. Metal concentrations vary from threefold to ninefold among fundamental geologic materials of the earth's crust, such as shales, granites, and sandstones (Figure 1).[12] Concentrations in undisturbed estuarine and coastal environments also may vary by as much as an order of magnitude (typified by Cu, Pb, and Zn in Figure 1). The range of reported metal concentrations is greatest in large, complex environments affected by human activities (Figure 1).[22-27] Large variations at the upper end of the concentration range typify such contaminated environments,

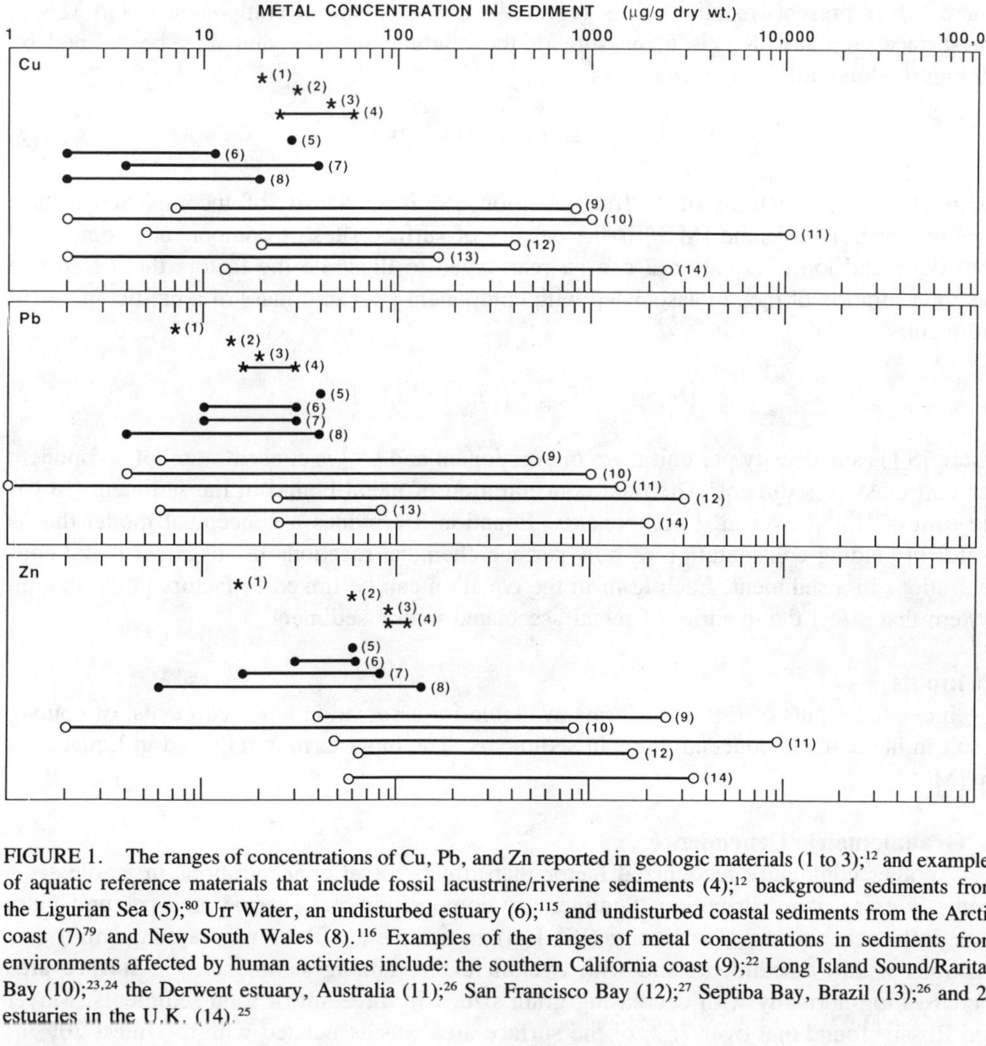

FIGURE 1. The ranges of concentrations of Cu, Pb, and Zn reported in geologic materials (1 to 3);[12] and examples of aquatic reference materials that include fossil lacustrine/riverine sediments (4);[12] background sediments from the Ligurian Sea (5);[80] Urr Water, an undisturbed estuary (6);[115] and undisturbed coastal sediments from the Arctic coast (7)[79] and New South Wales (8).[116] Examples of the ranges of metal concentrations in sediments from environments affected by human activities include: the southern California coast (9);[22] Long Island Sound/Raritan Bay (10);[23,24] the Derwent estuary, Australia (11);[26] San Francisco Bay (12);[27] Septiba Bay, Brazil (13);[26] and 27 estuaries in the U.K. (14).[25]

but localities with concentrations as low as found in reference or crustal materials also may be found within many such systems.

III. NATURAL PROCESSES AFFECTING METAL CONCENTRATIONS

Metal concentrations in marine and estuarine sediments are not only determined by metal inputs but also are affected by reactions at particle surfaces that influence the quantity of metal adsorbed, reduction/oxidation reactions, and differences in solution chemistry that affect particle formation and adsorption or desorption of metals.

A. SURFACE CHEMISTRY

The processes that control the association of metals with particle surfaces may cause variability in metal concentrations. In oxic sediments, metals are controlled by adsorption and complexation processes that may be empirically expressed by a simple complexation reaction:[8,28-30]

$$M_i^{z+} + S_j \overset{K_{M_iS_j}}{\longleftrightarrow} M_iS_j \qquad (1)$$

where M_i^{z+} represents metal ion i, S_j represents a surface site of component j, and M_iS_j is the surface complex. $K_{M_iS_j}$ is a measure of the binding intensity and may be defined by analogy with equilibrium constants as

$$K_{M_iS_j} = [M_iS_j]/[M_i^{z+}][S_j] \tag{2}$$

where $[M_i^{Z+}]$ is the activity of the free metal ion, $[S_j]$ is the density of uncomplexed surface sites on component j, and $[M_iS_j]$ is the density of surface sites of component j complexed with the metal ion i. Equation 2 can be rearranged to illustrate the factors that determine the concentration of metal i associated with component j per unit mass of a multicomponent sediment:[20]

$$[M_iS_j] = K_{M_iS_j} \cdot [M_i^{z+}] \cdot [S_j] \cdot [A_j] \tag{3}$$

where $[S_j]$ is site density per unit mass of component and $[A_j]$ is concentration of component per unit mass of sediment. The total concentration of metal bound in the sediment $[M_i]$ is the sum of $[M_iS_j]$ over all j components. Equation 3 provides a conceptual model that is useful in guiding consideration of how surface chemical reactions might affect metal concentrations in a sediment. Each term in the equation can be linked to factors in the natural system that affect the quantity of metal associated with a sediment.

1. Inputs

Increased inputs of metals in forms available for association with sediments, of course, result in increases in concentrations in sediments. The input term is reflected in Equation 3 by $[M_i^{Z+}]$.

2. Granulometric Dependence

Another commonly recognized factor that affects metal concentrations in whole sediments is grain size distribution. Sedimentary composition and surface area per unit mass both influence binding site density ([S] [A]) in sediments. Both vary among grain size fractions within a sediment, and with overall texture among sediments.[31-33] Surface area increases log-normally with decreasing grain size.[34] In three silt or sand sediments, Mayer and Rossi[34] found that over 75% of the surface area was associated with the finest 20% of the grains in silt and sand sediments.

The relationship of metal concentration to granulometric factors is not always simple. For example, silt/clay sized particles in estuarine/marine sediments are typically aggregates (see review in Reference 8), and particle aggregation may add complexity to the relationship of metal concentration and surface area.[33] Possibly as a result of aggregation, lower concentrations of Fe oxide and organic material coatings, as well as metal concentrations, have been found in the <2-μm fraction of sediments than in fractions 2 to 10 μm.[35,36]

Exceptional circumstances also may affect the relationship between sediment texture and metal concentrations. In sediments that are predominantly silt/clay, metal concentrations in fractions >63 μm occasionally may slightly exceed concentrations in the <63-μm fraction.[37,38] Physical or chemical processes may affect the degree of accumulation of metal-reactive sediment components on a surface (e.g., movement of water across a redox boundary may facilitate accumulation of iron or manganese oxides). Anomalous, conservative, metal-rich particles of a specific size also may be generated by some types of human activities, and enrich certain sediment size fractions of the sediments.[39] Despite the above, numerous studies show that finer fractions within sediments and finer textured sediments in estuaries and marine environments consistently have higher concentrations of Fe oxides, Mn oxides, or organic materials and higher concentrations of metals than do coarse materials.[32,40-45]

3. Concentrations of Sediment Components

The concentrations of the most important metal-reactive components of sediments (Fe oxides, Mn oxides, organic materials)[46,47] may vary widely among estuarine and marine environments.[8,20,18] Metal concentrations are affected by differences among sediments in [S], [A], and/or K for individual components, or differences in the aggregate values of these variables.[20,49] The nature of Fe oxides, organic materials, or Mn oxides will determine site densities for those components;[1,8] and the relative concentrations of Fe, Mn, and organic materials will determine aggregate differences in binding-site density or binding intensity.[8] Separating the effects of changing surface area from the effects of changing composition can be difficult in natural sediments, because many of the metal-reactive components of the sediment accumulate on particle surfaces.[33,46,47,50]

B. REDUCTION/OXIDATION REACTIONS

In anoxic environments, solubility equilibria appear to be much more important than surface chemical reactions in controlling metals in solid phases.[5] Complexation with organic materials may occur in some anoxic sediments, but insoluble sulfides appear to dominate metal chemistry at equilibrium in most.[46,47,51,52] It is thought that most metals occur coprecipitated with abundant sulfide species, such as FeS_2, or as specific metal sulfides.[46,47] Because of domination by solubility rather than surface chemistry, texture and composition could have more minor effects on metal concentrations in anoxic than in oxic sediments, although this hypothesis has not been adequately tested.

The effects of diagenesis on the concentrations of different metals at the interface of reduction/oxidation zones and with depth in the sediments are not completely understood.[9] Studies of dated cores show metal distributions with depth in many sediments that appear to reflect the pollution history of the sediment.[53-56] Such results suggest a minimal influence of diagenesis after deposition, a result of the stability of metal sulfide associations once the sediments are buried. In other studies, some influence of reduction/oxidation reactions on metal concentrations is suggested. Diagenetic reduction and mobilization clearly affect Mn concentrations in vertical sediment profiles.[9,57,58] In organic-rich fjord sediments, that the authors suggested were undisturbed by human activities, metal concentrations followed the same distribution as I/organic C, an indicator of diagenesis.[52] The profile of metal enrichment in surface sediments and declining concentration with depth, similar to that seen in contaminated sediments, appeared to have a diagenetic origin. Precipitation may enrich metal concentrations in sediments associated with anoxic, sulfide-rich marine waters.[59] Reducing conditions in bottom waters of Corpus Christi Bay, Texas caused precipitation of Cd and Zn sulfides, creating such metal-rich bottom sediments.[60] When bottom waters were reoxygenated, depletion of some metals occurred. Thus seasonal fluctuation in concentrations of some elements in sediments was linked to climatically driven seasonal oxygen depletion.

C. DESORPTION

Because of the high ionic strength of seawater and potential competition for binding sites with Ca and Mg, release of metals from oxic sediments is possible as salinities increase or reduction/oxidation state changes. Elevated metal concentrations in pore waters and mass balance calculations suggested release of Cu, Zn, and Cd occurred from the oxic surface of salt marsh sediments undersaturated with sulfides.[51] Release of Cu, Cd, Ni, and Zn from Swedish coastal sediments was demonstrated by *in situ* determinations of metal flux at the sediment/water interface.[61] The flux of oxygen from the water to the sediments controlled reactions at the interface.[62] Depletion of oxygen halted release of the metals. The extent of metal loss from oxic sediments may be reduced by readsorption to precipitating Fe and Mn oxides,[8,51,63-65] but the balance between the desorption and readsorption at the redox interface is not well understood.

Desorption of metals from particulates to solution when freshwater and seawater mix can be demonstrated in laboratory experiments and could contribute to metal depletion of estuarine sediments.[66] However, readsorption also occurs at the freshwater/seawater interface in estuaries due to the formation of new particles, increased turbidity, increased pH, and the occurrence of complex recirculation patterns.[66,67] Net desorption of Cd appeared to occur on both suspended particulates and in bottom sediments of the Elbe estuary, but desorption of Zn was not evident.[67] A net seasonal decrease in Cu concentrations in sediments of the Tamar estuary was attributed partly to desorption.[58] An empirical analysis of sediment mixing employing isotope techniques also suggested only desorption could explain reduced Pb concentrations in the sediments of the Gironde estuary.[68] Nevertheless, increased metal concentrations in solution in areas of suspected desorption are difficult to identify in estuaries (perhaps partly due to the complexities of estuarine circulation[68]); and in many estuaries, linking large changes in metal sediment concentrations to desorption is difficult.

IV. DETECTING ANOMALOUS METAL CONCENTRATIONS

Separating the natural concentrations of metals from anomalies imposed by human inputs to marine and estuarine sediments necessitates defining a relevant baseline or background as a reference, establishing comparability among sediments, and defining contamination by deviation from a comparable reference. Detecting anomalies also may be enhanced by identifying the pollutant phase.

A. IDENTIFYING THE POLLUTANT PHASE

Procedures that extract the nonresidual fraction of the total metal from sediments may increase sensitivity of detection of anomalies and estimate "potentially" reactive metal, both in terms of sediment/water mobility and bioavailability.[16,69-72] Although the concentration of metal removed by such procedures is a better estimate of reactivity than total metal,[11] it probably seldom is precisely that involved in geochemical and biological reactions.

Nonresidual metals are defined by a variety of approaches. Weak acids that are commonly employed include 25% acetic acid,[16] 0.4 to 0.6 N HCl,[48,69] hydroxylamine hydrochloride in acetic acid,[72] or warm 8 N nitric acid.[70,73] The sum of metal extracted by oxidizing and reducing steps in sequential techniques also is employed to estimate nonresidual metal.[71] Among more than 200 samples from San Francisco Bay, concentrations of Cu and Zn extracted by the HCl procedure correlated strongly with concentrations extracted by the acetic acid procedure ($r = 0.85$ and slope $= 0.80$ to 0.83 for both metals).[74] Thus, only small differences among some interpretations result when different types of weak acid are employed.

The proportion of total metal in the nonresidual fraction increases with enrichment.[72] However, other environmental factors also affect extractability. For example, nonresidual Zn and Cu were 20 to 23% and 15 to 18% of the total (respectively) in the lower St. Lawrence estuary and Saguenay Fjord compared to 10 and 5% of total in the Bay of Fundy, although levels of enrichment were comparable.[40] Nonresidual Pb was similar in the three environments. Nonresidual Zn, Pb, and Cu were 59, 35, and 14% extractable, respectively, in the contaminated Mersey and 25, 27, and 14% extractable in the less contaminated Ravenglass estuary.[71] The proportion of nonresidual Zn and Pb increased with contamination, but Cu did not. In oxidized sediments from San Francisco Bay, Cu averaged 65% extractable in HCl and 53% extractable in acetic acid at a site twofold enriched over the average for Bay sediments.[74] Mean extractability among sites in the Bay was 55% for HCl and 51% for acetic acid. In this case, nonresidual Cu increased with contamination in HCl, but not in acetic acid. Inhibition of the weak acid extractability of metals that form strong sulfides also occurs in anoxic sediments, relative to oxic sediments.[48,74,75]

The relationship between nonresidual metal and total metal is hyperbolic in some studies with some metals,[70] and linear in some studies.[48] When the relationship is linear the slope is less than 1.0.[48,76] Thus all "contaminant" metal added to a sediment may not be extracted as nonresidual. To determine the proportion of contaminant metal extracted as nonresidual, background metal concentrations must be subtracted from all values.[76]

B. DEFINING BACKGROUND

Establishing the metal concentration that preceded human activities is the first prerequisite to determining the extent of anthropogenic metal contamination in a sediment. Fossil lacustrine and riverine sediments,[12] or representative values from the earth's crust[77] may be employed as general references (Figure 1). "Precivilization" sediments from dated sediment cores provide more specific information about an environment, but require assumptions about sediment stability and lack of diagenetic redistribution of metals. Metal concentrations from environments or localities that appear to be relatively free from human disturbance and are hydrologically analogous to those under study are valuable as operational baselines. Similar reduction/oxidation or adsorption/desorption conditions in the baseline and test environments will improve accuracy and sensitivity in detection of contamination. However, proving that such environments represent an historic background is usually difficult.[12] Deep-sea sediments are enriched in concentrations of metals such as Cu, Ni, Pb, Zn, and Cd, and are unacceptable as a background reference for coastal and estuarine sediments.[79,80]

Many definitions assume background is a static value.[12,80] As shown previously, however, natural geologic and geochemical heterogeneity probably characterize most environments. Metal concentrations in whole sediments from "undisturbed" estuaries and coastal systems are within the same range as different rock types and fossil sediments (Figure 1), but values for any individual metal differ by sixfold or more within the same system.[32,40,42] Sensitive determinations of background or baseline conditions must account for such natural variability.

Salomons and Forstner[9] suggest that uniformity in grain size, composition, and origin is important to "obtain an ideal comparative basis for environmental studies." Employment of methods that improve comparability between reference and sample sediments is a more flexible alternative where sediments are diverse in character.

C. ESTABLISHING COMPARABILITY

The insensitivity of comparing metal concentrations among whole sediments of different textures or different composition is well established. Absolute metal concentrations in whole sediments have been useful in interpreting dated cores dominated by homogenous, anoxic sediments[53-55] or in comparing surface sediments of similar texture. In most studies, however, methods for improving the comparability of sediment metal concentrations are necessary. Unfortunately, all available approaches have limitations that must be recognized in application. The optimal approach should be determined by the goals of the study.

1. Identifying Homogenous Populations of Sediment

One approach to comparing heterogeneous sediments is to identify homogenous subgroups of sediments within the data set, and establish interpretations from commonalities within the subgroups. Determining the cumulative frequency of metal concentrations is a soil science technique that separates metal values into log-normally distributed populations.[81] The method assumes that environments with similar degrees of enrichment will fall into the same population distributions. Graphical representation is employed to determine if one or more populations of data are present. Rule[76] found that separation of data from Chesapeake Bay into three groups with cumulative frequency curves aided division of data by geographic area and sediment characteristics, and facilitated data analysis within similar sediments. The difficulty of achieving distinct separations of populations has limited the use of cumulative frequency in many natural sediments, however.[82]

R-mode factor analysis also is employed to classify sediments into statistically homogenous groups. This method establishes subgroups of sediments that have highly intercor related variables in common. Determination of geographically distributed subpopulations of sediments on the basis of metal concentrations and an array of textural and chemical characteristics helped explain metal distributions in the sediments of the Saguenay Fjord,[83] the Bay of Fundy,[40] the Gulf of St. Lawrence,[84] and Delaware Bay.[85] Classification of sediments into statistically homogenous groups, employing R-mode factor analysis, also was suggested as a preliminary step before attempting detailed statistical analysis of geochemical associations.[49] Correlations between independent variables within such groups may hinder interpretation of additional statistical associations, however.

2. Separation of Fine-Grained Sediments

If surveillance of the distribution of metal contamination is the goal, then separation of the sediment fraction <63 μm and direct analysis of metal concentrations in that fraction is economical and effective.[9] Such procedures reduce biases introduced by large particles with low surface areas and increase sensitivity in detecting contamination (i.e., anomalies are accentuated in fine fractions).[45] Size fractions smaller than 63 μm (e.g., <20 μm) have higher metal concentrations, may be more homogenous in composition,[32,45] and are better predictors of overall surface area in a sediment.[34] However, fractions smaller than 63 μm also are more difficult to separate in adequate quantity for direct analysis. Determination of the proportion of the finest particles within the <63-μm fraction (the cutoff should be <20 μm or finer)[34] allows compensation for inhomogeneities, may aid intercalibration of alternative separation approaches,[45,48] and is a reasonable surrogate for surface area.[34] If the study goals include considerations other than simply surveillance (e.g., determination of transport or mass loadings), then consideration of a single separated size fraction is an inadequate strategy.

3. Normalization

Normalizing metal concentrations to a textural or compositional characteristic may improve the sensitivity of comparisons of metal concentrations in sediments. Normalization assumes a baseline regression between a variable and metal concentrations, and the influence of the baseline is removed by creating a ratio. The most common choices for normalizing metal concentrations include the weight of a fine particle size fraction,[33] an empirical measure of surface area,[33] a conservative constituent of the sediment,[56,86] an inert constituent of the sediment,[87] or a metal-reactive component of the sediment that is unaffected by human activities.[37,88,89] Salomons and Forstner[9] suggested that normalization to a conservative element (e.g., Al, Sc, and Cs are common choices) combined with separation and direct analysis of <63-μm sediments may be the most powerful available surveillance strategy.

The choice of a normalizing parameter can influence interpretations. Ackerman[42] showed that normalization with Cs was a better choice (i.e., the strongest correlation with metal concentrations and the greatest difference between sandy and fine sediments) than several other conservative elements (Eu, Fe, Rb, Sc, Sm, and Th) in correcting for particle size biases of Zn concentrations in a core. Data from Williams et al.[41] indicate that interpretation of the difference in metal concentrations between fine and sandy sediments may change two- to threefold if the data were normalized by the weight fraction of <63-μm particles rather than normalized to iron concentrations. Correction for one inert constituent may be inadequate if others also are abundant enough to dilute metal concentrations. Possible diluents include quartz,[87] carbonates,[48] and feldspars.[9]

Although normalization does not require it, empirical demonstration of the underlying regression is valuable. The confidence limits of the regression will define significant differences among normalized values. Empirical demonstration also may aid interpretation of biases that occur when the intercept of the relationship is not zero. If the baseline regression is strong,[42] normalization is effective; but if baseline regressions are variable or nonlinear, care must be employed in the use of normalizing techniques. Nonlinear relationships of metal concentrations with surface area and average mineralogic composition have been observed.[32,90] Finally, special care must be employed when normalizing with variables whose fundamental relationship with metal concentrations may be affected by more than one process. For example, concentrations of Fe in surficial bed sediments may differ spatially or temporally with changes in the reduction/oxidation state of subsurface sediments.[91]

Normalization to background of metal concentrations in sediments of different character may be accomplished by calculation of an enrichment factor relative to the earth's crust. In the equation

$$EF = (M_x/Al_x)/(M_c/Al_c) \tag{4}$$

M_x and Al_x are metal and aluminum concentrations in the sample and M_c and Al_c are metal and aluminum concentrations in the crust.[9,56,92] This approach has been employed with both surface sediments and dated cores, and with normalizing factors other than Al (e.g., Fe).[37,76] Interpretation of enrichment factors does not seem especially sensitive to the choice of reference material.[37] However, comparisons of enrichment factors among studies will be affected by the decomposition procedure because strong decompositions may underestimate Fe or Al concentrations compared to actual total concentrations.[93]

4. Regression Techniques

Techniques that directly employ regression analysis also may be effective in improving comparisons among sediments of variable texture or composition, especially when employed with separation of fine particulate materials.[76] Quantitatively linking textural or compositional factors to fluctuating baseline metal concentrations may be accomplished by (1) definition of correlations between metal concentrations and conservative or reactive constituents *within a baseline data set where the influence of natural factors should be strongest* (in most studies, this would be some subset of less contaminated sediments representative of the study area); (2) determination of confidence limits for the regression (e.g., from standard deviations of the slope and intercept); and (3) determination of metal anomalies as values that lie outside the confidence limits.[73,94]

Step 1 is critical to this approach, but is seldom explicitly considered. Equation 3 suggests that demonstration by regression of the relationship between metal concentration ($[\bar{M}_i]$) and an independent variable represented in the equation will be most clear *if all other terms are constant.* For example, the hypothesis that metal concentrations are linked to Fe oxides may be tested by regression of $[\bar{M}_i]$ against $[A_{Fe}]$. The significance of the relationship may be influenced by variations in any term in the equation that can affect $[\bar{M}_i]$. Thus, linkages between metals and texture/composition variables will be most evident among sediments with similar degrees of contamination, because the regression implicitly assumes that metal inputs (M_i^{Z+}) to the sediments are constant. The effect of varying metal inputs is illustrated by the different slopes that characterize such regressions when sediments from the same system but with different levels of contamination are considered.[42,76,80,94,95] Alternatively, if the range of $[A_j]$ is great, variations in $[M_i^{Z+}]$ will be less disruptive.[48]

Perhaps because differences in metal inputs add variance, relationships between metal concentrations and sediment texture, organic material, or Fe concentrations are commonly weak where multiple sources of strong metal input are known to occur.[96,97] Stronger relationships are observed where metal concentrations are not excessively enriched.[32,45,83,84] In

San Francisco Bay, the correlation of Cu with organic carbon in fine-grained sediments was weak ($r^2 = 0.16$) among 16 stations of varying metal enrichment sampled 20 to 25 times each over 3 years. However, among the four least enriched of these stations, the correlation was very strong ($n = 84$; $r^2 = 0.70$).[98] Schults et al.[97] found that percent clay and organic materials together correlated well with the least enriched metals in Commencement Bay sediments (Cr and Cd), but correlated poorly with metals that showed strong enrichment and 50-fold variability in the data set (Pb and Zn). Concentrations of Cu also were variable, and only showed significant correlation if "outliers" (i.e., metal contaminated sediments) were removed from the data set. Rule[76] found stronger relationships between metal concentrations and Fe within groups of sediments separated by the cumulative frequency distribution of the metals than among all sediments sampled in Chesapeake Bay.

Strict criteria are essential for reducing subjectivity if a baseline data set is to be defined for regression analysis. At a minimum, the baseline stations must be representative of the environment(s) under study, the data set must be large enough to provide confidence that the regression is representative, and some concentration criteria must be consistently employed. For example, time series data from four stations were employed to establish an operational baseline for Cu in San Francisco Bay based upon the previously described relationship with organic carbon concentrations.[27,98] The 4 baseline stations had the lowest maximum Cu concentration among the 16 stations studied; 84 observations were included in the regression; organic carbon concentrations spanned the full range observed in the study; and the 4 stations were from different regions of the Bay. Temporal fluctuations in total Cu as large as 6-fold were shown to be largely reflective of changing concentrations of organic carbon (or some covarying factor) in the sediments at these baseline stations.

Although baseline has seldom been rigorously defined, residuals or outliers from regressions between metal concentrations and a sediment component or characteristic have been employed to distinguish sites or times of sediment contamination.[32,45,73,94,99] If a strong correlation can be established in a carefully defined baseline data set, and confidence limits are calculated, contamination may be quantitatively determined in any individual sample or set of samples by this method.[32,98]

Correlation methods have fundamental hazards that are well established.[100] A few worth special emphasis with regard to studies of metal contamination in sediments include:

1. Recognition that correlative relationships do not necessarily reflect cause and effect is important. Statistical studies are most effective in suggesting mechanisms or hypotheses that should be studied more directly.
2. Metal concentrations are affected by a complex array of processes in nature. Regression analyses may be too simplistic to describe the dominant processes in some environments.
3. Occurrence of confounding variables is common in complex natural systems. Coincidence among independent variables (some of which may not be considered) is difficult to avoid in data with narrow ranges, and especially when such data are from a single environment.
4. Multiple regression procedures are most effective in studies that consider broad ranges in the values of independent variables and/or comparisons across several systems. Problems with cross-correlations among independent variables are less likely in such studies.
5. Correlations observed in broad data sets depict broad phenomena that may be less evident within more narrow ranges of conditions.[5]
6. A clear statement of the hypothesis being tested is an obvious prerequisite to any regression analysis.

In summary, rigorous attention to the implicit hypotheses, assumptions, and limitations of statistical approaches are essential. Many of the statistical studies of metals in marine and estuarine sediments to date are difficult to interpret because of an inadequate degree of such attention.

V. CHARACTERISTICS OF METAL CONTAMINATION IN ESTUARINE AND MARINE SEDIMENTS

Spatial and temporal distributions of metal concentrations in sediments reflect inputs from natural and human sources, superimposed upon natural geochemical influences, and physical processes that dilute and redistribute the particles with which metals associate. Thus anthropogenic metal contamination may influence metal distributions observed in an environment, but distributions also will reflect processes other than simply metal inputs.

A. DISTRIBUTION NEAR HUMAN SOURCES OF METAL INPUT

Numerous studies have demonstrated that metal inputs from human activities are reflected by metal concentrations in sediments.[26,53,60,70,80,94,97] Concentrations decline, in the simplest cases, with distance away from the source of input as determined by net water and particle movement and the extent of dilution with less contaminated sediments.[101,102] Dilution away from large sources of contamination follows an exponential function in sediments near New Bedford Harbor[103] and offshore from southern California.[56] Where single sources dominate, dilution trends may be characterized by ratios between metals or correlations among metals.[22]

Many marine and estuarine systems are subject to multiple metal inputs from human activities that make dilution trends difficult to identify. In San Francisco Bay, metal distributions in sediments are complex, reflecting numerous localities of enrichment, zones of dilution around each, and localities outside the influence of point source inputs.[27] Because of these complexities, regional trends in contamination of sediments of the Bay are difficult to characterize.

Sediments also integrate the temporal variability that commonly characterizes metal inputs from human sources.[102] Concentrations in sediments will be maintained near a source by continuous input, but will change as inputs change. The general trend in Cu concentrations in sediments near a sewage outfall in South San Francisco Bay followed Cu discharges from the sewage works over a 10-year period.[27,102] Aperiodic episodes of contamination, lasting from 1 month to 1 year, were observed at other stations in the Bay.[27] Temporal variability of metal concentrations in Liverpool Bay sediments followed changes in metal inputs by sludge dumping into the Bay.[104]

B. PHYSICAL TRANSPORT: BAYS AND COASTAL ENVIRONMENTS

Concentration gradients determined by metal inputs may be strongly affected by physical processes that redistribute sediments. Physical transport and sorting of sediments is the most important factor controlling the distribution of metals in environments such as Delaware Bay,[85] the St. Lawrence bay and estuary,[86] the Bay of Fundy,[40] False Bay in South Africa,[105] the Pagassitikos Gulf in Greece,[106] and the coastal zone of Texas.[60] High concentrations of metals in whole sediments in such environments are found where the finest sediments accumulate.

Fine particles seek their preferred sites of deposition within coastal zones regardless of the point of introduction.[107] Thus, even where sediments are transported into a system by humans (e.g., in the disposal of dredge spoils), the ultimate distribution of the particulate material and its associated metal load is determined by natural depositional processes. In the Hudson-Raritan estuarine system, localized areas serve as fine-particle and contaminant traps, collecting material that has been added over a considerably larger area.[108] Contamination of depositional areas also was observed in several European estuaries[96] and in Liverpool Bay[104] and should be expected in most estuarine and coastal marine systems. Thus if metal inputs to a location are continuous, the source of that input may be responsible for both consistent contamination in the locality of the input and secondary contamination in areas of deposition.[108]

The influence of physical processes on metal concentrations is most complex in estuaries. When concentrations are corrected for texture differences among sediments a declining downstream trend from the head to the mouth is observed in many estuaries.[68,83,109,110] The head-to-mouth decline in metal concentration in the Rhine, Elbe, and St. Lawrence estuaries was attributed to mixing of metal-enriched riverine sediments with metal-poor sediments of marine origin.[83,109-111] Concentrations of Mn and Zn were depleted in Chesapeake Bay sediments relative to average values for suspended sediments in the incoming Susquehanna River because most sediments in the estuary originated from infrequent major flood events during which incoming particulates are particularly metal-poor.[88,113] In other estuaries, the distribution of metals is more complex. Monthly sampling of sediments in the Tamar estuary showed a complex seasonal variability in the downstream profile of Fe, Cu, and Zn.[58] Internal processes (such as landward sediment transport) appeared to govern metal distributions in the Tamar during low flow, and riverine inputs were dominant during high flow.[58,112] Direct contaminant inputs to the estuary from human activities create complex metal distributions in San Francisco Bay,[27] in the Hudson-Raritan system,[108] and in developed estuaries in Brazil.[114]

VI. CONCLUSIONS

Determination of metal concentrations in sediments have been useful in numerous studies in assessing metal contamination of marine and estuarine environments. However, concentrations that reflect metal inputs are superimposed upon variations in texture, composition, reduction/oxidation reactions, adsorption/desorption reactions and physical transport, sorting or dilution of sediments. The above influences will cause fluctuations in sediment metal concentrations in time and space that occur independent of metal inputs. Consideration of the effect of such influences is essential to realistic interpretation of the significance of inputs. Multiple strategies for improving comparability of sediments and defining deviations from background may be the best approach to interpretation of contamination,[76] given the limitations of most individual approaches. Surveillance or monitoring studies also must consider temporal variability, spatial heterogeneity, and redistribution of contamination by physical transport in their design.

REFERENCES

1. **Davis, J. A. and Leckie, J. O.,** Surface ionization and complexation at the oxide/water interface, *J. Colloidal Interface Sci.*, 67, 90, 1978.
2. **Benjamin, M. M. and Leckie, J. O.,** Conceptual model for metal-ligand-surface interactions during adsorption, *Env. Sci. Technol.*, 15, 1050, 1981.
3. **Balistrieri, L. S. and Murray, J. W.,** The adsorption of Cu, Pb, Zn and Cd on goethite from major ion seawater, *Geochim. Cosmochim. Acta*, 46, 1253, 1982.
4. **Altmann, R. S. and Leckie, J. O.,** Metal binding in heterogeneous multicomponent systems: mathematical and experimental modelling, in *Oceanic Processes in Marine Pollution: Vol. 2, Physicochemical Processes and Waste in the Ocean*, O'Connor, T. P., Burt, W. V., and Duedall, I. W., Eds., Robert E. Krieger, Malabar, FL, 1987, 145.

5. **Campbell, P.G.C., Lewis, A. G., Chapman, P. M., Crowder, A. A., Fletcher, W. K., Imber, B., Luoma, S. N., Stokes, P. M., and Winfrey, M.,** *Biologically Available Metals in Sediments,* National Research Council of Canada, Division of Chemistry Publication, Ottawa, 296pp, in press.
6. **Turekian, K. K.,** The fate of metals in the oceans, *Geochim. Cosmochim. Acta,* 41, 1139, 1977.
7. **Olsen, C. R., Cutshall, N. H., and Larsen, I. L.,** Pollutant-particle associations and dynamics in coastal marine environments: a review, *Mar. Chem.,* 11, 501, 1982.
8. **Luoma, S. N. and Davis, J. A.,** Requirements for modeling trace metal partitioning in oxidized estuarine sediments, *Mar. Chem.,* 12, 159, 1983.
9. **Salomons, W. and Forstner, U.,** *Metals in the Hydrocycle,* Springer-Verlag, Berlin, 1984.
10. **Luoma, S. N.,** Bioavailability of trace metals to aquatic organisms — a review, *Sci. Total Environ.,* 28, 1, 1983.
11. **Luoma, S. N.,** Can we determine the biological availability of sediment bound trace elements?, *Hydrobiologia,* 176/177, 379, 1989.
12. **Forstner, U. and Wittman, G. T. W.,** *Metal Pollution in the Aquatic Environment,* Springer-Verlag, Berlin, 1983.
13. **Jones, B. and Bowser, C.,** The mineralogy and related chemistry of lake sediments, in *Lakes: Chemistry, Geology and Physics,* Lerman, A., Ed., Springer-Verlag, New York, 1978, 179.
14. **Fletcher, W. K.,** *Analytical Methods in Geochemical Prospecting,* Elsevier, Amsterdam, 1981.
15. **Tessier, A., Campbell, P. G. C., and Bisson, M.,** Sequential extraction procedure for the speciation of particulate trace metals, *Anal. Chem.,* 51, 844, 1979.
16. **Loring, D. H.,** Potential bioavailability of metals in eastern Canadian estuarine and coastal sediments, *Rapp. P.V. Reun. Cons. Explor. Mer,* 181, 93, 1981.
17. **Rendell, P. S., Bately, G. E., and Cameron, A. J.,** Adsorption as a control of metal concentrations in sediment extracts, *Environ. Sci. Technol.,* 14, 314, 1980.
18. **Jouanneau, J. M., Latouche, C., and Pautrizel, R.,** Critical analysis of sequential extractions through the study of several attack constituent residues, *Environ. Technol. Lett.,* 4, 509, 1983.
19. **Tipping, E., Hetherington, N. B., Hilton, J., Thompson, D. W., Bowles, E., and Hamilton-Taylor, J.,** Artifacts in the use of selective chemical extraction to determine distributions of metals between oxides of manganese and iron, *Anal. Chem.,* 57, 1944, 1985.
20. **Luoma, S. N.,** A comparison of two methods for determining copper partitioning in oxidized sediments, *Mar. Chem.,* 20, 45, 1986.
21. **Kheboian, C. and Bauer, C. F.,** Accuracy of selective extraction procedures for metal speciation in model aquatic sediments, *Anal. Chem.,* 59, 1417, 1987.
22. **Katz, A. and Kaplan, I. R.,** Heavy metals behavior in coastal sediments of Southern California: a critical review and synthesis, *Mar. Chem.,* 10, 261, 1981.
23. **Greig, R. A., Reid, R. N., and Wenzloff, D. R.,** Trace metal concentrations in sediments from Long Island Sound, *Mar. Pollut. Bull.,* 8, 183, 1977.
24. **Greig, R. A. and McGrath, R. A.,** Trace metals in sediments of Raritan Bay, *Mar. Pollut. Bull.,* 8, 188, 1977.
25. **Bryan, G. W., Langston, W. J., and Hummerstone, L. G.,** *The Use of Biological Indicators of Heavy Metal Contamination in Estuaries,* Marine Biological Assoc. U. K. Occasional Publ. #1, Plymouth, U.K., 1980.
26. **De Lacerda, L. D., Pfeiffer, W. C., and Fiszman, M.,** Heavy metal distribution, availability and fate in Sepetiba Bay, Brazil, *Sci. Total Environ.,* 65, 163, 1987.
27. **Luoma, S. N. and Phillips, D. J. H.,** Distribution, variability, and impacts of trace elements in San Francisco Bay, *Mar. Pollut. Bull.,* in press.
28. **Oakley, S. M., Nelson, P. O., and Williamson, K. J.,** Model of trace metal partitioning in marine sediments, *Environ. Sci. Technol.,* 15, 474, 1981.
29. **Davies-Colley, R. J., Nelson, P. O., and Williamson, K. J.,** Copper and cadmium uptake by estuarine sedimentary phases, *Environ. Sci. Technol.,* 18, 491, 1984.
30. **Tessier, A., Rapin, R., and Carignan, R.,** Trace metals in oxic lake sediments: possible adsorption onto iron oxyhydroxides, *Geochim. Cosmochim. Acta,* 49, 183, 1985.
31. **Gibbs, R. J.,** Mechanisms of trace metal transport in rivers, *Science,* 180, 71, 1973.
32. **Mayer, L. M. and Fink, L. K.,** Granulometric dependence of chromium accumulation in estuarine sediments in Maine, *Estuarine Coastal Mar. Sci.,* 11, 491, 1980.
33. **Horowitz, A. J. and Elrick, K. A.,** The relation of stream sediment surface area, grain size and composition to trace element chemistry, *Appl. Geochem.,* 2, 437, 1987.
34. **Mayer, L. M. and Rossi, P. M.,** Specific surface areas in coastal sediments: relationships with other textural factors, *Mar. Geol.,* 45, 241, 1982.
35. **DeFlaun, M. F. and Mayer, L. M.,** Relationships between bacteria and grain surfaces in intertidal sediments, *Limnol. Oceanogr.,* 28, 873, 1983.

36. **Schoer, J.,** Iron-oxo-hydroxides and their significance to the behavior of heavy metals in estuaries, *Environ. Technol. Letters,* 6, 189, 1985.
37. **Sinex, S. A. and Helz, G. R.,** Regional geochemistry of trace elements in Chesapeake Bay sediments, *Environ. Geol.,* 3, 315, 1981.
38. **Axtmann, E. and Luoma, S. N.,** unpublished data, 1986.
39. **Thorne, L. T. and Nickless, G.,** The relationship between heavy metals and particle size fractions within the Severn estuary (U.K.) inter-tidal sediments, *Sci. Total Environ.,* 19, 207, 1981.
40. **Loring, D. H.,** Geochemical factors controlling the accumulation and dispersal of heavy metals in the Bay of Fundy sediments, *Can. J. Earth Sci.,* 19, 930, 1982.
41. **Williams, S. C., Simpson, H. J., Olsen, C. R., and Bopp, R. F.,** Sources of heavy metals in sediments of the Hudson River estuary, *Mar. Chem.,* 6, 195, 1978.
42. **Ackerman, F.,** A procedure for correcting the grain size effect in heavy metal analyses of estuarine and coastal sediments, *Environ. Technol. Lett.,* 1, 518, 1980.
43. **Forstner, U. and Salomons, W.,** Trace metal analysis on polluted sediments. I. Assessment of sources and intensities, *Environ. Technol. Lett.,* 1, 494, 1980.
44. **Wheeler, R. B., Anderson, J. B., Schwarzer, R. R., and Hokanson, C. L.,** Sedimentary processes and trace metal contaminants in the Buccaneer oil/gas field, Northwestern Gulf of Mexico, *Environ. Geol.,* 3, 163, 1980.
45. **Ackerman, F., Bergmann, H. and Schleichert, U.,** Monitoring of heavy metals in coastal and estuarine sediments — a question of grain size: <20μm vs. <60μm, *Environ. Technol. Lett.,* 4, 317, 1983.
46. **Jenne, E. A.,** Controls on Mn, Fe, Co, Ni, Cu, and Zn concentrations in soils and water: the significant role of hydrous Mn and Fe oxides, in *Trace Inorganics in Water,* Gould, R. F., Ed., American Chemical Society, Washington, D.C., 1968, 337.
47. **Jenne, E. A.,** Trace element sorption by sediments and soils — sites and processes, in *Symposium on Molybdenum in the Environment,* Chappel, W. and Peterson, K., Eds., Marcel Dekker, New York, 1977, 425.
48. **Luoma, S. N. and Bryan, G. W.,** Statistical assessment of the forms of trace metals in oxidized estuarine sediments employing chemical extractants, *Sci. Total Environ.,* 17, 165, 1981.
49. **Jaquet, J. M., Davaud, F. R., and Vernet, J. P.,** Basic concept and associated statistical methodology in the geochemical study of lake sediments, *Hydrobiol.,* 91, 139, 1982.
50. **Sundby, B., Silverberg, N., and Chesselet, R.,** Pathways of manganese in an open estuarine system, *Geochim. Cosmochim. Acta,* 45, 298, 1981.
51. **Giblin, A. E., Luther, G. W., III, and Valiela, I.,** Trace metal solubility in salt marsh sediments contaminated with sewage sludge, *Estuarine Coastal Shelf Sci.,* 23, 477, 1986.
52. **Ridgway, I. M. and Price, N. B.,** Geochemical associations and post-depositional mobility of heavy metals in coastal sediments: Loch Etive, Scotland, *Mar. Chem.,* 21, 229, 1987.
53. **Goldberg, E. D., Gamble, E., Griffin, J. J., Koide, M.,** Pollution history of Narragansett Bay as recorded in its sediments, *Estuarine Coastal Mar. Sci.,* 5, 549, 1977.
54. **Goldberg, E. D., Hodge, V., Koide, M., Griffin, J., and Gamble, E.,** Pollution history of Chesapeake Bay, *Geochim. Cosmochim. Acta,* 42, 1413, 1978.
55. **Goldberg, E. D., Griffin, J. J., Hodge, V., Koide, M., and Windom, H.,** Pollution history of the Savannah river estuary, *Environ. Sci. Technol.,* 13, 588, 1979.
56. **Bertine, K. K. and Goldberg, E. D.,** History of heavy metal pollution in Southern California coastal zone — reprise, *Environ. Sci. Technol.,* 11, 297, 1977.
57. **Trefrey, J. H. and Presley, B. J.,** Manganese fluxes from Mississippi delta sediments, *Geochim. Cosmochim. Acta,* 46, 1715, 1982.
58. **Ackroyd, D. R., Millward, G. E., and Morris, A. W.,** Periodicity in the trace metal content of estuarine sediments, *Oceanol. Acta,* 10, 161, 1987.
59. **Jacobs, L., Emerson, S., and Skei, J.,** Partitioning and transport of metals across the O_2/H_2S interface in a permanently anoxic basin: Framvaren Fjord, Norway, *Geochim. Cosmochim. Acta,* 49, 1433, 1985.
60. **Holmes, C. W.,** Trace metal seasonal variations in Texas marine sediments, *Mar. Chem.,* 20, 13, 1986.
61. **Westerlund, S. F. G., Anderson, L. G., Hall, P. O., Iverfeldt, A., van der Loeff, M. M. R., and Sundby, B.,** Benthic fluxes of cadmium, copper, nickel, zinc, and lead in the coastal environment, *Geochim. Cosmochim. Acta,* 50, 1289, 1986.
62. **Sundby, G., Anderson, L. G., Hall, P. O. J., Iverfeldt, A., van der Loeff, M. M. R., and Westerlund, S. F. G.,** The effect of oxygen on release and uptake of cobalt, manganese, iron and phosphate at the sediment-water interface, *Geochim. Cosmochim. Acta,* 50, 1281, 1986.
63. **Gobiel, D., Sundby, B., and Silverberg, N.,** Factors influencing particulate matter geochemistry in the St. Lawrence estuary turbidity maximum, *Mar. Chem.,* 10, 123, 1981.
64. **Hem, J. D.,** Redox processes at surfaces of manganese oxide and their effects on aqueous metal ions, *Chem. Geol.,* 21, 199, 1978.
65. **Salomons, W.,** Sediments and water quality, *Environ. Technology Lett.,* 6, 315, 1985.

66. **Salomons, W.,** Adsorption processes and hydrodynamic conditions in estuaries, *Environ. Technol. Lett.*, 1, 356, 1980.
67. **Ahlf, W.,** The River Elbe: behavior of Cd and Zn during estuarine mixing, *Environ. Technol. Lett.*, 4, 405, 1983.
68. **Elbaz-Ooulichet, F., Holliger, P., Huang, W. W., and Martin, J.-M.,** Lead cycling in estuaries, illustrated by the Gironde estuary, France, *Nature,* 308, 409, 1984.
69. **Chester, R. and Voutsinou, F. G.,** The initial assessment of trace metal pollution in coastal sediments, *Mar. Pollut. Bull.*, 12, 84, 1981.
70. **Cosma, B., Drago, M., Piccazzo, M., Scarponi, G., and Tucci, S.,** Heavy metals in Ligurian Sea sediments: distribution of Cr, Cu, Ni, and Mn in superficial sediments, *Mar. Chem.*, 8, 125, 1979.
71. **Badri, M. A. and Aston, S. R.,** Observations on heavy metal geochemical associations in polluted and non-polluted sediments, *Environ. Pollut. Ser. B,* 6, 181, 1983.
72. **Calamano, W. and Forstner, U.,** Chemical extraction of heavy metals in polluted river sediments in central Europe, *Sci. Total Environ.*, 28, 77, 1983.
73. **Cosma, B., Frache, R., Baffi, F., and Dadone, A.,** Trace metals in sediments from the Ligurian Coast, Italy, *Mar. Pollut. Bull.*, 13, 127, 1982.
74. **Luoma, S. N., Cain, D., and Axtmann, E.,** unpublished data, 1987.
75. **Thomson, E. A., Luoma, S. N., Cain, C. J., and Johansson, C. E.,** The effect of sample storage on the extraction of Cu, Zn, Fe, Mn and organic material from oxidized estuarine sediments, *Water Air Soil Pollut.*, 14, 215, 1980.
76. **Rule, J. H.,** Assessment of trace element geochemistry of Hampton Roads Harbor and Lower Chesapeake Bay area sediments, *Environ. Geol. Water Sci.*, 8, 209, 1986.
77. **Turekian, K. K. and Wedepohl, K. H.,** Distribution of the elements in some major units of the earth's crust, *Bull. Geol. Soc. Am.*, 72, 175, 1961.
78. **Chester, R., Aston, S. R., and Bruty, D.,** The trace-element partition geochemistry in an ancient deep-sea sediment core from the Bermuda Rise, *Mar. Geol.*, 21, 271, 1976.
79. **Campbell, J. A. and Loring, D. H.,** Baseline levels of heavy metals in the waters and sediments of Baffin Bay, *Mar. Pollut. Bull.*, 11, 257, 1980.
80. **Donazzolo, R., Merlin, O. H., Vitturi, L. M., Orio, A. A., Pavoni, B., Perin, G., and Rabitti, S.,** Heavy metal contamination in surface sediments from the Gulf of Venice, Italy, *Mar. Pollut. Bull.*, 12, 417, 1981.
81. **Siegel, F. R.,** *Applied Geochemistry,* Wiley-Interscience, New York, 202, 1974.
82. **Chester, R., Kudoja, W. M., Thomas, A., and Towner, J.,** Pollution reconnaissance in stream sediments using non-residual trace metals, *Environ. Pollut. Ser. B.*, 10, 213, 1985.
83. **Loring, D. H.,** The distribution and partition of zinc, copper, and lead in the sediments of the Saguenay fjord, *Can. J. Earth Sci.*, 13, 960, 1976.
84. **Loring, D. H.,** Geochemistry of zinc, copper, and lead in the sediments of the estuary and Gulf of St. Lawrence, *Can. J. Earth Sci.*, 15, 757, 1978.
85. **Bopp, F., III and Biggs, R. B.,** Metals in estuarine sediments: factor analysis and its environmental significance, *Science,* 214, 441, 1981.
86. **Bruland, K. W., Bertine, K., Koide, M., Goldberg, E. D.,** History of metal pollution in southern California coastal zone, *Environ. Sci. Technol.*, 8, 425, 1974.
87. **Helmke, P. A., Koons, R. D., Schomberg, P. J., and Iskandar, I. K.,** Determination of trace element contamination of sediments by multielement analysis of clay-size fraction, *Environ. Sci. Technol.*, 11, 985, 1977.
88. **Helz, G. R., Sinex, S. A., Ferri, K. L., and Nichols, M.,** Processes controlling Fe, Mn and Zn in sediments of Northern Chesapeake Bay, *Estuarine Coastal Shelf Sci.*, 21, 1, 1985.
89. **Presley, B. J., Trefrey, J. H., and Shokes, R. F.,** Heavy metal inputs to Mississippi delta sediments, *Water Air Soil Pollut.*, 13, 481, 1980.
90. **Oliver, B. G.,** Heavy metal levels of Ottawa and Rideau River sediments, *Environ. Sci. Technol.*, 7, 135, 1973.
91. **Thomson-Becker, E. A. and Luoma, S. N.,** Temporal fluctuations in grain size, organic materials and iron concentrations in intertidal surface sediments of San Francisco Bay, *Hydrobiol.*, 129, 91, 1985.
92. **Sinex, S. A. and Helz, G. R.,** Entrapment of zinc and other trace elements in a rapidly flushed industrialized harbor, *Environ. Sci. Technol.*, 16, 820, 1982.
93. **Loring, D. H.,** personal communication, 1987.
94. **Edgren, M.,** Sediment as indicator of pollution, *Ambio*, 5, 133, 1977.
95. **Langston, W.,** Metals in sediments and benthic organisms in the Mersey estuary, *Estuarine Coastal Shelf Sci.*, 23, 239, 1986.
96. **Figueres, G., Martin, J. M., Meybeck, M., and Seyler, P.,** A comparative study of mercury contamination in the Tagus estuary (Portugal) and major French estuaries (Gironde, Loire, Rhone), *Estuarine Coastal Shelf Sci.*, 20, 183, 1985.

97. **Schults, D. W., Ferraro, S. P., Ditsworth, G. R., and Sercu, K. A.,** Selected chemical contaminants in surface sediments of Commencement Bay and the Tacoma Waterways, Washington, USA, *Mar. Environ. Res.*, 22, 271, 1987.
98. **Luoma, S. N. and Thomson-Becker, E. A.,** Temporal variability of Cu, Zn, and Ag in surface sediments of San Francisco Bay, in prep.
99. **Trefrey, J. H. and Presley, B. J.,** Heavy metal transport from the Mississippi River to the Gulf of Mexico, in *Marine Pollutant Transfer,* Windom, H. L. and Duce, R. A., Eds., Heath and Co., Lexington, 39, 1976.
100. **Draper, N. and Smith, H.,** *Applied Regression Analysis,* John Wiley & Sons, New York, 1981.
101. **Moore, J. W.,** Influence of water movements and other factors on distribution and transport of heavy metals in a shallow bay (Canada), *Arch. Environ. Contam. Toxicol.*, 10, 715, 1981.
102. **Thomson, E. A., Luoma, S. N., Johansson, C. E. and Cain, D. J.,** Comparison of sediments and organisms in identifying sources of biologically available trace metal contamination, *Water Res.*, 18, 755, 1984.
103. **Stoffers, P., Summerhayes, C., Forstner, U., and Patchineelam, S. R.,** Copper and other heavy metal contamination in sediments from New Bedford Harbor, Massachusetts: a preliminary note, *Environ. Sci. Technol.*, 11, 819, 1977.
104. **Norton, M. G., Rowlatt, S. M., and Nunny, R. S.,** Sewage sludge dumping and contamination of Liverpool Bay sediments, *Estuarine Coastal Shelf Sci.*, 19, 69, 1984.
105. **Rosental, R., Eagle, G. A., and Orren, M. J.,** Trace metal distribution in different chemical fractions of nearshore marine sediments, *Estuarine Coastal Shelf Sci.*, 22, 303, 1986.
106. **Voutsinou-Taliadouri, F. and Satsmadjis, J.,** Trace metals in the Pagassitikos Gulf, Greece, *Estuarine Coastal Shelf Sci.*, 15, 221, 1982.
107. **Young, R. A., Swift, D. J. P., Clarke, T. L., Harvey, G. R., and Betzer, P. R.,** Dispersal pathways for particle-associated pollutants, *Science,* 229, 431, 1985.
108. **Olsen, C. R., Larsen, I. L., Brewster, R. H., Cutshall, N. H., Bopp, R. F., and Simpson, H. J.,** *A geochemical assessment of sedimentation and contaminant distributions in the Hudson-Raritan estuary,* NOAA Technical Rep. NOS OMS 2, National Oceanographic and Atmospheric Administration, Seattle, Washington, 1984.
109. **Muller, G. and Forstner, U.,** Heavy metals in sediments of the Rhine and Elbe estuaries: mobilization or mixing effect?, *Environ. Geol.*, 1, 33, 1975.
110. **Duinker, J. C. and Nolting, R. F.,** Mixing, removal and mobilization of trace elements in the Rhine estuary, *Neth. J. Sea Res.*, 12, 205, 1978.
111. **Salomons, W. and Mook, W. G.,** Trace metal concentrations in estuarine sediments: mobilization, mixing or precipitation, *Neth. J. Sea Res.*, 11, 119, 1977.
112. **Loring, D. H., Rantala, R. T. T., Morris, A. W., Bale, A. J., and Howland, R. J. M.,** Chemical composition of suspended particles in an estuarine turbidity maximum zone, *Can. J. Fish. Aquat. Sci.*, 40, 201, 1983.
113. **Helz, G. R. and Sinex, S. A.,** Influence of infrequent floods on the trace metal composition of estuarine sediments, *Mar. Chem.*, 20, 1, 1986.
114. **De Souza, C. M. M., Pestana, M. H. D., and Lacerda, L. D.,** Geochemical partitioning of heavy metals in sediments of three estuaries along the Rio De Janeira coast (Brazil), *Sci. Total Environ.*, 58, 63, 1986.
115. **Taylor, D.,** Distribution of heavy metals in the sediment of an unpolluted estuarine environment, *Sci. Total Environ.*, 6, 259, 1976.
116. **de Forest, A., Murphy, S. P., and Pettis, R. W.,** Heavy metals in sediments from the Central New South Wales coastal region, *Aust. J. Mar. Freshwater Res.*, 29, 777, 1978.

Chapter 5

HEAVY METAL LEVELS IN MARINE INVERTEBRATES

P. S. Rainbow

TABLE OF CONTENTS

I. INTRODUCTION

Essential starting points for a study of heavy metal levels in marine invertebrates would be the seminal work of Vinogradov[1] *The Elementary Chemical Composition of Marine Organisms* and the later welcome compilation by Eisler[2] *Trace Metal Concentrations in Marine Organisms*. A brief glance at either would convince any reader of the impossibility of the task of producing a comprehensive review of heavy metal levels in marine invertebrates in a single chapter. No such attempt is made here. Rather, the aim of this chapter is to explore aspects that might assist in any understanding and appreciation of the significance of heavy metal concentrations in marine invertebrates — the significance with respect to the biology of the organism as opposed to their significance to man, the theme of a later chapter[3] in this volume.

The body concentration (typically μg/g) of a heavy metal in a marine invertebrate is a result of the net difference in the absolute amounts (μg) of metal taken up and lost by the body, as diluted by the body tissue present. Any change in metal uptake or loss, or in body weight, therefore, has the potential to change the body metal concentration. In this chapter, metal concentrations are expressed in terms of dry weight, usually more reproducible than fresh weight which depends on the degree of desiccation of the invertebrate analyzed.

II. UPTAKE OF HEAVY METALS

Routes for the uptake of heavy metals by marine invertebrates include uptake from solution and uptake from food. The uptake of heavy metals from solution into the bodies of marine invertebrates at permeable surfaces is generally considered to be a passive process not requiring expenditure of energy,[4-8] a situation contrasting with that for alkali metals such as sodium, potassium and calcium.[7] Heavy metals dissolved in the sea occur as different chemical species (ions, complexes, etc.) in equilibrium with inorganic and organic complexing agents including chlorides and carbonates.[9-12] The results of laboratory uptake experiments[13-19] involving heavy metals are explained by many (but not all)[20,21] authors by implicating the free metal ion (e.g., Cd^{++}, Zn^{++}) as the bioavailable dissolved form of the metal, although at equilibrium in the sea the free metal ion often represents only a small percentage of the total dissolved metal concentration.[9-12] A possible scenario for passive uptake would be a process of facilitated diffusion followed by passive sequestration of the metal to intracellular metal-binding ligands.[6,7,22,23] The free metal ion might bind with a membrane-associated transport protein and subsequently to a cascade of endogenous metal-binding ligands according to the stability constants of the ligands for the metal and their relative concentrations.[8] The sequestration of the metal in the cell would maintain a free ion activity within the cell below that of the surrounding medium (even if the total internal metal concentration greatly exceeds that externally) and the metal would continue to enter passively along the diffusion gradient.[8,24] Externally absolute concentrations of free metal ion would be restocked from the equilibria with other dissolved forms of the metal.

The free metal ion model also explains much of the data concerning the changed bioavailabilities of many heavy metals (reflected for example in altered toxicities and uptake rates) with variations in the physicochemical characteristics of media, such as changes in salinity[25] or in concentrations of organic chelating agents.[26] The percentages of total dissolved zinc and cadmium present in the form of the free metal ion for example have been modeled[9,10] to increase with decreased salinity, and zinc and cadmium are more bioavailable to invertebrates at lower salinities.[25,27-30] It should be stressed that it remains to be clarified whether there is any link between the uptake of particular heavy metals and that of calcium,[20] and whether physiological responses of estuarine invertebrates to reduced salinities are more significant to heavy metal uptake than associated physicochemical changes in the surrounding media.[28]

Water-soluble chelating agents able to complex a heavy metal reduce the uptake of that metal from solution,[26] as in the case of EDTA chelation with zinc and cadmium,[8,15,18,26] the presence of the chelating agent altering equilibria to reduce the concentration of the free metal ion. Typical affinities of water-soluble ligands for metals will probably differ in the reducing conditions which prevail interstitially in anoxic sediments. Lipid-soluble complexing agents in contrast often increase the bioavailability of heavy metals to marine invertebrates.[26] In such situations, the chelated metal may be taken up by an atypical route through the lipophilic membrane, bypassing the more usual route modeled above.[19,26] Similarly organo-metallic compounds, such as methyl mercury or tributyl tin,[32] whether synthesized naturally or artifactually, probably owe their high toxicities to their rapid accumulation in marine invertebrates[33,34] via such atypical routes.

Uptake of metals into the body via permeable surfaces might occur all over the body particularly in small and/or soft bodied invertebrates or mostly at sites of high permeability such as the gills. Uptake from solution will also take place in the alimentary tract when any of the medium is swallowed during 'drinking' or food ingestion.[35] The uptake of metal into the body across permeable surfaces is to be distinguished from the passive adsorption of metals onto binding sites on the outside of an invertebrate — particularly the cuticle of an aquatic arthropod.[23] Such metal never enters the organism and is available for passive exchange with the media. Nevertheless, this adsorbed metal does contribute to the total body concentration of a metal, often to a negligible extent[19,36,37] but occasionally significantly, as in small crustaceans of high surface area to volume ratio[38] or under abnormally high ambient concentrations of metals[39] especially iron and manganese.[40] In crustaceans, this adsorbed metal is not the only metal associated with the cuticle, for internally accumulated metal may be transported into the cuticle, subsequently either being lost with the molt or being reabsorbed internally.[40-42] The molting of crustaceans temporarily increases the rate of uptake of metals into the body, presumably in correlation with a temporary increase in body surface permeability prior to the tanning and/or calcification of new cuticle.[19,36,37,42,43]

Uptake of heavy metals from solution can take place in unexpected ways. Mangum[44] showed that in large marine gastropods like *Busycon carica,* seawater is taken into the foot as the animal expands after retraction into the shell, and this seawater mixes freely with the blood. Depledge and Phillips,[45] confirming the presence of the same process in *Hemifusus tuba,* point out its important implications as a route of heavy metal entry. Mason *et al.*[8] have identified another potential route of heavy metal entry in the polychaete worm *Neanthes arenaceodentata,* albeit under conditions of metabolic inhibition. The polychaete depends on the beating of cilia lining metanephridial canals to ensure that there is a constant flow of fluid from the coelom through the nephridia to the external medium. If these cilia are inhibited, for example by a xenobiotic compound, the external medium (with dissolved metals) may leak directly into the coelom without traversing gut or epidermis.[8]

Metals associated with particles usually become available to marine invertebrates only after ingestion and digestion in the alimentary tract. Burrowing invertebrates may be bathed in the interstitial water of the sediment or by their own irrigation currents interacting to varying degrees with interstitial water. The interstitial water contains dissolved metals in equilibria with sediment-associated metals, often under redox and other physicochemical conditions very different from those in overlying oxygenated waters. Dissolved interstitial metals may be available to infauna, especially to soft bodied invertebrates including members of most worm phyla.

Mucus sheets used in the ciliary feeding of many filter-feeding invertebrates including bivalve mollusks may adsorb heavy metals, these metals then being made available for assimilation in the gut. In some ascidian tunicates, vanadium adsorbed onto mucus may be taken up by pinocytosis into the cells of the pharynx.[46] Similarly, lamellibranch bivalves may also take up small metal-rich particles by pinocytosis in the gills.[47,48]

The food supply itself is a major route of heavy metal uptake into marine invertebrates, the availability of the constituent metals varying greatly with metal type and concentration, nature of food, digestive physiology of the consumer, etc.[49] The total metal concentration of the food ingested will not necessarily itself be a measure of metal availability. For example, the bioavailabilities in food of both lead and arsenic to the deposit feeding bivalve *Scrobicularia plana* depend on the concentration of iron present,[50,51] and silver concentrations in *S. plana* are reduced when copper concentrations in the sediment are high.[52] There is little evidence of the biomagnification of heavy metal concentrations along food chains,[49,53,54] although amplification of a metal concentration can occur if food intake is high, the metal is assimilated efficiently and is not excreted significantly.[49]

III. EXCRETION OF HEAVY METALS

Excretion of heavy metals from the bodies of marine invertebrates may also take place by a variety of routes. Passive desorption will reduce the body concentration of externally adsorbed metal if external physicochemical conditions change appropriately. Defecation will also remove the contribution to the total body metal concentration of the nonavailable metal contained in the gut contents. Internally accumulated metal may be lost in solution or in particulate form.

Metals may be excreted through permeable surfaces including the gills. For example, the crabs *Carcinus maenas* and *Cancer magister*[24,55] and the littoral prawn *Palaemon elegans*[41] excrete zinc across the gills, whereas zinc loss in the urine from the antennary glands is more important for the lobster *Homarus gammarus*.[24,55-57] The latter species loses manganese in three ways — by urinary excretion (20 to 40%), across the body surface (40 to 80%), and via the alimentary tract (up to 20%).[40] Cells lining the digestive tract of invertebrates (e.g., midgut diverticula, caeca or hepatopancreas) may release metal-rich granules into the gut lumen as in the bivalve *Cerastoderma edule,* the cockle,[58] or in the case of iron-rich ferritin crystals in stegocephalid amphipod crustaceans.[59] Metals detoxified in granules in the hepatopancreas or caeca of malacostracan crustaceans, for example lead in *Carcinus maenas,*[60] or copper in the amphipod *Corophium volutator,*[61] will be released into the lumen of the alimentary tract when the epithelial cells complete their cell cycle.

The cells of the kidneys of bivalve mollusks also have the ability to excrete metal-rich granules.[49] For example, the major excretory route of zinc from the mussel *Mytilus edulis* is in granular form from the kidneys into the urine (albeit slowly),[62] and the kidneys of many bivalves[48,62-65] similarly accumulate metal-rich granules with the potential for excretion.[66]

Other routes of metal loss include the excretion of iron into the byssus by the mussel *M. edulis*,[48,67] and the loss of metals (adsorbed and absorbed) with the cast molts of crustaceans, although some of both the essential metal zinc and the nonessential metal cadmium are resorbed into the body of the prawn *P. elegans* before ecdysis.[41,42] If an aquatic arthropod consumes its cast molt, metal excreted with the cast may become available for uptake again,[40] as well as any metal newly adsorbed onto the many binding sites of the fresh cast.[41,42]

IV. ACCUMULATION OF HEAVY METALS

Heavy metals are taken up passively by marine invertebrates even when external total metal concentrations are low. Thus marine invertebrates have the capacity to accumulate the metals to high body concentrations, depending on the relationship between metal uptake and excretion (net metal content) and the rate of diluting body growth. This relationship brings about a particular metal accumulation strategy in an organism, the most usual being that of net body accumulation over time in correlation with ambient metal bioavailability.

In the case of such net accumulators, metal excretion does not balance metal uptake. In some marine invertebrates, however, metal excretion rates can be matched to metal uptake rates such that the total body metal concentration does not change. This accumulation strategy is that of regulation — strictly the maintenance of an approximately constant body metal concentration over a wide range of ambient metal bioavailabilities. Metal uptake rates typically are beyond the short-term physiological control of individual invertebrates carrying out metal regulation, although these can vary metal excretion rates. Long-term differences in permeability to metals between populations or between species[28,68] will, however, assist in the adoption of a regulation strategy. Evidence for the presence of regulation strategies may come from laboratory experiments,[19,24,28,36,37,40,41,43,56,57,68-70] or from comparative analyses of metal concentrations of an invertebrate from a wide range of sites of known (suspected) different metal bioavailabilities.[55,69-73]

Decapod crustaceans for example regulate body concentrations of the essential metals zinc, copper, and manganese, balancing metal uptake rates by varied rates of metal excretion.[19,23,24,28,36,37,40,41,43,49,55-57,69,70] Specifically the littoral prawn *Palaemon elegans* regulates body concentrations of zinc to about 80 μg/g in dissolved zinc concentrations up to about 100 μg/l under defined physicochemical conditions, until under extreme conditions of high zinc bioavailability, regulation breaks down and net accumulation of body zinc begins.[19,28,36,43,70] The rate of zinc uptake by the prawn is affected by physicochemical factors beyond the control of the prawn,[19,28,36,43] but at least until the threshold of regulation breakdown, zinc excretion can be increased to match the rate of zinc uptake increasing with increased dissolved zinc concentration. On regulation breakdown, body zinc concentrations only need to double before lethality,[68] suggesting that the accumulated zinc is maintained in a metabolically available form with little significant resort to detoxification. Other crustaceans including barnacles[69,74,75] and amphipods[75] do not regulate body concentrations of these trace metals to a constant level over a range of dissolved metal bioavailabilities.

Other invertebrates show evidence of an ability to maintain relatively low body concentrations of a metal as the ambient metal bioavailability is increased. Such an ability may not be so definite as in decapods and is often referred to as partial regulation. In effect, there is no clear distinction between perfect regulators and net accumulators, but there exists a spectrum of invertebrate examples with decreasing net accumulation rates. An invertebrate may accumulate a metal at such a low rate that it closely approaches the strategy of regulation. For example, the amphipod *Echinogammarus pirloti* shows little (although real) change in net accumulation of body zinc with increased zinc bioavailability,[75] and the zinc accumulation strategy of another amphipod *Gammarus zaddachi* appears similar.[76]

Although regulation or partial regulation is not a typical metal accumulation strategy, examples have been reported from several invertebrate phyla in addition to crustaceans, usually involving zinc and to a lesser extent copper (both essential metals). In the bivalve mollusks, certain mytilids show evidence of regulation of body zinc concentrations, for example *Mytilus edulis*[76] and the tropical species *Septifer virgatus,*[71] *Perna viridis,*[72] and *Trichomya hirsuta.*[77] This strategy of zinc accumulation contrasts markedly with that of ostreid bivalves which are notoriously strong accumulators of zinc.[2,71] Among the gastropod mollusks, there is evidence of degrees of zinc regulation for *Haliotis tuberculata,*[73] *Littorina obtusata,*[54] *Littorina littorea,*[78,79] and *Gibbula umbilicalis,*[79] and for some copper regulation in *L. littorea.*[78,79] Of the polychaetes, the estuarine *Nereis diversicolor* can regulate body zinc concentrations.[68,76,80] A zinc-tolerant population of *N. diversicolor* had a reduced permeability to zinc in comparison to 'normal' worms and probably more effective zinc excretion,[70] both assisting in the regulation of body zinc levels.

The adoption of a regulation strategy in the accumulation of a metal does seem to be a feature of physiologically 'advanced' examples of invertebrate taxa, often those living in estuaries. Perhaps familiar adaptations of invertebrates to estuarine life, including reduced permeability and efficient excretory systems, can be considered as preadaptations to the adoption of metal regulating accumulation strategy. Furthermore, among the crustaceans, a decapod characteristic, such as a generally impermeable tanned and calcified exoskeleton (with the specific exception of the gills) and a macrophagous habit might also be preadaptations, reducing metal uptake and enhancing the possibility of matching metal excretion to uptake.[74,81]

Most marine invertebrates fail to match rates of heavy metal excretion to rates of metal uptake and will necessarily be net accumulators of heavy metals. Since all heavy metals are toxic at some concentration, such accumulation will usually involve detoxification of the metal to render it unavailable metabolically and restrict potential toxic effects. Examples of such invertebrates will cover a gradient from strong to weak accumulators of heavy metals. Notoriously strong accumulators include barnacles[74] and oysters[71] in the case of zinc, and such invertebrates are good candidates for use as biomonitors of available heavy metals.[3]

Detoxification of accumulated heavy metals by marine invertebrates takes a variety of forms[4,82] including the formation of metal-rich granules and binding by ligands (often proteins) with high metal affinities.[82-86] The detoxified product may be stored permanently as in the case of zinc phosphate granules beneath the midgut epithelia of barnacles, accumulating in vast numbers under conditions of high zinc bioavailability to produce body concentrations up to 150,000 μg/g.[74] Bivalve kidneys exemplify the transition from permanent to temporary storage of detoxified metal, accumulating large quantities of detoxified metaliferous granules with varying degrees and rates of release into the urine according to metal and bivalve species.[62-65,82] Similar temporary storage of detoxified metals for varying lengths of time takes place in the midgut and its derivatives in many invertebrate groups.[58-61] Relative rates of accumulation and release of detoxified metal will determine the metal concentration of the tissue concerned, and thence of the body.

Such tissue changes may even take place independently of the uptake of new metal. Copper, for example, is an essential component of hemocyanin, the respiratory pigment of certain mollusks and crustaceans. In decapod crustaceans, the concentration of hemocyanin in the blood falls at the time of the molt.[87,88] This fall is usually interpreted[23,87,88] to indicate that hemocyanin is broken down releasing copper, some of which is stored temporarily in the hepatopancreas,[87] possibly detoxified by binding as copperthionein.[88] Any resynthesis of hemocyanin may involve use of copper from this store,[87] or of copper newly accumulated.[87,88] The relationship between hemocyanin, copper, blood and hepatopancreas are complicated by changes in the water content of the hemolymph at the time of the molt, and are under critical review.[89]

Thus the nature of the accumulation strategy adopted, itself dependent on the detoxification process involved, will determine the final metal content of the marine invertebrate. The metal concentration is also affected by the weight of body tissue present. The rate of growth of an invertebrate may be so high that growth 'dilutes' the metal content, itself increasing with time, preventing any increase in metal concentration.

Decapod crustaceans, for example, accumulate the nonessential metal cadmium at all dissolved cadmium exposures,[39,42,69,70,75] with little, if any, significant cadmium excretion.[42,75] In most temperate coastal waters, cadmium concentrations and hence bioavailabilities are such that growth rates cause the dilution of the increasing accumulated cadmium content, and body concentrations of decapods fall into a typical range (about 1 to 2 μg/g).[70] Raised body cadmium concentrations in British coastal decapods (about 10 to 15 μg/g) are indicative of locally raised cadmium bioavailability.[70] The tropical decapod *Metapenaeopsis palmensis* from Hong Kong waters with a high absolute growth rate had body cadmium concentrations below detection limits (<0.06 to <0.32 μg/g),[90] providing a probable example of the dilution of body metal content by rapid growth. At the other extreme, the deep sea decapod *Systellaspis debilis* has exceedingly high body cadmium concentrations (up to

43 μg/g)[91,92] although the dissolved cadmium levels in the ocean are very low in comparison with coastal waters.[12] These decapods may have atypically low growth rates causing high cadmium concentrations, but body cadmium concentrations decrease, not increase, with size (taken to be equivalent to age).[91,92] *S. debilis* may, therefore, be accumulating cadmium from a particularly cadmium-rich diet.[91]

Examination of the relationship between metal concentration and size (weight) in a population of invertebrates can, therefore, be of assistance in investigating the biology of accumulated metals in marine invertebrates.[93] Metal concentrations decreasing with body size may indicate that a significant proportion of the metal content is surface adsorbed,[38,92] smaller specimens having a larger surface area to volume ratio. In the case of the mesopelagic decapod *S. debilis,* the concentration of iron, manganese, zinc, and cadmium all decreased with size but that of copper increased.[92] Follow up work has shown that juvenile *S. debilis* lack hemocyanin, perhaps in reflection of possible copper deficiency in the deep sea.[92,94]

V. ESSENTIAL REQUIREMENTS FOR HEAVY METALS

The body concentration of a metal in an invertebrate will depend on the accumulation strategy in operation, particularly on the method of detoxification employed, but in the case of essential metals, body concentrations of metabolically available metal must also reach a minimum concentration. Essential heavy metals include iron, zinc, molybdenum, selenium, copper, vanadium, manganese, nickel, tin, cobalt, and chromium which play a variety of roles in biochemistry often as enzyme cofactors.[95]

It is possible to attain theoretical estimates of the metabolic requirements of marine invertebrates.[92,96,97] Estimates of enzymatic requirements for a particular metal can be made using the weight of enzymes in tissue and the proportion of enzymes known to be associated with that particular metal, to calculate the contribution of enzyme-associated metal atoms to the total tissue dry weight.[97] There are many assumptions in these calculations[97] which are at best an approximation of the correct order of magnitude. Recognizing these assumptions, White and Rainbow calculated enzyme requirements of 34.5 μg/g Zn, 26.3 μg/g Cu, 27.4 μg/g Fe, and 3.9 μg/g Mn for marine invertebrates.[92,97] Certain invertebrates also have clear further requirements for particular metals. For example, decapod crustaceans and gastropod mollusks need extra copper to meet the copper requirements of typical hemocyanin loads.[97] Such calculations can be checked against body concentrations in invertebrates from habitats believed to be nonpolluted, and given the assumptions made, agreement is good.[92,97] The assumptions made in such calculations[97] need considerable refinement, but the calculation of theoretical metabolic requirements does appear to be an approach worth pursuing when attempting to evaluate the significance of essential trace metal concentrations in invertebrates.

The regulated body concentrations of essential metals in decapod crustaceans also provide guidelines as to the total metal requirements of marine invertebrates. As discussed earlier, decapod crustaceans maintain relatively constant body zinc concentrations (about 80 μg/g Zn in *Palaemon elegans*) and yet are killed if the body zinc concentration only doubles.[60] It appears that the accumulated body zinc is kept in a metabolically available form,[23] in contrast for example to zinc accumulated in barnacles (up to 150,000 μg/g Zn).[74] The regulated body concentrations of decapods are of the same order as the theoretical calculations of the metabolic requirements.[97] Changes in regulated body concentrations of zinc, for example intraspecifically with temperature in the case of *P. elegans,*[43] or interspecifically between closely related species (for example the palaemonids *P. elegans* and *Palaemonetes varians* and the pandalid decapod *Pandalus montagui*[28,37]) probably reflect subtle differences in metabolic requirements for zinc.[23]

Essential metals may also play structural roles. Zinc is a major inorganic component of the jaws of nereid polychaete worms, apparently as a structural characteristic.[98,99] The zinc in the jaws of *Nereis virens* accounts for over 40% of the total body burden of zinc in small worms but 30% or less in large worms.[99] (The constancy of the high zinc concentration in the jaw only partially accounts for the constancy of body metal levels of zinc in *N. diversicolor* in sediments of different zinc concentration.[68,99]) Manganese in nereid jaws can contribute over 50% to the total manganese body burden of small *N. virens,* but less in *Nereis diversicolor.*[99] In contrast, the jaws of glycerid polychaetes contain little zinc but have copper as the principal structural metal component (up to 67% total body burden of copper).[100]

VI. OTHER ROLES FOR ACCUMULATED HEAVY METALS

The extraordinary accumulation of vanadium by seas quirts has entered the folklore of marine biologists. Not all ascidian tunicates concentrate vanadium to the same extent, highest concentrations being found in the Ascidiidae and closely related Perophoridae (suborder Phlebobranchiata),[101] and in the suborder Aplousobranchia.[102] The vanadium in these tunicates is found in vacuoles in blood cells (vanadocytes), which also contain sulfate, probably as sulfuric acid.[101-105] The vanadium (in an organic complex) has a high reducing action.[101] Other ascidians (for example in the suborder Stolidobranchia) do contain vanadium in concentrations higher than in other marine invertebrates, though not as high as those in the sea squirts of the suborder Phlebobranchiata.[101,102] These other tunicates also concentrate a number of other metals including titanium, chromium, manganese, iron, niobium, and tantalum, also in blood cells and associated with acid in which they exhibit a powerful reducing action.[101] In the Phlebobranchiata (especially Ascidiidae), therefore, the 'primitive' concentration of these other metals has been superseded by the increased concentration of vanadium,[101] presumably in association with an increased efficiency of action. Carlisle[101] considered the function of the V-containing blood pigments to be that of powerful reducing agents concerned primarily in the synthesis of tunicin, a form of cellulose, constituting the tunic of sea squirts. There is evidence, for example, that the pigment may be involved in this role of ascidian embryos.[106] A possible secondary role in Ascidiidae, that of a carrier of molecular oxygen in conditions of low oxygen tension,[101] has been discounted by Mascara *et al.*[107] Rowley[108] has suggested that the vanadium may play an antimicrobial role in ascidian blood, and vanadium in the tunic may have antifouling and antipredator properties.[109,110]

Certain marine invertebrates may accumulate a heavy metal to a very high concentration for apparently specific reasons. The ampharetid polychaete *Melinna palmata* has copper concentrations atypically high in comparison with related ampharetids, much of this copper being present in the branchiae (about 10,000 μg/g Cu).[111] The function of the high copper concentration may be to reduce the palatability of the tissue, for the branchiae are distasteful to certain small fish.[111] The cirratulid polychaete *Tharyx marioni* accumulates remarkable concentrations of the metalloid arsenic (usually exceeding 2000 μg/g As) even under low ambient arsenic conditions.[112] Other cirratulids from the same habitats contained considerably less arsenic ($<$100 μg/g As).[112] Much of the arsenic in *T. marioni* is in the palps (6,000 to 13,000 μg/g As), but its function remains obscure.[112] The presence of the arsenic for example did not appear to influence predator behavior.[112] Other anomalously strong accumulators of arsenic are the gastropod mollusks *Hemifusus ternatanus* and *H. tuba* (spindle shells).[113,114]

VII. CONCLUSIONS

The concentrations of heavy metals in the bodies (and tissues) of marine invertebrates, therefore, depend on the accumulation strategy adopted by each species for each metal. The accumulation strategy results from the net difference between rates of uptake and excretion of metal, as affected by changes in body tissue. These changes may be increases due to growth, storage of metabolic reserves or gametes, or decreases resulting from starvation, gamete release, etc. The accumulation of heavy metals typically involves detoxification and storage to some degree, either temporary or permanent. Essential metals need to be present in metabolically available form at concentrations sufficient to meet physiological demands which may be biochemical roles widespread in all invertebrates or structural or metabolic functions specific to the organism itself. Relative rates of metal uptake and excretion are affected by features of the biology of the organism, including the permeability of external surfaces, the nature of the food and its method of capture (small planktonic and detritus particles are often metal-rich and filtering involves the passage of large volumes of water across often permeable surfaces), and the efficiency of osmoregulatory systems present, for example, in estuarine invertebrates.

REFERENCES

1. **Vinogradov, A. P.,** *The Elementary Chemical Composition of Marine Organisms,* Sears Foundation for Marine Research, Memoir 2, Yale University, New Haven, CT, 1953, 1.
2. **Eisler, R.,** *Trace Metal Concentrations in Marine Organisms,* Pergamon Press, New York, 1981, 1.
3. **Phillips, D. J. H.,** Use of macroalgae and invertebrates as monitors of metal levels in estuaries and coastal waters, in this volume, chap. 6.
4. **Combs, T. L. and George, S. G.,** Mechanisms of immobilisation and detoxification of metals in marine organisms, in *Proc. 12th Eur. Symp. Marine Biology, Physiology and Behaviour of Marine Organisms,* McLusky, D. S. and Berry, A. J., Eds., Pergamon Press, Oxford, 1978, 179.
5. **Carpene, E. and George, S. G.,** Absorption of cadmium by gills of *Mytilus edulis, Mol. Physiol.*, 1, 23, 1981.
6. **Williams, R. J. P.,** Physico-chemical aspects of inorganic element transfer through membranes, *Philos. Trans. R. Soc. Ser. B,* 294, 57, 1981.
7. **Williams, R. J. P.,** Natural selection of the chemical elements, *Proc. R. Soc. Ser. B,* 213, 361, 1981.
8. **Mason, A. Z., Jenkins, K. D., and Sullivan, P. A.,** Mechanisms of trace metal accumulation in the polychaete *Neanthes arenaceodentata, J. Mar. Biol. Assoc. U.K.*, 68, 61, 1988.
9. **Zirino, A. and Yamamoto, S.,** A pH-dependent model for the chemical speciation of copper, zinc, cadmium and lead in seawater, *Limnol. Oceanogr.*, 17, 661, 1972.
10. **Mantoura, R. F. C., Dickson, A., and Riley, J. P.,** The complexation of metals with humic materials in natural waters, *Estuarine Coastal Mar. Sci.*, 6, 387, 1978.
11. **Van den Berg, C. M. G.,** Organic and inorganic speciation of copper in the Irish Sea, *Mar. Chem.*, 14, 201, 1984.
12. **Bruland, K. W.,** Trace elements in seawater, in *Chemical Oceanography,* Vol. 8, Riley, J. P. and Chester, R., Eds., Academic Press, London, 1983, 157.
13. **Sunda, W. G., Engel, D. W., and Thuotte, R. M.,** Effects of chemical speciation on toxicity of cadmium to grass shrimp, *Palaemonetes pugio:* importance of free cadmium ion, *Environ. Sci. Technol.*, 12, 409, 1978.
14. **Engel, D. W. and Fowler, B. A.,** Factors influencing cadmium accumulation and its toxicity to marine organisms, *Environ. Health Perspect.*, 28, 81, 1979.
15. **Rainbow, P. S., Scott, A. G., Wiggins, E. A., and Jackson, R. W.,** Effect of chelating agents on the accumulation of cadmium by the barnacle *Semibalanus balanoides,* and complexation of soluble Cd, Zn and Cu, *Mar. Ecol. Prog. Ser.*, 2, 143, 1980.
16. **Fowler, S. W. and Heyraud, M.,** Biologically transformed zinc and its availability for bioaccumulation by marine organisms, in *Management of Environment,* Patel, B., Ed., Wiley Eastern, New Delhi, 1980, 389.
17. **Zamuda, C. D. and Sunda, W. G.,** Bioavailability of dissolved copper to the American oyster *Crassostrea virginica.* I. Importance of chemical speciation, *Mar. Biol. Berlin,* 66, 77, 1982.
18. **Jenkins, K. D. and Sanders, B. M.,** Relationships between free cadmium ion activity in seawater, cadmium accumulation and subcellular distribution and growth in polychaetes, *Environ. Health Perspect.*, 65, 205, 1986.
19. **Nugegoda, D. and Rainbow, P. S.,** Effect of a chelating agent (EDTA) on zinc uptake and regulation by *Palaemon elegans* (Crustacea:Decapoda), *J. Mar. Biol. Assoc. U.K.*, 68, 25, 1988.

20. **Wright, D. A.,** The effect of calcium on cadmium uptake by the shore crab, *Carcinus maenas, J. Exp. Biol.*, 67, 163, 1977.
21. **Simkiss, K.,** Lipid solubility of heavy metals in saline solutions, *J. Mar. Biol. Assoc. U.K.*, 63, 1, 1983.
22. **Simkiss, K., Taylor, M., and Mason, A. Z.,** Metal detoxification and bioaccumulation in molluscs, *Mar. Biol. Lett.*, 3, 187, 1982.
23. **Rainbow, P. S.,** The significance of trace metal concentrations in decapods, in *Symp. Zool. Soc. London, 59, Aspects of Decapod Crustacean Biology,* Fincham, A. A. and Rainbow, P. S., Eds., Oxford University Press, Oxford, 1988, 291.
24. **Bryan, G. W.,** The metabolism of Zn and ^{65}Zn in crabs, lobsters and freshwater crayfish, in *Radioecological Concentration Processes,* Aberg, G. and Hungate, F. P., Eds., Pergamon Press, Oxford, 1966, 1005.
25. **McLusky, D. S., Bryant, V., and Campbell, R.,** The effects of temperature and salinity on the toxicity of heavy metals to marine and estuarine invertebrates, *Oceanogr. Mar. Biol. Annu. Rev.*, 24, 481, 1986.
26. **Ahsanullah, M. and Florence, T. M.,** Toxicity of copper to the marine amphipod *Allorchestes compressa* in the presence of water- and lipid-soluble ligands, *Mar. Biol. Berlin,* 84, 41, 1984.
27. **Wright, D. A.,** The effect of salinity on cadmium uptake by the tissues of the shore crab *Carcinus maenas, J. Exp. Biol.*, 67, 137, 1977.
28. **Nugegoda, D. and Rainbow, P. S.,** Effects of salinity changes on zinc uptake and regulation by the decapod crustaceans *Palaemon elegans* and *Palaemonetes varians, Mar. Ecol. Prog. Ser.*, 51, 57, 1989.
29. **Phillips, D. J. H.,** The common mussel *Mytilus edulis* as an indicator of pollution by zinc, cadmium, lead and copper. I. Effects of environmental variables on uptake of metals, *Mar. Biol. Berlin,* 38, 59, 1976.
30. **George, S. G., Carpene, E., and Coombs, T. L.,** The effect of salinity on the uptake of cadmium by the common mussel, *Mytilus edulis* (L.), in *Proc. 12th Eur. Symp. Marine Biology, Physiology and Behaviour of Marine Organisms,* McLusky, D. S. and Berry, A. J., Eds., Pergamon Press, Oxford, 1978, 189.
31. **Jensen, S. and Jernelov, A.,** Biological methylation of mercury in aquatic organisms, *Nature (London),* 223, 753, 1969.
32. **Bryan, G. W., Gibbs, P. E., Burt, G. R., and Hummerstone, L. G.,** The effects of tributyltin (TBT) accumulation on adult dog-whelks, *Nucella lapillus:* long term field and laboratory experiments, *J. Mar. Biol. Assoc. U.K.*, 67, 525, 1987.
33. **Corner, E. D. S. and Sparrow, B. W.,** The modes of action of toxic agents. II. Factors influencing the toxicities of mercury compounds to certain crustacea, *J. Mar. Biol. Assoc. U. K.*, 36, 459, 1957.
34. **Corner, E. D. S. and Rigler, F. H.,** The modes of action of toxic agents. III. Mercuric chloride and n-amyl mercuric chloride on crustaceans, *J. Mar. Biol. Assoc. U.K.*, 37, 85, 1958.
35. **Jennings, J. R., and Rainbow, P. S.,** Accumulation of cadmium by *Artemia salina, Mar. Biol. Berlin,* 51, 47, 1979.
36. **White, S. L. and Rainbow, P. S.,** Regulation of zinc concentration by *Palaemon elegans* (Crustacea:Decapoda): zinc flux and effects of temperature, zinc concentration and moulting, *Mar. Ecol. Prog. Ser.*, 16, 135, 1984.
37. **Nugegoda, D. and Rainbow, P. S.,** Zinc uptake and regulation by the sublittoral prawn *Pandalus montagui* (Crustacea:Decapoda), *Estuarine Coastal Shelf Sci.*, 26, 619, 1988.
38. **Rainbow, P. S. and Moore, P. G.,** Comparative metal analyses in amphipod crustaceans, *Hydrobiologia,* 141, 273, 1986.
39. **Jennings, J. R. and Rainbow, P. S.,** Studies on the uptake of cadmium by the crab *Carcinus maenas* in the laboratory. I. Accumulation from seawater and a food source, *Mar. Biol. Berlin,* 50, 131, 1979.
40. **Bryan, G. W., and Ward, E.,** The absorption and loss of radioactive and non radioactive manganese by the lobster, *Homarus vulgaris, J. Mar. Biol. Assoc. U.K.*, 45, 65, 1965.
41. **White, S. L. and Rainbow, P. S.,** Zinc flux in *Palaemon elegans* (Crustacea:Decapoda): moulting, individual variation and tissue distribution, *Mar. Ecol. Prog. Ser.*, 19, 153, 1984.
42. **White, S. L. and Rainbow, P. S.,** Accumulation of cadmium by *Palaemon elegans* (Crustacea:Decapoda), *Mar. Ecol. Prog. Ser.*, 32, 17, 1986.
43. **Nugegoda, D. and Rainbow, P. S.,** The effect of temperature on zinc regulation by the decapod crustacean *Palaemon elegans* Rathke, *Ophelia,* 27, 17, 1987.
44. **Mangum, C. P.,** A note on blood and water mixing in large marine gastropods, *Comp. Biochem. Physiol.*, 63A, 389, 1979.
45. **Depledge, M. H. and Phillips, D. J. H.,** Circulation, respiration and fluid dynamics in the gastropod mollusc, *Hemifusus tuba* (Gmelin), *J. Exp. Mar. Biol. Ecol.*, 95, 1, 1986.
46. **Kalk, M.,** Absorption of vanadium by tunicates, *Nature (London),* 198, 1010, 1963.
47. **Bevelander, G. and Nakahara, H.,** Correlation of lysosomal activity and ingestion by mantle epithelium, *Biol. Bull.*, 131, 76, 1966.
48. **George, S. G., Pirie, B. J. S., and Coombs, T. L.,** The kinetics of accumulation and excretion of ferric hydroxide in *Mytilus edulis* (L.) and its distribution in the tissues, *J. Exp. Mar. Biol. Ecol.*, 23, 71, 1976.

49. **Bryan, G. W.,** Heavy metal contamination in the sea, in *Marine Pollution,* Johnston, R., Ed., Academic Press, London, 1976, 185.
50. **Luoma, S. N. and Bryan, G. W.,** Factors controlling the availability of sediment-bound lead to the estuarine bivalve *Scrobicularia plana, J. Mar. Biol. Assoc. U.K.,* 58, 793, 1978.
51. **Langston, W. J.,** Arsenic in U.K. estuaries and its availability to benthic organisms, *J. Mar. Biol. Assoc. U.K.,* 60, 869, 1980.
52. **Luoma, S. N. and Bryan, G. W.,** A statistical study of environmental factors controlling concentrations of heavy metals in the burrowing bivalve *Scrobicularia plana* and the polychaete *Nereis diversicolor, Estuarine Coastal Mar. Sci.,* 15, 95, 1982.
53. **Moore, M. N.,** Elemental accumulation in organisms and food chains, in *Analysis of Marine Ecosystems,* Longhurst, A. R., Ed., Academic Press, London, 535, 18, 1981.
54. **Young, M. L.,** The transfer of ^{65}Zn and ^{59}Fe along a *Fucus serratus* (L.) → *Littorina obtusata* (L.) food chain, *J. Mar. Biol. Assoc. U.K.,* 55, 583, 1975.
55. **Bryan, G. W.,** Concentrations of zinc and copper in the tissues of decapod crustaceans, *J. Mar. Biol. Assoc. U.K.,* 48, 303, 1968.
56. **Bryan, G. W.,** Zinc regulation in the lobster *Homarus vulgaris.* I. Tissue zinc and copper concentrations, *J. Mar. Biol. Assoc. U.K.,* 44, 549, 1964.
57. **Bryan, G. W., Hummerstone, L. G., and Ward, E.,** Zinc regulation in the lobster *Hommarus gammarus:* importance of different pathways of absorption and excretion, *J. Mar. Biol. Assoc. U.K.,* 66, 175, 1986.
58. **Owen, G.,** Observations on the stomach and digestive diverticula of the Lamellibranchia. I. The Anisomyaria and Eulamellibranchia, *Q. J. Microsc. Sci.,* 96, 517, 1955.
59. **Moore, P. G. and Rainbow, P. S.,** Ferritin crystals in the gut caeca of *Stegocephaloides christianiensis* Boeck and other Stegocephalidae (Amphipoda:Gammaridea): a functional interpretation, *Philos. Trans. R. Soc. Ser. B,* 306, 219, 1984.
60. **Hopkins, S. P. and Nott, J. A.,** Some observations on concentrically structured, intracellular granules in the hepatopancreas of the shore crab *Carcinus maenas* (L.), *J. Mar. Biol. Assoc. U.K.,* 59, 867, 1979.
61. **Icely, J. D. and Nott, J. A.,** Accumulation of copper in the 'hepatopancreatic' caeca of *Corophium volutator* (Crustacea:Amphipoda), *Mar. Biol. Berlin,* 57, 193, 1980.
62. **George, S. G. and Pirie, B. J. S.,** Metabolism of zinc in the mussel, *Mytilus edulis* (L.): a combined ultrastructural and biochemical study, *J. Mar. Biol. Assoc. U.K.,* 60, 575, 1980.
63. **George, S. G. and Pirie, B. J. S.,** The occurrence of cadmium in the sub-cellular particles in the kidney of Cd-exposed marine mussel, *Mytilus edulis:* the use of electron probe microprobe analysis, *Biochim. Biophys. Acta,* 580, 234, 1979.
64. **George, S. G., Pirie, B. J. S., and Coombs, T. L.,** Isolation and elemental analysis of metal-rich granules from the kidney of the scallop, *Pecten maximus* (L.), *J. Exp. Mar. Biol. Ecol.,* 23, 71, 1980.
65. **Carmichael, N. G., Squibb, K. S., and Fowler, B. A.,** Metals in the molluscan kidney. A comparison of two closely related bivalve species *(Argopecten),* using X-ray microanalysis and atomic absorption spectroscopy, *J. Fish. Res. Board Can.,* 36, 1149, 1979.
66. **Bryan, G. W.,** The occurrence and seasonal variation of trace metals in the scallops *Pecten maximus.* (L.) and *Chlamys opercularis* (L.), *J. Mar. Biol. Assoc. U.K.,* 53, 145, 1973.
67. **Pentreath, R. J.,** The accumulation and retention of ^{65}Zn, ^{54}Mn, ^{58}Co and ^{59}Fe by the mussel *Mytilus edulis, J. Mar. Biol. Assoc. U.K.,* 53, 127, 1973.
68. **Bryan, G. W. and Hummerstone, L. G.,** Adaptation of the polychaete *Nereis diversicolor* to estuarine sediments containing high concentrations of zinc and cadmium, *J. Mar. Biol. Assoc. U.K.,* 53, 839, 1973.
69. **Rainbow, P. S.,** Accumulation of Zn, Cu and Cd by crabs and barnacles, *Estuarine Coastal Shelf Sci.,* 21, 669, 1985.
70. **White, S. L. and Rainbow, P. S.,** Regulation and accumulation of copper, zinc and cadmium by the shrimp *Palaemon elegans, Mar. Ecol. Prog. Ser.,* 8, 95, 1982.
71. **Phillips, D. J. H. and Yim, W. W.-S.,** A comparative evaluation of oysters, mussels and sediments as indicators of trace metals in Hong Kong waters, *Mar. Ecol. Prog. Ser.,* 6, 285, 1981.
72. **Phillips, D. J. H.,** Organochlorines and trace metals in green-lipped mussels *Perna viridis* from Hong Kong waters: a test of indicator ability, *Mar. Ecol. Prog. Ser.,* 21, 251, 1985.
73. **Bryan, G. W., Potts, G. W., and Forster, G. R.,** Heavy metals in the gastropod mollusc *Haliotis tuberculata* (L.), *J. Mar. Biol. Assoc. U.K.,* 57, 379, 1977.
74. **Rainbow, P. S.,** Heavy metals in barnacles, in *Barnacle Biology,* Southward, A. J., Ed., A. A. Balkema, Rotterdam, 1987, 405.
75. **Rainbow, P. S. and White, S. L.,** Comparative strategies of heavy metal accumulation by crustaceans: zinc, copper and cadmium in a decapod, an amphipod and a barnacle, *Hydrobiologia,* 174, 245, 1989.
76. **Amiard, J. C., Amiard-Triquet, C., Berthet, B., and Métayer, C.,** Comparative study of the patterns of bioaccumulation of essential (Cu, Zn) and non-essential (Cd, Pb) trace metals in various estuarine and coastal organisms, *J. Exp. Mar. Biol. Ecol.,* 106, 73, 1987.

77. **Klumpp, D. W. and Burdon-Jones, C.,** Investigations of the potential of bivalve molluscs as indicators of heavy metal levels in tropical marine waters, *Aust. J. Mar. Freshwater Res.,* 33, 285, 1982.
78. **Bryan, G. W., Langston, W. J., Hummerstone, L. G., Burt, G. R., and Ho, Y. B.,** An assessment of the gastropod, *Littorina littorea,* as an indicator of heavy-metal contamination in United Kingdom estuaries, *J. Mar. Biol. Assoc. U.K.,* 63, 327, 1983.
79. **Amiard-Triquet, C., Métayer, C., and Amiard, J. C.,** Études *in situ* et expérimentales de l'écotoxicologie de quatre métaux (Cd, Pb, Cu, Zn) chez des algues et des mollusques gastéropodes brouteurs, *Water Air Soil Pollut.,* 34, 11, 1987.
80. **Bryan, G. W., Langston, W. J., and Hummerstone, L. G.,** The use of biological indicators of heavy metal contamination in estuaries, *Occas. Publ. Mar. Biol. Assoc. U.K.,* 1, 1, 1980.
81. **Rainbow, P. S.,** The biology of heavy metals in the sea, *Int. J. Environ. Stud.,* 25, 195, 1985.
82. **George, S. G.,** Biochemical and cytological assessments of metal toxicity in marine animals, this volume, chap. 8.
83. **Simkiss, K.,** Biomineralization and detoxification, *Calcif. Tissue Res.,* 24, 199, 1979.
84. **Simkiss, K.,** Calcium, pyrophosphate and cellular pollution, *Trends Biochem. Sci.,* 6, 3, 1981.
85. **Mason, A. Z. and Nott, J. A.,** The role of intracellular biomineralised granules in the regulation and detoxification of metals in gastropods with special reference to the marine prosobranch *Littorina littorea, Aquat. Toxicol.,* 1, 239, 1981.
86. **George, S. G.,** Subcellular accumulation and detoxification of metals in aquatic animals, in *Physiological Mechanisms of Marine Pollutant Toxicity,* Vernberg, W. B., Calabrese, A., Thurberg, F. P., and Vernberg, F. J., Eds., Academic Press, New York, 1982, 3.
87. **Zuckerkandl, E.,** Hémocyanine et cuivre chez un crustacé décapode, dans leur rapports avec le cycle d'intermue, *Ann. Inst. Oceanogr. Monaco (N.S.),* 38, 1, 1960.
88. **Engel, D. W.,** Metal regulation and molting in the blue crab, *Callinectes sapidus:* copper, zinc and metallothionein, *Biol. Bull.,* 172, 69, 1987.
89. **Depledge, M. H. and Bjerregaard, P.,** Haemolymph protein composition and copper levels in decapod crustaceans, *Helgoländer Meeresuntors,* 439, 1989.
90. **Rainbow, P. S.,** Trace metal concentrations in a Hong Kong penaeid prawn, *Metapenaeopis palmensis* (Haswell), in *Proc. 2nd Int. Workshop Marine Flora and Fauna of Hong Kong and Southern China,* Hong Kong 1986, Morton, B.S., Ed., Hong Kong University Press, Hong Kong, in press.
91. **Ridout, P. S., Willcocks, A. D., Morris, R. J., White, S. L., and Rainbow, P. S.,** Concentrations of Mn, Fe, Cu, Zn, and Cd in the mesopelagic decapod *Systellaspis debilis* from the East Atlantic Ocean, *Mar. Biol. Berlin,* 87, 285, 1985.
92. **White, S. L. and Rainbow, P. S.,** Heavy metal concentrations and size effects in the mesopelagic decapod crustacean *Systellaspis debilis, Mar. Ecol. Prog. Ser.,* 37, 147, 1987.
93. **Boyden, C. R.,** Effect of size on metal content of shellfish, *J. Mar. Biol. Assoc. U.K.,* 57, 675, 1975.
94. **Rainbow, P. S. and Abdennour, C.,** Copper and haemocyanin in the mesopelagic decapod crustacean *Systellaspis debilis, Oceanol. Acta,* 12, 91, 1989.
95. **Harrison, P. M. and Hoare, R. J.,** *Metals in Biochemistry,* Chapman and Hall, London, 1980.
96. **Pequegnat, J. E., Fowler, S. W., and Small, L. F.,** Estimates of the zinc requirements of marine organisms, *J. Fish. Res. Board Can.,* 26, 145, 1969.
97. **White, S. L. and Rainbow, P. S.,** On the metabolic requirements for copper and zinc in molluscs and crustaceans, *Mar. Environ. Res.,* 16, 215, 1985.
98. **Bryan, G. W. and Gibbs, P. E.,** Zinc — a major inorganic component of nereid polychaete jaws, *J. Mar. Biol. Assoc. U. K.,* 59, 969, 1979.
99. **Bryan, G. W. and Gibbs, P. E.,** Metals in nereid polychaetes: the contribution of metals in the jaws to the total body burden, *J. Mar. Biol. Assoc. U.K.,* 60, 641, 1980.
100. **Gibbs, P. E. and Bryan, G. W.,** Copper — the major component of glycerid polychaete jaws, *J. Mar. Biol. Assoc. U.K.,* 60, 205, 1980.
101. **Carlisle, D. B.,** Vanadium and other metals in ascidians, *Proc. R. Soc. Ser. B,* 171, 31, 1968.
102. **Michibata, H., Terada, T., Anada, N., Yamakara, K., and Namakurai, T.,** The accumulation and distribution of vanadium, iron and manganese in some solitary ascidians, *Biol. Bull,* 171, 672, 1986.
103. **Rowley, A. F.,** The blood cells of *Ciona intestinalis:* an electron probe X-ray microanalytical study, *J. Mar. Biol. Assoc. U.K.,* 62, 607, 1982.
104. **Bell, M. V., Pirie, B. J. S., McPhail, D. B., Goodman, B. A., Falk-Petersen, I.-B., and Sargent, J. R.,** Contents of vanadium and sulphur in the blood cells of *Ascidia mentula* and *Ascidiella aspersa, J. Mar. Biol. Assoc. U.K.,* 62, 709, 1982.
105. **Pirie, B. J. S. and Bell, M. V.,** The localisation of inorganic elements, particularly vanadium and sulphur, in haemolymph from the ascidians *Ascidia mentula* (Muller) and *Ascidiella aspersa* (Muller), *J. Exp. Mar. Biol. Ecol.,* 74, 187, 1984.
106. **Robinson, W. E., Kustin, K., and Cloney, R. A.,** The influence of tunichrome and other reducing compounds on tunic and fin formation in embryonic *Ascidia callosa* Stimpson, *J. Exp. Zool.,* 237, 63, 1986.

107. **Mascara, I. G., McLeod, G. C., and Kustin, K.,** Isolation, properties and structural studies on a compound from tunicate blood cells that may be involved in vanadium accumulation, *Biochem. J.*, 181, 457, 1979.
108. **Rowley, A.,** Preliminary investigations on the possible antimocrobial properties of tunicate blood cell vanadium, *J. Exp. Zool.*, 227, 319, 1983.
109. **Stoecker, D.,** Resistance of a tunicate to fouling, *Biol. Bull.*, 155, 615, 1978.
110. **Stoecker, D.,** Relationships between chemical defense and ecology in benthic ascidians, *Mar. Ecol. Prog. Ser.*, 3, 257, 1980.
111. **Gibbs, P. E., Bryan, G. W., and Ryan, K. P.,** Copper accumulation by the polychaete *Melinna palmata:* an antipredation mechanism?, *J. Mar. Biol. Assoc. U.K.*, 61, 707, 1981.
112. **Gibbs, P. E., Langston, W. J., Burt, G. R. and Pascoe, P. L.,** *Tharyx marioni* (Polychaeta): a remarkable accumulator of arsenic, *J. Mar. Biol. Assoc. U.K.*, 63, 313, 1983.
113. **Phillips, D. J. H. and Depledge, M. H.,** Chemical forms of arsenic in marine organisms, with emphasis on *Hemifusus* species, *Water Sci. Technol.*, 18, 213, 1986.
114. **Phillips, D. J. H. and Depledge, M. H.,** Distribution of inorganic and total arsenic in tissues of the marine gastropod *Hemifusus ternatanus, Mar. Ecol. Prog. Ser.*, 34, 261, 1986.

Chapter 6

USE OF MACROALGAE AND INVERTEBRATES AS MONITORS OF METAL LEVELS IN ESTUARIES AND COASTAL WATERS

David J. H. Phillips

TABLE OF CONTENTS

I. INTRODUCTION

Trace metals are directly toxic to aquatic organisms and are also significantly accumulated by many marine and estuarine species. This gives rise to concerns both with respect to the possible detrimental effects of metals on coastal resources, and in terms of the potential impacts of metals on human health. Monitoring programs are, therefore, required to establish both spatial and temporal trends in metal abundance and bioavailability in estuaries and other coastal waters.

Such monitoring could rely upon the quantification of metals in the waters themselves, in the underlying sediments, or in organisms.[1-5] It is now generally held that the analysis of water for metals suffers from several disadvantages. Thus, metal concentrations are very low in natural waters and are difficult to quantify while avoiding extraneous contamination.[6-8] Temporal variability is also a problem, rendering mean levels of pollutants difficult to determine.[9-11] There is also no definite relationship between metals in "soluble" and "particulate" fractions of natural waters and their availabilities to organisms.[3,12]

The monitoring of metals in coastal environments by the analysis of sediments (see Chapter 4) offers several advantages over the use of water, particularly in terms of providing time-averaged values for pollutant abundance. However, it is difficult to account for the effects of particle size and organic carbon content on metal levels.[1-3,13] In addition, no simple method exists to determine the bioavailabilities of metals in sediments.[14-17]

As a result of these problems, the use of organisms (which by definition reflect bioavailable levels of metals) is now the most widely employed method to monitor trace elements in coastal waters.[1,3,18,19] Organisms employed to quantify pollutant abundance or bioavailability by virtue of their tissue concentrations of contaminants have been referred to as *bioindicators* (also employed for species whose presence or absence in certain environments is indicative of the impacts of pollution or other factors), *sentinel organisms,* and *bio-monitors.* The latter term is employed in the present review.

II. THE EARLY DEVELOPMENT OF BIO-MONITORING TECHNIQUES

Organisms were employed in the early 1960s to monitor radionuclides released from the Hanford reactors to the Columbia River, and those in coastal waters in California. The analysis of water samples demanded huge sample volumes, and yielded barely detectable amounts of the isotopes of interest (^{54}Mn, ^{60}Co, and ^{65}Zn). However, the analysis of phytoplankton and euphausiids[20,21] and of macroalgae, mussels, barnacles, and other species including fish[22,23] permitted the quantification of these radionuclides fairly simply. Later studies[24] favored the use of mussels (*Mytilus edulis* and *M. californianus*) and barnacles (*Lepas anatifera* and *Pollicipes polymerus*) in particular, as these species heavily accumulated the isotopes of interest.

The mid- to late 1960s saw an increase in the use of biota in monitoring programs. Butler[25-27] employed bivalves to monitor pesticide levels in many estuaries of the U.S. In the U.K., monitoring programs using shellfish and finfish species to quantify both pesticides and trace metals commenced in the late 1960s and have continued to date.[28-30]

However, very little was then known about the mechanisms of uptake and excretion of metals by species used as bio-monitors. Such data became available slowly through the 1970s, and research continues to the present, serving to hone and improve the techniques employed, both in the sampling and analytical arenas.

III. IMPROVEMENTS AND IMPORTANT CONCEPTS

The use of an aquatic organism as bio-monitor of metals in coastal waters is defensible only when the resulting picture of environmental contamination (either in a spatial or temporal sense) truly reflects ambient metal bioavailabilities. Prerequisites for a species to be a useful bio-monitor may be summarized as follows:[1,3,31,32]

- Bio-monitors should be sessile or sedentary, thus being representative of the study area.
- Bio-monitors should be hardy, tolerating high levels of pollutants and large ranges in salinity, and permitting laboratory studies of pollutant kinetics.
- Bio-monitors should be abundant in study areas, easy to identify and sample, and should provide sufficient tissue for analysis of the contaminants of interest.
- A simple correlation should exist between the pollutant concentration found in the tissues of a bio-monitor and the average ambient bioavailable pollutant concentration. This correlation should be the same at all study sites.

It is unfortunate that many studies to date have lost sight of these basic essential prerequisites. It is not sufficient to simply collect and analyse previously unstudied species, or to analyse any and all organisms available at a study site. Only certain organisms (conforming well to the above prerequisites) should be employed in biomonitoring studies.

Perhaps the most important property of a bio-monitor is that it must not regulate the accumulated levels of pollutants in its body or tissues (see Chapter 5). The propensity of species to regulate metals varies with the element concerned, and both between and within phyla. For example, decapod crustaceans are known to regulate their internal tissue concentrations of the essential metals copper, manganese, and zinc;[33-41] however, other nonessential elements, such as cadmium are not subject to regulation.[37,39,42,43] By contrast, barnacles do not appear to metabolically regulate the concentrations of any metals in their bodies,[44-46] accumulating many elements to very high absolute concentrations. Macroalgae are thought to be passive accumulators of metals, and most bivalve mollusks are also poor regulators of their body load of elements.[1,3,35,47-50] However, some bivalves have been found to regulate their net uptake of zinc into soft tissues, although this is restricted to only a few species.[48,49]

Phillips[1,3] has also emphasized the importance of the *rates* of uptake and excretion of contaminants by bio-monitors, as these kinetics define the period over which such species reflect the ambient pollution conditions. If uptake and excretion rates are slow, the contaminant concentrations present in the tissues of bio-monitoring species will be indicative of the average bioavailability of those contaminants over a long period. By contrast, species which exhibit rapid turnover of contaminants will reflect only recent pollution conditions. Surprisingly little information is available on contaminant kinetics in most bio-monitors, and those data which are published are frequently based on laboratory studies of pollutant uptake (mostly from solution). These often do not adequately mimic contaminant kinetics in the field,[3,51,52] presumably because the contaminants are taken up into different tissues and/or cellular sequestration sites (from which they are lost at different rates) in the two situations.

Despite this dearth of high-quality information, it is generally held true that metals exhibit relatively long half-lives in most bio-monitors compared to pollutants such as organochlorines or hydrocarbons.[3,53,54] This relatively slow excretion of metals by bio-monitors simplifies the design of monitoring studies for metals, and reduces the need for multiple sampling with time,[54] although the impacts of seasonal variations in metal levels of bio-monitors must be carefully considered in the design of monitoring programs.[55-59]

The movement or migration of a bio-monitor causes great problems for the interpretation of data in monitoring programs. Thus, for example, contaminant levels in fish are a complex function of their exposure in space and time to pollutants, and these variables cannot be separated.[1,3] The use of species which exhibit significant movement or migration as bio-monitors is generally not defensible,[3,54] and sedentary or sessile invertebrate species are preferred as bio-monitors by most authors.

Finally here, it should be noted that programs employing bio-monitors to measure metal bioavailabilities in aquatic environments generally seek to identify significant differences in metal abundance between samples from different sites, or between those taken at the same location at different times. The identification of such spatial and/or temporal differences is only efficiently accomplished if the variation between metal concentrations exhibited by the individuals sampled at any site and time is minimized. The sources of variability and the residual "inherent variability" of metals within populations of bio-monitors have been analyzed and discussed by several authors.[1,3,53,59-62] Sampling regimes to reduce within-site variability in metal concentrations in bio-monitors have been proposed,[63] and the number of individuals required in any one sample for analysis has been discussed.[53,60,61] The data of Gordon et al.[60] are particularly useful in the design of sampling programs, although it should be noted that the inherent variability of metals in bio-monitors varies between species, sites, and with the element concerned.[3,59,61]

IV. THE PRESENT STATUS

A. MACROALGAE

Marine and estuarine macroalgae have long been known to concentrate metals to levels many times those found in the surrounding waters.[64,65] This is believed now to be due to their accumulation of metals by both passive and active processes, involving ion-exchange types of binding to surface polysaccharides as well as metal sequestration within membrane-limited vacuoles rich in polyphenols.[66,67]

The binding of metals by macroalgae was shown by Bryan[68] to be strong, with only minimal exchange between bound metals and ambient waters. Most metals were thought to be taken up in proportion to their external concentration,[35,36,68] giving rise to the possibility of the use of these seaweeds as bio-monitors.

Early studies of macroalgae as monitors of metals in coastal environments were mainly restricted to the U.K.[69-74] Direct comparisons between metal levels in waters and those in macroalgae generally provided encouraging results.[75,76] Seeliger and Edwards[75] found close correlations between dissolved levels of copper and lead in the waters of Raritan Bay near New York and concentrations of these elements in four species of algae. Morris and Bale[76] produced similar data for cadmium, copper, and zinc in solution and in *Fucus vesiculosus* from the Bristol Channel. However, the relationship for manganese in the latter studies was poor, and the alga was thought to partially regulate this element.[76]

Later studies provided further insight into the extent of the usefulness of macroalgae to monitor metals in aquatic environments. These species essentially respond to metals present in solution in the ambient waters.[1,68] As a result, spatial trends in contamination provided by data on metal concentrations in macroalgae may not match those found for other species, such as bivalve mollusks, which respond not only to metals in solution but also to those in suspension (adsorbed to inorganic particulates, or in phytoplankton food).[77-80]

The details of sampling of macroalgae for metal studies are also important. Metal concentrations vary in different portions of the plant, generally being higher in older parts of the alga and least in fast-growing tips.[32,35,65,68,72-74] This is believed to be a function of the strong binding of metals by macroalgae, giving rise to increased element concentrations with age.[3,35,81] Some authors have suggested that this may be advantageous, permitting the contamination history to be identified at any given location through the analysis of different portions of the alga, each of different age.[82,83]

However, other authors[1,3,84] have proposed that the impacts of growth on metal levels in macroalgae are detrimental to the use of these species as efficient bio-monitors. Phillips[1,3] suggested that the known impacts of many variables (portion of the plant analyzed, season, light intensity, temperature, possibly shore position [related to degree of submergence[85]]) on metal concentrations in macroalgae were likely to be mediated at least in part through changes in the growth rates of the plants. Given this, any factor which tends to alter growth rates of macroalgae (those cited above, turbidity, nutrient availability, etc.) would interfere with the accuracy of inter-site comparisons of metal levels in these species.[86] Rice and Lapointe[84] reached the same conclusion, based on laboratory studies of the net uptake of several trace elements by *Ulva fasciata*. This problem has not been solved to date, but is a potential major interference in the use of macroalgae as bio-monitors in coastal environments.

A further problem in employing macroalgae to monitor trace metals concerns the surface contamination of algae by metals associated with fine particulate material. Such material is difficult to eliminate by washing prior to analysis and interferes with the quantification of metals actually present in the plant tissues.[72,87,88] This problem is particularly prevalent for metals which are heavily associated with particulates in coastal waters, such as iron, lead, and chromium.[87,88]

Phillips[1,3] has reviewed data on interactions between metals in macroalgae, and also discussed the impacts of such effects on the use of bio-monitors. Bryan[68] showed that the uptake of ^{65}Zn by the brown alga *Laminaria digitata* in the laboratory could be reduced by additions of cadmium, copper, or manganese to the external medium. Foster[89] found that the exceptionally heavy contamination of waters of Dulas Bay in Anglesey, Wales was not reflected by metal levels in *Fucus vesiculosus* or *Ascophyllum nodosum* for all elements. The uptake of large amounts of copper and zinc by these species appeared to reduce the accumulation of other elements, perhaps through the saturation of available binding sites.[89] These two examples involve unusually high concentrations of metals in solution, much greater than those normally found in coastal waters. It is not known whether such metal interactions occur at lower ambient exposure concentrations.

The toxic effects of metals on macroalgae may also influence their ability to act as efficient bio-monitors in certain instances. Most elements are not significantly toxic to macroalgae at concentrations likely to be present in coastal waters; however, copper is an exception to this general rule.[68,90] In certain instances, species have evolved adaptive mechanisms to deal with this, including the extracellular release of compounds which chelate copper ions, the use of particular storage sites within the tissues to detoxify accumulated copper, and the active exclusion of copper by the plant. The last of these mechanisms would clearly interfere with any attempt to employ these species as bio-monitors of copper in aquatic environments, and is effectively a case of metal regulation. Hall et al.[91] have provided evidence for such an exclusion of copper in copper-tolerant strains of the marine fouling alga *Ectocarpus siliculosus*. By contrast, Reed and Moffat[92] found that copper-tolerant isolates of *Enteromorpha compressa* did not rely on exclusion of the metal, but on internal detoxification of copper subsequent to its normal uptake. These examples are extreme, in that the species concerned are fouling organisms found on vessels employing copper-based antifouling paints. There is no evidence that such adaptive mechanisms have evolved in macroalgae in more normal coastal environments.

It may be concluded that, despite the existence of significant unsolved problems concerning their use, macroalgae continue to be of benefit as bio-monitors of metals in estuaries and other coastal waters. Studies of shallow and even deep-water species continue to be reported, both for stable metals[93-97] and rare radionuclides.[98-100] The latter application is reminiscent of the early development of bio-monitors (see Section II above), and the driving forces for such studies are no different today from those existing over two decades ago.

B. CRUSTACEANS

1. Decapod Crustaceans

Early studies by Bryan,[33-36] extended by Rainbow and co-workers[37-43,101-105] have demonstrated that the metals copper, manganese, and zinc are subject to metabolic regulation in the tissues and whole bodies of decapod crustaceans. Several factors (such as molting, temperature, and animal size) may interfere with the concentrations of these three elements in decapod crustaceans, but there is no correlation between their tissue concentrations and their levels in ambient seawater.[3,35,36,39,106,107] While the regulation of zinc can be overcome under some conditions by exposure to very high concentrations of zinc in solution (100 μg/l or greater), this is most unlikely to occur in the environment. As a result, decapod crustaceans cannot be employed as bio-monitors of these metals with useful results, despite attempts to the contrary.[108-110]

By contrast, some metals are accumulated by decapod crustaceans in proportion to environmental exposure. Cadmium is one of these,[37,39,42,43,101,102,111-115] and the levels of this element in some species are sufficiently high to be potentially hazardous to human health.[116-118] Other nonessential metals accumulated by decapods in proportion to ambient concentrations include lead,[119] mercury,[120] and vanadium.[121,122]

Despite this, decapods have been little used as bio-monitors. This is largely due to the impacts of external variables on their uptake of metals. For example, the accumulation of cadmium by these crustaceans[3,111-113] is dictated by salinity. Much of the metal taken up by decapods is adsorbed to or bound in the exoskeleton;[3,33,34,42] molting, therefore, affects the amounts of some metals present in the tissues.[38,104,121] Animal size is also known to determine metal concentrations in at least some species.[105,123] In addition, few decapods can be considered to be truly sedentary species which would accurately reflect the element bioavailability at the location of their capture. It is concluded that decapod crustaceans are not suitable as bio-monitors of metals in coastal waters, at least in most instances.

2. Barnacles

The accumulation and sequestration of metals by barnacles offers a direct contrast to the processes described above for decapod crustaceans.[3,39,124] Early studies of radionuclide uptake by barnacles[22-24] showed these organisms to accumulate isotopes to high concentration, and later investigations of stable metals have confirmed this tendency. Ireland[125,126] published some of the initial work on metals in barnacles, using *Balanus* (now *Semibalanus) balanoides*. It was shown that this species accumulates greater amounts of zinc than those found in mussel (*Mytilus edulis*) tissues by at least an order of magnitude,[125] and that concentrations of copper, lead, manganese and zinc vary considerably with season in *B. balanoides*.[126]

Further studies by Walker and co-workers[44-46] confirmed the unusually high accumulation of metals by barnacles, and showed that both copper and zinc are sequestered in granules in parenchyma cells below the gut epithelium.[107,124,127,128] It was suggested on the basis of comparisons of contamination levels of barnacles at various sites with those of other species that barnacles might be useful bio-monitors of zinc in coastal environments.[45]

While later studies have also generally concluded that barnacles exhibit many of the prerequisites to act as efficient bio-monitors of metals in aquatic ecosystems, surprisingly little use has been made of these species for such purposes. Similarly, although some data are available on the uptake and depuration of metals in barnacles,[39,129,130] they have been largely ignored in laboratory investigations. This is unfortunate, and there is scope for further work on the use of barnacles as bio-monitors.

It is relevant here that Ireland[125] reported differences between the profiles of contamination for zinc at several locations, based on either analyses of *Balanus balanoides* or mussels (*M. edulis*). This suggests that the two bio-monitors respond to different portions of the total metal load in the environment, i.e., that the bioavailabilities of zinc in distinct forms in coastal waters differ from the two species. Recent studies[131,132] of three species of barnacles and the mussel *Perna viridis* in Hong Kong waters (in which well-defined gradients in metal pollution have been demonstrated) have generally not shown such subtleties; the contamination profiles exhibited by all four species matched well for most metals studied. Barnacles are thus thought to have considerable potential as biomonitors of metals in coastal waters and deserve further attention.

C. GASTROPOD MOLLUSKS

Gastropod mullusks have been much less extensively studied as potential bio-monitors in aquatic ecosystems than have bivalve mollusks. Nevertheless, many gastropods conform well to at least some of the prerequisites for a species to be a useful bio-monitor.

The earliest monitoring studies using gastropods were undertaken in the Bristol Channel, and employed both transplanted and native dogwhelks (*Nucella lapillus*) and limpets (*Patella vulgata*).[133-135] These species were shown to accumulate and excrete metals slowly, and both animal size and the shore level at which the samples were taken were found to affect metal concentrations in the soft tissues.[3,133-135] Later studies have confirmed size to be an important variable influencing metal levels in gastropods.[136-138]

While several authors have simply assumed that gastropod mollusks are efficient bio-monitors of metals in aquatic environments,[139,140] there is evidence for at least partial regulation of certain metals in some species. Young[81] postulated that the winkle *Littorina obtusata* was able to partially regulate its tissue content of zinc when fed macroalgae of differing degrees of contamination. Both Anderlini[141] and Bryan et al.[138] reported data suggesting that zinc is regulated in abalone of the genus *Haliotis*. Evidence for the regulation of copper by gastropods is equivocal. Betzer and Pilson[142] considered that the channeled whelk *Busycon canaliculatum* regulated copper poorly, if at all, while Martoja et al.[143] reported that the winkle *Littorina littorea* accumulated copper as its sulfide with age in the tissues and that this was independent of environmental exposure to the metal. There is little evidence for the regulation of nonessential metals in gastropods.

Studies of the uptake of zinc and iron by *L. obtusata* concluded that the accumulation of metals from food is important in determining their concentrations in gastropods.[81] Gastropod species differ considerably in their preferred diet, some being herbivorous (e.g., *L. littorea,* feeding mainly on species of *Fucus* and other algae) and others carnivorous (e.g., *N. lapillus,* which feeds on barnacles and mussels). In situations where comparative levels of metals in solution do not bear a consistent relationship to those associated with particulates at all locations, this introduces differences into the contamination profiles displayed by such species.[3,144] Thus, Ireland and Wootton[144] noted that contaminant profiles generated from the analysis of *L.littorea* and *N. lapillus* at nine sites on the coast of Wales did not match. Other authors have reported similar data for other species.[77,95] This is reminiscent of the differences between spatial trends in metal levels produced by studies of macroalgae and bivalve mollusks,[77-80] discussed in Section IV. A above. As noted by Phillips,[3] such differences do not negate the possibility of using these species as bio-monitors; rather, spatial trends in contamination should be compared with a knowledge of the dietary habits of species as a basis for interpretation.

Several extraneous factors (in addition to animal size, discussed above) are known to influence the concentrations of metals present in gastropod mollusks, and these should be considered if such species are to be employed as bio-monitors. Flegal and Martin[145] have emphasized the importance of depuration of some species prior to analysis, as metals present in ingested sediment (or in particulates trapped in the mantle cavity[81]) may interfere significantly with analytical data. Seasonal changes in metal concentration may also be considerable, and these variations may be due either to changes in the ambient concentrations of metals or to the effects of gametogenesis and spawning.[142] Finally, a case of metal interactions has been reported by Nelson et al.,[146] in which the uptake of copper by slipper limpets (*Crepidula fornicata*) was enhanced by the co-presence of silver.

It may be concluded that the presently available information on metal kinetics in gastropod mollusks is sparse, and is generally insufficient to recommend enthusiastically such species for use as bio-monitors of metals in coastal waters. The existence of at least partial regulation of some of the essential metals is particularly notable, and to some extent parallels the situation in decapod crustaceans.

D. BIVALVE MOLLUSKS

Bivalve mollusks have been more frequently employed as bio-monitors of metals in aquatic environments than have species of any other family or phylum and the available literature on their use for such purposes is considerable. Most studies have been carried out in temperate rather than tropical waters, although the latter areas have received greater attention recently, concomitant to the increased urbanization and industrialization of tropical nations. Reviews of the use of bivalves as bio-monitors of aquatic pollutants have been produced by several authors;[3,147,148] the present paper largely concentrates upon more recent data, using these reviews as a basis to describe the older material.

Studies of metal uptake, sequestration, and excretion in bivalves have been very extensive,[3,47,127,149-161] and these have provided a firm basis for the evaluation of the usefulness of bivalves as bio-monitors of metals in aquatic ecosystems. In general, the accumulation of metals has been shown to be relatively rapid in bivalves and to reflect ambient exposure levels closely. Metals are sequestered both in granular form and bound to metallothioneins; the prevalence of each of these mechanisms of metal storage varies between elements and by species. The excretion of metals is thought to be generally slow,[52,150,154,162-164] at least by comparison to most organochlorines or hydrocarbons.[53] It should be repeated here, however, that estimates of the half-lives of metals in biota generated from laboratory experiments are frequently unreliable indications of the situation in the field.[51,52] The best estimates of metal depuration rates are provided by transplantation of field populations of bivalves. Okazaki and Panietz[162] provided an excellent example of this technique, and reported half-lives of 12 metals in *Crassostrea gigas* to range from 23 to 60 d, while the range for *C. virginica* was from 70 to 180 d. The rates of metal uptake and excretion by bivalves are important, as these define the period over which a species time-averages ambient concentrations of metals.[1,3,53,54] Certain authors have also provided insights into the effects of the chemical speciation of metals on their uptake by bivalves;[165,166] however, these studies are in their infancy and much remains to be discovered.

Most of the studies cited above on metal kinetics in bivalves have concluded that these species do not regulate element concentrations in their tissues, but respond faithfully to changes in the ambient levels of metals. However, a few isolated exceptions to this general rule are known; interestingly, these all involve zinc. Phillips and Yim[167] reported that the mytilid *Septifer virgatus* (incorrectly named as *S. bilocularis* in the original paper[48]) partially regulates its tissue concentrations of zinc in Hong Kong waters. Similar findings were later published for zinc in the hairy mussel *Trichomya hirsuta* from Australia,[49] and in green-lipped mussels (*Perna viridis*) from Hong Kong.[48,131,132] Unpublished data on *Perna canaliculus* (which is closely related to *P. viridis*) from New Zealand waters also suggest the partial regulation of zinc.[237] It is not known why these few species have developed such a capacity for regulating their tissue levels of zinc; however, it is clear that this is a most exceptional attribute among bivalves.

Several species of bivalve mollusks exhibit broad geographical ranges (e.g., species of the genera *Mytilus, Perna,* and *Crassostrea*), which permits their use in large-scale studies of metal distribution,[55,78,79,168] embodied by the "Mussel Watch" concept.[53,169] In addition, many bivalves are euryhaline in nature (allowing investigations of estuaries and purely marine environments using the same species), and most are highly tolerant of pollutant-induced stress.

Thus, few examples exist where pollutants are thought to potentially reduce the capacity of bivalves to reflect ambient metal levels. However, an important exception to this general rule concerns the toxic effects of copper on the blue mussel *Mytilus edulis* (probably the most widely used bio-monitor to date). Early indications that *M. edulis* might not accurately reflect copper bioavailabilities in all situations[56,63] were followed by detailed investigations of the sublethal effects of copper on this species.[170-178] Copper has been documented to be of unusual toxicity to *M. edulis*, affecting filtration rates and growth rates, and eliciting a valve closure response at concentrations not dissimilar to those found in contaminated environments. The valve closure response effectively isolates the internal tissues of the mussel from the ambient seawater. In locations which are highly contaminated by copper, this may reduce the effectiveness of *M. edulis* as a bio-monitor. Other studies have suggested that the tolerance of *M. edulis* and other bivalves to copper may be either genetically based or adaptive.[179,180]

Variables which affect the concentrations of metals in bivalves have been reviewed in detail elsewhere.[1,3] These include animal size,[136,137,155,181-191] season (sexual condition),[5,58,59,155,184,189,192-194] shore level of sampling,[3,187,195] sex of the individuals sampled,[3,192,194] and water salinity and temperature.[3,190,196] Most of the effects of such parameters can be eliminated or accounted for during sampling[3,53,63] or in the interpretation of monitoring data.

However, interactions between pollutants are difficult to account for, or even to recognize, in bio-monitoring studies.[1,3] The existence of such interactions casts doubt upon the reliability of a species to accurately reflect the bioavailabilities of individual toxicants, and indeed, upon the very concept of bio-monitoring.[3] Luckily, interactions between metals (or between metals and other contaminants) are rare among bivalves, although a few examples are known. Fowler et al.[197] found that mercury displaced iron in the mantle fringes of the quahog, *Mercenaria mercenaria*. The uptake of copper by *Mytilus edulis* may be affected by the presence of other elements,[56] and that of cadmium may be reduced by the co-presence of zinc,[196] although these laboratory findings are based on high doses of metals and are of uncertain validity in field conditions. Luoma and Bryan[16] showed that lead uptake by the burrowing bivalve *Scrobicularia plana* was influenced by the co-presence of iron. However, this may be mediated by the scavenging of lead by hydrous oxides of iron in sediments externally to the animal, rather than through a bona fide interaction between the elements in the tissues of *S. plana*.[3] It should also be noted that speculation by Young and Jan[198] that the uptake of cadmium by the rock scallop *Hinites multirugosus* was reduced by the co-presence of high levels of DDE and PCBs in the heavily contaminated Palos Verdes area of California has not been resolved to date.

While several authors have reported that no interactions between elements could be observed in studies of metal uptake by various species of bivalves,[157,199-201] two recent reports have added further instances of metal interactions to those discussed above; coincidentally, these both involve silver. Luoma and Bryan[202] found that levels of silver in *S. plana* are low when copper concentrations are high in local sediments. However, as noted above for lead and iron in *S. plana*, this does not necessarily imply the existence of interactions within the tissues of the bivalve. Finally, Calabrese et al.[176] found that mussels (*M. edulis*) exposed to silver accumulated greater amounts of copper than did controls not exposed to silver. These data are reminiscent of the results cited previously for silver and copper uptake by slipper limpets, *Crepidula fornicata*.[146]

Several authors have also considered the use of bivalve shells or even the byssus for bio-monitoring purposes.[182,203-207] However, metals vary in concentration in different regions of the shell and also with depth (and chemical composition) in the shell. In addition, the factors influencing the deposition of metals in shells or byssal threads are poorly understood. Although some elements (especially the transuranics) are heavily deposited in these tissues, this is no guarantee of their adequacy for bio-monitoring purposes. There is scope for additional work in this field.

In conclusion, most species of bivalves conform admirably to the requirements for an organism to be an effective bio-monitor of pollution. The use of these animals in national and regional monitoring programs, such as those in the U.S.[55,208-213] is clearly of great benefit, and there can be no doubt that this is an improvement over more traditional monitoring techniques employing water or sediment analysis. Where native bivalves of the required species are not present, transplantation techniques are readily available.[214,215] Some authors have stated that there are definite benefits to the use of transplants,[216] and certain programs employ these routinely rather than using native animals, although occasional problems have been encountered with the interpretation of data from transplanted samples.[217] Whether native organisms or transplants are used, however, bivalve mollusks are in most cases a highly reliable and efficient tool for estimating the bioavailability of metals in coastal environments. This is particularly the case where the main sources of variance[3] are accounted for, and the necessary attention is paid to the choice of species used, the sampling regime, and the analytical methodology.

E. OTHER INVERTEBRATE SPECIES

Several aquatic species other than those described above have been considered for use as bio-monitors, and some have actually been employed in monitoring programs. However, in general, much less is known of the kinetics of metal uptake and excretion in these species, and their use as bio-monitors has, therefore, not been shown to be legitimate as yet.

Henze[218] first noted that ascidians accumulate large quantities of certain metals. Their uptake and sequestration of vanadium in particular has excited further work more recently,[219-221] and Papadopoulou and Kanias[222] have suggested that certain ascidian species may be useful as bio-monitors. However, although species such as *Ciona intestinalis* exhibit widespread geographical distributions and may have potential for this purpose,[3] almost nothing is known of element kinetics in ascidians or the impacts of external variables upon these.

By contrast to the extensive studies on gastropod and (particularly) bivalve mollusks discussed above, relatively few data are available for metals in cephalopods. It is known that both squid and octopus accumulate elements to high levels,[223-225] but no attempts to employ these species as bio-monitors per se have been reported. Many cephalopod species exhibit significant movement, and this detracts from their likely usefulness as bio-monitors.[3]

More data are available on annelids, but these species do not appear to always be well suited to a role in bio-monitoring programs. At least some polychaete species are known to regulate certain elements, such as zinc,[36,226,227] and the acquisition of tolerance to other metals (e.g., cadmium and copper) may be based upon a reduced net accumulation,[227,228] which severely restricts the usefulness of such species as bio-monitors. The uptake of metals by polychaetes and other species is also affected by animal size,[36,227,229] by salinity in certain instances,[230] and by the reproductive cycle of the worm.[231] Nevertheless, polychaetes, such as *Nereis diversicolor* and *N. succinea* have been employed to monitor metals in coastal ecosystems by some authors,[95,232] and studies of chromium uptake by *Neanthes arenaceodentata* suggest that this element is accumulated in proportion to environmental exposure levels.[233] Investigations of metals in the feather duster tube worm *Eudistylia vancouveri*[234,235] and in nematode species[236] are also informative, although little is known of element kinetics in these organisms.

V. SUMMARY AND CONCLUSIONS

There can be no doubt that the use of biota to monitor metal levels in coastal waters has greatly improved our knowledge of the impacts of these contaminants in such areas. It has been argued that metals are of no consequence in aquatic ecosystems unless they are bioavailable (or at least of direct toxicity to biota of such ecosystems, which probably equates to bioavailability in most cases). There is no viable alternative at present to the use of bio-monitors, if the bioavailability of metals is the primary concern.[18]

Nevertheless, it is certainly the case that this technique has been abused by many researchers. A lack of attention in establishing that a species possesses the necessary attributes to perform efficiently as a bio-monitor results in the compilation of data which are either meaningless or of very limited use in defining water quality. Such data are unfortunately not uncommon in the literature, and this debases the value of this powerful monitoring technique.[54]

The most basic requirement for a species to act as an accurate bio-monitor is that its tissue levels of contaminants should faithfully reflect the ambient bioavailability of those pollutants, which implies a lack of regulation of contaminant uptake. On this basis, the most appropriate bio-monitors include macroalgae, bivalve mollusks, and perhaps barnacles. This does not imply, however, that all species of these families are equally useful bio-monitors in all situations. There is a need for continued research (which should emphasize the use of these types of organisms in particular) to further improve bio-monitoring techniques; this will require both laboratory and field investigations. It cannot be denied that great strides have been made in the past two decades through the use of bio-monitors; however, much remains to be learned before we can be confident that the pollution threat from metals to coastal waters is fully understood.

REFERENCES

1. **Phillips, D. J. H.,** The use of biological indicator organisms to monitor trace metal pollution in marine and estuarine environments — a review, *Environ. Pollut.,* 13, 281, 1977.
2. **Förstner, U.,** Inorganic pollutants, particularly heavy metals in estuaries, in *Chemistry and Biogeochemistry of Estuaries,* Olausson E. and Cato, I., Eds., John Wiley & Sons, New York, 1980, 307.
3. **Phillips, D. J. H.,** *Quantitative Aquatic Biological Indicators: Their Use to Monitor Trace Metal and Organochlorine Pollution,* Applied Science Publishers, London, 1980.
4. **Prosi, F.,** Heavy metals in aquatic organisms, in *Metal Pollution in the Aquatic Environment,* Förstner, U. and Wittmann, G. T. W., Eds., Springer-Verlag, Berlin, 1981, 271.
5. **Thompson, E. A., Luoma, S. N., Johansson, C. E., and Cain, D. J.,** Comparison of sediments and organisms in identifying sources of biologically available trace metal contamination, *Water Res.,* 18, 755, 1984.
6. **Bruland, K. W., Knauer, G. A., and Martin, J. H.,** Zinc in north-east Pacific water, *Nature,* 271, 741, 1978.
7. **Bruland, K. W., Knauer, G. A., and Martin, J. H.,** Cadmium in northeast Pacific waters, *Limnol. Oceanogr.,* 23, 618, 1978.
8. **Burnett, M. and Patterson, C. C.,** Analysis of natural and industrial lead in marine ecosystems, in *Lead in the Marine Environment,* Branica, M., and Konrad, Z., Eds., Pergamon Press, Oxford, 1980, 15.
9. **Fukai, R., Murray, C. N., and Huynh-Ngoc, L.,** Variations of soluble zinc in the Var River and its estuary, *Estuarine Coastal Mar. Sci.,* 3, 177, 1975.
10. **Grimshaw, D. L., Lewin, J., and Fuge, R.,** Seasonal and short-term variations in the concentration and supply of dissolved zinc to polluted aquatic environments, *Environ. Pollut.,* 11, 1, 1976.
11. **Boyden, C. R., Aston, S. R., and Thornton, I.,** Tidal and seasonal variations of trace elements in two Cornish estuaries, *Estuarine Coastal Mar. Sci.,* 9, 303, 1979.
12. **Florence, T. M.,** The speciation of trace elements in waters, *Talanta,* 29, 345, 1982.

13. **Mayer, L. M. and Fink, L. K.,** Granulometric dependence of chromium accumulation in estuarine sediments in Maine, *Estuarine Coastal Mar. Sci.*, 11, 491, 1980.
14. **Ayling, G. M.,** Uptake of cadmium, zinc, copper, lead and chromium in the Pacific oyster, *Crassostrea gigas,* grown in the Tamar River, Tasmania, *Water Res.*, 8, 729, 1974.
15. **Knezovich, J. P., Harrison, F. L., and Wilhelm, R. G.,** The bioavailability of sediment-sorbed organic chemicals: a review, *Water Air Soil Pollut.*, 32, 233, 1987.
16. **Luoma, S. N. and Bryan, G. W.,** Factors controlling the availability of sediment-bound lead to the estuarine bivalve *Scrobicularia plana, J. Mar. Biol. Assoc. U.K.*, 58, 793, 1978.
17. **Loring, D. H.,** Potential availability of metals in eastern Canadian estuarine and coastal sediments, *Rapp. P. V. Reun. Cons. Int. Explor. Mer.*, 181, 93, 1981.
18. **Waldichuk, M.,** Biological availability of metals to marine organisms, *Mar. Pollut. Bull.*, 16, 7, 1985.
19. **Nelson, A. and Donkin, P.,** Processes of bioaccumulation: the importance of chemical speciation, *Mar. Pollut. Bull.*, 16, 164, 1985.
20. **Osterberg, C., Small, L., and Hubbard, L.,** Radioactivity in large marine plankton as a function of surface area, *Nature,* 197, 883, 1963.
21. **Osterberg, C., Pattullo, J., and Pearcy, W.,** Zinc-65 in euphausiids as related to Columbia River water off the Oregon coast, *Limnol. Oceanogr.*, 9, 249, 1964.
22. **Folsom, T. R., Young, D. R., Johnson, J. N., and Pillai, K. C.,** Manganese-54 and zinc-65 in coastal organisms of California, *Nature,* 200, 327, 1963.
23. **Folsom, T. R. and Young, D. R.,** Silver-110*m* and cobalt-60 in oceanic and coastal organisms, *Nature,* 206, 803, 1965.
24. **Young, D. R. and Folsom, T.R.,** Mussels and barnacles as indicators of the variation of ^{54}Mn, ^{60}Co and ^{65}Zn in the marine environment, in *Radioactive Contamination of the Marine Environment,* International Atomic Energy Agency, Vienna, 1973, 633.
25. **Butler, P. A.,** The problem of pesticides in estuaries, in *A Symp. Estuarine Fisheries, Trans. Amer. Fish. Soc. Special Publ.*, No, 3, 110, 1966.
26. **Butler, P. A.,** Monitoring pesticide pollution, *BioScience,* 19, 889, 1969.
27. **Butler, P. A.,** Organochlorine residues in estuarine molluscs, 1965—1972 — National Pesticides Monitoring Program, *Pestic. Monit. J.*, 6, 238, 1973.
28. **Portmann, J. E.,** Monitoring metals in marine animals, *Mar. Pollut. Bull.*, 2, 157, 1971.
29. **Holden, A. V.** International cooperative study of organochlorine and mercury residues in wildlife, 1969—71, *Pestic. Monit. J.*, 7, 37, 1973.
30. **Murray, A. J. and Portmann, J. E.,** Management of conservative pollutants in the marine environment: some advantages and a few problems of using fish and shellfish, *Mar. Pollut. Bull.*, 13, 300, 1982.
31. **Butler, P. A., Andrén, L., Bonde, G. J., Jernelov, A., and Reish, D. J.,** Monitoring organisms, in FAO Technical Conference on Marine Pollution and its Effects on Living Resources and Fishing, Rome 1970, Supplement 1, Report of the Seminar on Methods of Detection, Measurement and Monitoring of Pollutants in the Marine Environment, FAO Fisheries Rep. No. 99, Food and Agriculture Organization, Rome, 1971, 101.
32. **Haug, A., Melsom, S., and Omang, S.,** Estimation of heavy metal pollution in two Norwegian fjord areas by analysis of the brown alga *Ascophyllum nodosum, Environ. Pollut.*, 7, 179, 1974.
33. **Bryan, G. W.,** Zinc regulation in the lobster *Homarus vulgaris.* I. Tissue zinc and copper concentrations, *J. Mar. Biol. Assoc. U.K.*, 44, 549, 1964.
34. **Bryan, G. W.,** Concentrations of zinc and copper in the tissues of decapod crustaceans, *J. Mar. Biol. Assoc. U.K.*, 48, 303, 1968.
35. **Bryan, G. W.,** The effects of heavy metals (other than mercury) on marine and estuarine organisms, *Proc. R. Soc. London Ser. B,* 177, 389, 1971.
36. **Bryan, G. W.,** Heavy metal contamination in the sea, in *Marine Pollution,* Johnston, R., Ed., Academic Press, London, 1976, 185.
37. **White, S. L. and Rainbow, P. S.,** Regulation and accumulation of copper, zinc and cadmium by the shrimp *Palaemon elegans, Mar. Ecol. Prog. Ser.*, 8, 95, 1982.
38. **White, S. L., and Rainbow, P. S.,** Regulation of zinc concentration by *Palaemon elegans* (Crustacea: Decapoda): zinc flux and effects of temperature, zinc concentration and moulting, *Mar. Ecol. Prog. Ser.*, 16, 135, 1984.
39. **Rainbow, P. S.,** Accumulation of Zn, Cu and Cd by crabs and barnacles, *Estuarine Coastal Mar. Sci.*, 21, 669, 1985.
40. **Nugegoda, D. and Rainbow, P. S.,** The effect of temperature on zinc regulation by the decapod crustacean *Palaemon elegans* Rathke, *Ophelia,* 27, 17, 1987.
41. **Nugegoda, D. and Rainbow, P. S.,** Effect of a chelating agent (EDTA) on zinc uptake and regulation by *Palaemon elegans* (Crustacea: Decapoda), *J. Mar. Biol. Assoc. U.K.*, 68, 25, 1988.
42. **Jennings, J. R. and Rainbow, P. S.,** Studies on the uptake of cadmium by the crab *Carcinus maenas* in the laboratory. I. Accumulation from seawater and a food source, *Mar. Biol.*, 50, 131, 1979.

43. **White, S. L. and Rainbow, P. S.,** Accumulation of cadmium by *Palaemon elegans* (Crustacea: Decapoda), *Mar. Ecol. Prog. Ser.*, 32, 17, 1986.
44. **Walker, G.,** "Copper" granules in the barnacle *Balanus balanoides, Mar. Biol.*, 39, 343, 1977.
45. **Walker, G., Rainbow, P. S., Foster, P., and Crisp, D. J.,** Barnacles: possible indicators of zinc pollution?, *Mar. Biol.*, 30, 57, 1975.
46. **Walker, G., Rainbow, P. S., Foster, P., and Holland, D. L.,** Zinc phosphate granules in tissue surrounding the midgut of the barnacle *Balanus balanoides, Mar. Biol.*, 33, 161, 1975.
47. **Simkiss, K., Taylor, M., and Mason, A. Z.,** Metal detoxification and bioaccumulation in molluscs, *Mar. Biol. Lett.*, 3, 187, 1982.
48. **Phillips, D. J. H.,** Organochlorines and trace metals in green-lipped mussels *Perna viridis* from Hong Kong waters: a test of indicator ability, *Mar. Ecol. Prog. Ser.*, 21, 251, 1985.
49. **Klumpp, D. W. and Burdon-Jones, C.,** Investigations of the potential of bivalve molluscs as indicators of heavy metal levels in tropical marine waters, *Aust. J. Mar. Freshwater Res.*, 33, 285, 1982.
50. **Roesijadi, G.,** The significance of low molecular weight, metallothionein-like proteins in marine invertebrates: current status, *Mar. Environ. Res.*, 4, 167, 1980-81.
51. **Fowler, S. W., La Rosa, J., Heyraud, M., and Renfro, W. C.,** Effect of different radiotracer labelling techniques on radionuclide excretion from marine organisms, *Mar. Biol.*, 30, 297, 1975.
52. **Fowler, S. W. Heyraud, M., and La Rosa, J.,** Factors affecting methyl and inorganic mercury dynamics in mussels and shrimp, *Mar. Biol.*, 46, 267, 1978.
53. NAS, *The International Mussel Watch,* National Academy of Sciences, Washington, D.C., 1980.
54. **Phillips, D. J. H. and Segar, D. A.,** Use of bio-indicators in monitoring conservative contaminants: programme design imperatives, *Mar. Pollut. Bull.*, 17, 10, 1986.
55. **Goldberg, E. D., Bowen, V. T., Farrington, J. W., Harvey, G., Martin, J. H., Parker, P. L., Risebrough, R. W., Robertson, E., Schneider, E., and Gamble, E.,** The mussel watch, *Environ. Conserv.*, 5, 101, 1978.
56. **Phillips, D. J. H.,** The common mussel *Mytilus edulis* as an indicator of pollution by zinc, cadmium, lead and copper. I. Effects of environmental variables on uptake of metals, *Mar. Biol.*, 38, 59, 1976.
57. **Simpson, R. D.,** Uptake and loss of zinc and lead by mussels *(Mytilus edulis)* and relationships with body weight and reproductive cycle, *Mar. Pollut. Bull.*, 10, 74, 1979.
58. **Luoma, S. N., Cain, D., and Johansson, C.,** Temporal fluctuations of silver, copper and zinc in the bivalve *Macoma balthica* at five stations in South San Francisco Bay, *Hydrobiologia,* 129, 109, 1985.
59. **Phillips, D. J. H. and Boyden, C. R.,** Seasonal variation and inherent variability of trace elements in oysters and their implications for indicator studies, *Mar. Ecol. Prog. Ser.*, 5, 29, 1981.
60. **Gordon, M., Knauer, G. A., and Martin, J. H.,** *Mytilus californianus* as a bioindicator of trace metal pollution: variability and statistical considerations, *Mar. Pollut. Bull.*, 11, 195, 1980.
61. **Wright, D. A, Mihursky, J. A., and Phelps, H. L.,** Trace metals in Chesapeake Bay oysters: intrasample variability and its implications for biomonitoring, *Mar. Environ. Res.*, 16, 181, 1985.
62. **Lobel, P. B.,** Role of the kidney in determining the whole soft tissue zinc concentration of individual mussels *(Mytilus edulis), Mar. Biol.*, 92, 355, 1986.
63. **Phillips, D. J. H.,** The common mussel *Mytilus edulis* as an indicator of pollution by zinc, cadmium, lead and copper. II. Relationship of metals in the mussel to those discharged by industry, *Mar. Biol.*, 38, 71, 1976.
64. **Jones, A. J.,** The arsenic content of some of the marine algae, *Pharm. J.*, 109, 86, 1922.
65. **Black, W. A. P. and Mitchell, R. L.,** Trace elements in the common brown algae and in sea water, *J. Mar. Biol. Assoc. U.K.*, 30, 575, 1952.
66. **Ragan, M. A., Smidsrød, O., and Larsen, B.,** Chelation of divalent metal ions by brown algal polyphenols, *Mar. Chem.*, 7, 265, 1979.
67. **Veroy, R. L., Montaño, N., de Guzman, M. L. B., Laserna, E. C., and Cajipe, G. J. B.,** Studies on the binding of heavy metals to algal polysaccharides from Phillippine seaweeds. I. Carrageenan and the binding of lead and cadmium, *Bot. Mar.*, 23, 59, 1980.
68. **Bryan, G. W.,** The absorption of zinc and other metals by the brown seaweed *Laminaria digitata, J. Mar. Biol. Assoc. U.K.*, 49, 225, 1969.
69. **Butterworth, J., Lester, P., and Nickless, G.,** Distribution of heavy metals in the Severn Estuary, *Mar. Pollut. Bull.*, 3, 72, 1972.
70. **Nickless, G., Stenner, R., and Terrile, N.,** Distribution of cadmium, lead and zinc in the Bristol Channel, *Mar. Pollut. Bull.*, 3, 188, 1972.
71. **Preston, A., Jefferies, D. J., Dutton, D. W. R., Harvey, B. R., and Steele, A. K.,** British Isles coastal waters: the concentrations of selected heavy metals in sea water, suspended matter and biological indicators — a pilot survey, *Environ. Res.*, 3, 69, 1972.
72. **Bryan, G. W., and Hummerstone, L. G.,** Brown seaweed as an indicator of heavy metals in estuaries in South-west England, *J. Mar. Biol. Assoc. U.K.*, 53, 705, 1973.

73. **Fuge, R. and James, K. H.,** Trace metal concentrations in brown seaweeds, Cardigan Bay, Wales, *Mar. Chem.,* 1, 281, 1973.
74. **Fuge, R. and James, K. H.,** Trace metal concentrations in *Fucus* from the Bristol Channel, *Mar. Pollut. Bull.,* 5, 9, 1974.
75. **Seeliger, U. and Edwards, P.,** Correlation coefficients and concentration factors of copper and lead in seawater and benthic algae, *Mar. Pollut. Bull.,* 8, 16, 1977.
76. **Morris, A. W., and Bale, A. J.,** The accumulation of cadmium, copper, manganese and zinc by *Fucus vesiculosus* in the Bristol Channel, *Estuarine Coastal Mar. Sci.,* 3, 153, 1975.
77. **Bryan, G. W. and Hummerstone, L. G.,** Indicators of heavy-metal contamination in the Looe Estuary (Cornwall) with particular regard to silver and lead, *J. Mar. Biol. Assoc. U.K.,* 57, 75, 1977.
78. **Phillips, D. J. H.,** The common mussel *Mytilus edulis* as an indicator of trace metals in Scandinavian waters. I. Zinc and cadmium, *Mar. Biol.,* 43, 283, 1977.
79. **Phillips, D. J. H.,** The common mussel *Mytilus edulis* as an indicator of trace metals in Scandinavian waters. II. Lead, iron and manganese, *Mar. Biol.,* 46, 147, 1978.
80. **Phillips, D. J. H.,** Trace metals in the common mussel, *Mytilus edulis* (L.), and in the alga *Fucus vesiculosus* (L.) from the region of the Sound (Öresund), *Environ. Pollut.* 18, 31, 1979.
81. **Young, M. L.,** The transfer of ^{65}Zn and ^{59}Fe along a *Fucus serratus* (L.) → *Littorina obtusata* (L.) food chain, *J. Mar. Biol. Assoc. U.K.,* 55, 583, 1975.
82. **Myklestad, S., Eide, I., and Melsom, S.,** Exchange of heavy metals in *Ascophyllum nodosum* (L.) Le Jol. *in situ* by means of transplanting experiments, *Environ. Pollut.,* 16, 277, 1978.
83. **Eide, I., Myklestad, S., and Melsom, S.,** Long-term uptake and release of heavy metals by *Ascophyllum nodosum* (L.) Le Jol. (Phaeophyceae) *in situ, Environ. Pollut. Ser. A,* 23, 19, 1980.
84. **Rice, D. L. and Lapointe, B. E.,** Experimental outdoor studies with *Ulva fasciata* Delile. II. Trace metal chemistry, *J. Exp. Mar. Biol. Ecol.,* 54, 1, 1981.
85. **Schonbeck, M. and Norton, T. A.,** The effects of brief periodic submergence on intertidal fucoid algae, *Estuarine Coastal Mar. Sci.,* 8, 205, 1979.
86. **Burdon-Jones, C., Denton, G. R. W., Jones, G. B., and McPhie, K. A.,** Regional and seasonal variations of trace metals in tropical Phaeophyceae from North Queensland, *Mar. Environ. Res.,* 7, 13, 1982.
87. **Aulio, K.,** Heavy metals in the green alga *Cladophora glomerata* as related to shore types in the Archipelago Sea, SW Finland, *Mar. Pollut. Bull.,* 14, 347, 1983.
88. **Barnett, B. E. and Ashcroft, C. R.,** Heavy metals in *Fucus vesiculosus* in the Humber Estuary, *Environ. Pollut. Ser. B,* 9, 193, 1985.
89. **Foster, P.,** Concentrations and concentration factors of heavy metals in brown algae, *Environ. Pollut.,* 10, 45, 1976.
90. **Stömgren, T.,** The effect of copper on length increase in *Ascophyllum nodosum* (L.) Le Jolis, *J. Exp. Mar. Biol. Ecol.,* 37, 153, 1979.
91. **Hall, A., Fielding, A. H., and Butler, M.** Mechanisms of copper tolerance in the marine fouling alga *Ectocarpus siliculosus* — evidence for an exclusion mechanism, *Mar. Biol.,* 54, 195, 1979.
92. **Reed, R. H. and Moffat, L.,** Copper toxicity and copper tolerance in *Enteromorpha compressa* (L.) Grev., *J. Exp. Mar. Biol. Ecol.,* 69, 85, 1983.
93. **Cullinane, J. P. and Whelan, P. M.,** Copper, cadmium and zinc in seaweeds from the South coast of Ireland, *Mar. Pollut. Bull.,* 13, 205, 1982.
94. **Woolston, M. E., Breck, W. G. and VanLoon, G. W.,** A sampling study of the brown seaweed, *Ascophyllum nodosum* as a marine monitor for trace metals, *Water Res.,* 16, 687, 1982.
95. **Langston, W. J.,** Metals in sediments and benthic organisms in the Mersey Estuary, *Estuarine Coastal Mar. Sci.,* 23, 239, 1986.
96. **Sawidis, T. and Voulgaropoulos, A. N.,** Seasonal bioaccumulation of iron, cobalt and copper in marine algae from Thermaikos Gulf of the northern Aegean Sea, Greece, *Mar. Environ. Res.,* 19, 39, 1986.
97. **Sears, J. R., Pecci, K. J., and Cooper, R. A.,** Trace metal concentrations in offshore, deep-water seaweeds in the western North Atlantic Ocean, *Mar. Pollut. Bull.,* 16, 325, 1985.
98. **Cross, J. E. and Day, J. P.,** Plutonium and americium in seaweed from the Channel Islands, *Environ. Pollut. Ser. B,* 2, 249, 1981.
99. **Topcuoğlu, S. and Fowler, S. W.,** Factors affecting the biokinetics of technetium (^{95m}Tc) in marine macroalgae, *Mar. Environ. Res.,* 12, 25, 1984.
100. **Guimarães, J. R. D. and Penna-Franca, E.,** ^{137}Cs, ^{60}Co and ^{125}I bioaccumulation by seaweeds from the Angra dos Reis nuclear power plant region, *Mar. Environ. Res.,* 16, 77, 1985.
101. **Jennings, J. R., Rainbow, P. S., and Scott, A. G.,** Studies on the uptake of cadmium by the crab *Carcinus maenas* in the laboratory. II. Preliminary investigation of cadmium-binding proteins, *Mar. Biol.,* 50, 141, 1979.
102. **Jennings, J. R. and Rainbow, P. S.,** Accumulation of cadmium by *Artemia salina, Mar. Biol.,* 51, 47, 1979.

103. **Rainbow, P. S. and Scott, A. G.,** Two heavy metal-binding proteins in the midgut gland of the crab *Carcinus maenas, Mar. Biol.*, 55, 143, 1979.
104. **White, S. L. and Rainbow, P. S.,** Zinc flux in *Palaemon elegans* (Crustacea: Decapoda): moulting, individual variation and tissue distribution, *Mar. Ecol. Prog. Ser.*, 19, 153, 1984.
105. **White, S. L. and Rainbow, P. S.,** Heavy metal concentrations and size effects in the mesopelagic decapod crustacean *Systellaspis debilis, Mar. Ecol. Prog. Ser.*, 37, 147, 1987.
106. **Bryan, G. W., Hummerstone, L. G., and Ward, E.,** Zinc regulation in the lobster *Homarus gammarus*: importance of different pathways of absorption and excretion, *J. Mar. Biol. Assoc. U.K.*, 66, 175, 1986.
107. **Rainbow, P. S.,** The biology of heavy metals in the sea, *Int. J. Environ. Stud.*, 25, 195, 1985.
108. **Balkas, T. Tuğrul, S., and Salihoğlu, I.,** Trace metal levels in fish and crustacea from Northeastern Mediterranean coastal waters, *Mar. Environ. Res.*, 6, 281, 1982.
109. **Sanders, M.,** Metals in crab, oyster and sediment in two South Carolina estuaries, *Mar. Pollut. Bull.*, 15, 159, 1984.
110. **Frenet, M. and Alliot, A.,** Comparative bioaccumulation of metals in *Palaemonetes varians* in polluted and non-polluted environments, *Mar. Environ. Res.*, 17, 19, 1985.
111. **O'Hara, J.,** Cadmium uptake by fiddler crabs exposed to temperature and salinity stress, *J. Fish. Res. Board Can.*, 30, 846, 1973.
112. **Hutcheson, M. S.,** The effect of temperature and salinity on cadmium uptake by the blue crab, *Callinectes sapidus, Chesapeake Sci.*, 15, 237, 1974.
113. **Wright, D. A.,** The effect of calcium on cadmium uptake by the shore crab *Carcinus maenas, J. Exp. Biol.*, 67, 163, 1977.
114. **Davies, I. M., Topping, G., Graham, W. C., Falconer, C. R., McIntosh, A. D., and Saward, D.,** Field and experimental studies on cadmium in the edible crab *Cancer pagurus, Mar. Biol.*, 64, 291, 1981.
115. **Brouwer, M. and Brouwer-Hoexum, T.,** Cadmium accumulation by the blue crab, *Callinectes sapidus:* involvement of haemocyanin and characterization of cadmium-binding proteins, *Mar. Environ. Res.*, 14, 71, 1984.
116. **Uthe, J. F., Chou, C. L., Loring, D. H., Rantala, R. T. T., Bewers, J. M., Dalziel, J., Yeats, P. A., and Charron, R. L.,** Effect of waste treatment at a lead smelter on cadmium levels in American lobster *(Homarus americanus),* sediments and seawater in the adjacent coastal zone, *Mar. Pollut. Bull.*, 17, 118, 1986.
117. **Maff, U. K.,** Working Party on the Monitoring of Foodstuffs for Heavy Metals: Fourth Report: Survey of Cadmium in Food, Her Majesty's Stationery Office, London, 1973.
118. **Friberg, L., Piscator, M., Nordberg, G. F., and Kjellström, T.,** *Cadmium in the Environment,* 2nd ed., CRC Press, Boca Raton, Fl, 1974.
119. **Hopkin, S. P. and Nott, J. A.,** Some observations on concentrically structured intracellular granules in the hepatopancreas of the shore crab *Carcinus maenas* (L.), *J. Mar. Biol. Assoc. U.K.*, 59, 867, 1979.
120. **Ray, G. L. and Tripp, M. R.,** The uptake of mercury from water by the grass shrimp *Palaemonetes vulgaris* (Say), *J. Environ. Qual.*, 5, 193, 1976.
121. **Miramand, P., Guary, J. C., and Fowler, S. W.,** Uptake, assimilation, and excretion of vanadium in the shrimp, *Lysmata seticaudata* (Risso), and the crab, *Carcinus maenas* (L.), *J. Exp. Mar. Biol. Ecol.*, 49, 267, 1981.
122. **Ünsal, M.,** Transfer pathways and accumulation of vanadium in the crab *Carcinus maenas, Mar. Biol.*, 72, 279, 1983.
123. **Rainbow, P. S. and Moore, P. G.,** Comparative metal analyses in amphipod crustaceans, *Hydrobiologia,* 141, 273, 1986.
124. **Rainbow, P. S.,** Heavy metals in barnacles, in *Barnacle Biology,* Southward, A. J., Ed., A.A. Balkema, Rotterdam, 1987, 405.
125. **Ireland, M. P.,** Result of fluvial zinc pollution on the zinc content of littoral and sublittoral organisms in Cardigan Bay, Wales, *Environ. Pollut.*, 4, 27, 1973.
126. **Ireland, M. P.,** Variations in the zinc, copper, manganese and lead content of *Balanus balanoides* in Cardigan Bay, Wales, *Environ. Pollut.*, 7, 65, 1974.
127. **Brown, B. E.,** The form and function of metal-containing 'granules' in invertebrate tissues, *Biol. Rev.*, 57, 621, 1982.
128. **White, K. N. and Walker, G.,** Uptake, accumulation, and excretion of zinc by the barnacle, *Balanus balanoides* (L.), *J. Exp. Mar. Biol. Ecol.*, 51, 285, 1981.
129. **Rainbow, P. S., Scott, A. G., Wiggins, E. A., and Jackson, R. W.,** Effect of chelating agents on the accumulation of cadmium by the barnacle *Semibalanus balanoides,* and complexation of soluble Cd, Zn and Cu, *Mar. Ecol. Prog. Ser.*, 2, 143, 1980.
130. **van Weerelt, M., Pfeiffer, W. C., and Fiszman, M.,** Uptake and release of ^{51}Cr(VI) and ^{51}Cr(III) by barnacles *(Balanus* sp.), *Mar. Environ. Res.*, 11, 201, 1984.
131. **Chan, H. M., Rainbow, P. S., and Phillips, D. J. H.,** Barnacles and mussels as monitors of trace metal bio-availability in Hong Kong waters, in *Proc. 2nd Int. Workshop on the Marine Flora and Fauna of Hong Kong and Southern China,* Morton, B. S., Ed., Hong Kong University Press, Hong Kong, (in press).

132. **Phillips, D. J. H. and Rainbow, P. S.,** Barnacles and mussels as biomonitors of trace elements: a comparative study, *Mar. Ecol. Prog. Ser.*, 49, 83, 1988.
133. **Nickless, G., Stenner, R., and Terrile, N.,** Distribution of cadmium, lead and zinc in the Bristol Channel, *Mar. Pollut. Bull.*, 3, 188, 1972.
134. **Peden, J. D., Crothers, J. H., Waterfall, C. E., and Beasley, J.,** Heavy metals in Somerset marine organisms, *Mar. Pollut. Bull.*, 4, 7, 1973.
135. **Stenner, R. D. and Nickless, G.,** Absorption of cadmium, copper and zinc by dog whelks in the Bristol Channel, *Nature*, 247, 198, 1974.
136. **Boyden, C. R.,** Trace element content and body size in molluscs, *Nature*, 251, 311, 1974.
137. **Boyden, C. R.,** Effect of size upon metal content of shellfish, *J. Mar. Biol. Assoc. U.K.*, 57, 675, 1977.
138. **Bryan, G. W., Potts, G. W., and Forster, G. R.,** Heavy metals in the gastropod mollusc *Haliotis tuberculata* (L.), *J. Mar. Biol. Assoc. U.K.*, 57, 379, 1977.
139. **Navrot, J., Amiel, A. J., and Kronfeld, J.,** *Patella vulgata*: a biological monitor of coastal metal pollution — a preliminary study, *Environ. Pollut.*, 7, 303, 1974.
140. **Ramelow, G. J.,** A study of heavy metals in limpets (*Patella* sp.) collected along a section of the Southeastern Turkish Mediterranean coast, *Mar. Environ. Res.*, 16, 243, 1985.
141. **Anderlini, V. C.,** The distribution of heavy metals in the red abalone, *Haliotis rufescens,* on the California coast, *Arch. Environ. Contam. Toxicol.*, 2, 253, 1974.
142. **Betzer, S. B. and Pilson, M. E. Q.,** Copper uptake and excretion by *Busycon canaliculatum* L., *Biol. Bull.*, 148, 1, 1975.
143. **Martoja, M., Tue, V. T., and Elkaim, B.,** Bioaccumulation de cuivre chez *Littorina littorea* (L.) (Gastéropode Prosobranche): signification physiologique et écologique, *J. Exp. Mar. Biol. Ecol.*, 43, 251, 1980.
144. **Ireland, M. P. and Wootton, R. J.,** Distribution of lead, zinc, copper and manganese in the marine gastropods, *Thais lapillus* and *Littorina littorea,* around the coast of Wales, *Environ. Pollut.*, 12, 27, 1977.
145. **Flegal, A. R. and Martin, J. H.,** Contamination of biological samples by ingested sediment, *Mar. Pollut. Bull.*, 8, 90, 1977.
146. **Nelson, D. A., Calabrese, A., Greig, R. A., Yevich, Y. Y., and Chang, S.,** Long-term silver effects on the marine gastropod *Crepidula fornicata, Mar. Ecol. Prog. Ser.*, 12, 155, 1983.
147. **Bryan, G. W., Langston, W. J., and Hummerstone, L. G.,** The use of biological indicators of heavy metal contamination in estuaries, *Occas. Publ. Mar. Biol. Assoc. U.K.*, 1, 1, 1980.
148. **Bryan, G. W., Langston, W. J., Hummerstone, L. G., and Burt, G. R.,** A guide to the assessment of heavy metal contamination in estuaries using biological indicators, *Occas. Publ. Mar. Biol. Assoc. U.K.*, 4, 1, 1985.
149. **George, S. G., Pirie, B. J. S., Cheyne, A. R., Coombs, T. L., and Grant, P. T.,** Detoxification of metals by marine bivalves: an ultrastructural study of the compartmentation of copper and zinc in the oyster *Ostrea edulis, Mar. Biol.*, 45, 147, 1978.
150. **Zaroogian, G. E., Morrison, G., and Heltshe, J. F.,** *Crassostrea virginica* as an indicator of lead pollution, *Mar. Biol.*, 52, 189, 1979.
151. **Carmichael, N. G., Squibb, K. S., Engel, D. W., and Fowler, B. A.,** Metals in the molluscan kidney: uptake and subcellular distribution of ^{109}Cd, ^{54}Mn and ^{65}Zn by the clam, *Mercenaria mercenaria, Comp. Biochem. Physiol.*, 65A, 203, 1980.
152. **George, S. G. and Pirie, B. J. S.,** Metabolism of zinc in the mussel *Mytilus edulis* (L.): a combined ultrastructural and biochemical study, *J. Mar. Biol. Assoc. U.K.*, 60, 575, 1980.
153. **George, S. G., Pirie, B. J. S., and Coombs, T. L.,** Isolation and elemental analysis of metal-rich granules from the kidney of the scallop, *Pecten maximus, J. Exp. Mar. Biol. Ecol.*, 42, 143, 1980.
154. **Scholz, N.,** Accumulation, loss and molecular distribution of cadmium in *Mytilus edulis, Helgol. Wiss. Meeresunters.*, 33, 68, 1980.
155. **Zaroogian, G. E.,** *Crassostrea virginica* as an indicator of cadmium pollution, *Mar. Biol.*, 58, 275, 1980.
156. **Roesijadi, G., Drum, A. S., Thomas, J. M., and Fellingham, G. W.,** Enhanced mercury tolerance in marine mussels and relationship to low molecular weight, mercury-binding proteins, *Mar. Pollut. Bull.*, 13, 250, 1982.
157. **Frazier, J. M. and George, S. G.,** Cadmium kinetics in oysters — a comparative study of *Crassostrea gigas* and *Ostrea edulis, Mar. Biol.*, 76, 55, 1983.
158. **Pirie, B. J. S., George, S. G., Lytton, D. G., and Thompson, J. D.,** Metal-containing blood cells of oysters: ultrastructure, histochemistry and X-ray microanalysis, *J. Mar. Biol. Assoc. U.K.*, 64, 115, 1984.
159. **Miller, W. L., Blake, N. J., and Byrne, R. H.,** Uptake of Zn^{65} and Mn^{54} into body tissues and renal concretions by the Southern quahog, *Mercenaria campechiensis* (Gmelin): effects of elevated phosphate and metal concentrations, *Mar. Environ. Res.*, 17, 167, 1985.
160. **Morse, M. P., Meyhofer, E., and Robinson, W. E.,** Accumulation of ^{109}cadmium in extracellular granules in the kidney of the bivalve mollusc *Mercenaria mercenaria* (L.), *Mar. Environ. Res.*, 17, 172, 1985.

161. **George, S. G., Pirie, B. J. S., Calabrese, A., and Nelson, D. A.,** Biochemical and ultrastructural observations of long-term silver accumulation in the mussel, *Mytilus edulis, Mar. Environ. Res.*, 18, 255, 1986.
162. **Okazaki, R. K. and Panietz, M. H.,** Depuration of twelve trace metals in tissues of the oysters *Crassostrea gigas* and *C. virginica, Mar. Biol.*, 63, 113, 1981.
163. **Beasley, T. M., Gonor, J. J., and Lorz, H. V.,** Technetium: uptake, organ distribution and loss in the mussel, *Mytilus californianus* (Conrad) and the oyster *Crassostrea gigas* (Thunberg), *Mar. Environ. Res.*, 7, 103, 1982.
164. **Widdows, J., Donkin, P., Salkeld, P. N., Cleary, J. J., Lowe, D. M., Evans, S. V., and Thomson, P. E.,** Relative importance of environmental factors in determining physiological differences between two populations of mussels *(Mytilus edulis), Mar. Ecol. Prog. Ser.*, 17, 33, 1984.
165. **Crecelius, E. A., Hardy, J. T., Gibson, C. I., Schmidt, R. L., Apts, C. W., Gurtisen, J. M., and Joyce, S. P.,** Copper bioavailability to marine bivalves and shrimp: relationship to cupric ion activity, *Mar. Environ. Res.*, 6, 13, 1982.
166. **Zamuda, C. D., Wright, D. A., and Smucker, R. A.,** The importance of dissolved organic compounds in the accumulation of copper by the American oyster, *Crassostrea virginica, Mar. Environ. Res.*, 16, 1, 1985.
167. **Phillips, D. J. H. and Yim, W. W.-S.,** A comparative evaluation of oysters, mussels and sediments as indicators of trace metals in Hong Kong waters, *Mar. Ecol. Prog. Ser.*, 6, 285, 1981.
168. **De Wolf, P.,** Mercury content of mussels from West European coasts, *Mar. Pollut. Bull.*, 6, 61, 1975.
169. **Goldberg, E. D.,** The mussel watch — a first step in global marine monitoring, *Mar. Pollut. Bull.*, 6, 111, 1975.
170. **Davenport, J.,** A study of the effects of copper applied continuously and discontinuously to specimens of *Mytilus edulis* (L.) exposed to steady and fluctuating salinity levels, *J. Mar. Biol. Assoc. U.K.*, 57, 63, 1977.
171. **Davenport, J. and Manley, A.,** The detection of heightened sea-water copper concentrations by the mussel *Mytilus edulis, J. Mar. Biol. Assoc. U.K.*, 58, 843, 1978.
172. **Davenport, J.,** The isolation response of mussels *(Mytilus edulis* L.) exposed to falling sea-water concentrations, *J. Mar. Biol. Assoc. U.K.*, 59, 123, 1979.
173. **Strömgren, T.,** Effect of heavy metals (Zn, Hg, Cu, Cd, Pb, Ni) on the length growth of *Mytilus edulis, Mar. Biol.*, 72, 69, 1982.
174. **Maung Myint, U. and Tyler, P. A.,** Effects of temperature, nutritive and metal stressors on the reproductive biology of *Mytilus edulis, Mar. Biol.*, 67, 209, 1982.
175. **Manley, A. R.,** The effects of copper on the behaviour, respiration, filtration and ventilation activity of *Mytilus edulis, J. Mar. Biol. Assoc. U.K.*, 63, 205, 1983.
176. **Calabrese, A., MacInnes, J. R., Nelson, D. A., Greig, R. A., and Yevich, P. P.,** Effects of long-term exposure to silver or copper on growth, bioaccumulation and histopathology in the blue mussel *Mytilus edulis, Mar. Environ. Res.*, 11, 253, 1984.
177. **Redpath, K. J.,** Growth inhibition and recovery in mussels *(Mytilus edulis)* exposed to low copper concentrations, *J. Mar. Biol. Assoc. U.K.*, 65, 421, 1985.
178. **Strömgren, T.,** The combined effect of copper and hydrocarbons on the length growth of *Mytilus edulis, Mar. Environ. Res.*, 19, 251, 1986.
179. **Hvilsom, M. M.,** Copper-induced differential mortality in the mussel *Mytilus edulis, Mar. Biol.*, 76, 291, 1983.
180. **Luoma, S. N., Cain, D. J., Ho, K., and Hutchinson, A.,** Variable tolerance to copper in two species from San Francisco Bay, *Mar. Environ. Res.*, 10, 209, 1983.
181. **Bryan, G. W., and Hummerstone, L. G.,** Heavy metals in the burrowing bivalve *Scrobicularia plana* from contaminated and uncontaminated estuaries, *J. Mar. Biol. Assoc. U.K.*, 58, 401, 1978.
182. **Bryan, G. W. and Uysal, H.,** Heavy metals in the burrowing bivalve *Scrobicularia plana* from the Tamar Estuary in relation to environmental levels, *J. Mar. Biol. Assoc. U.K.*, 58, 89, 1978.
183. **Davies, I. M. and Pirie, J. M.,** The mussel *Mytilus edulis* as a bio-assay organism for mercury in seawater, *Mar. Pollut. Bull.*, 9, 128, 1978.
184. **Cossa, D., Bourget, E., and Piuze, J.,** Sexual maturation as a source of variation in the relationship between cadmium concentration and body weight of *Mytilus edulis* L., *Mar. Pollut. Bull.*, 10, 174, 1979.
185. **Strong, C. R. and Luoma, S. N.,** Variations in the correlation of body size with concentrations of Cu and Ag in the bivalve *Macoma balthica, Can. J. Fish. Aquat. Sci.*, 38, 1059, 1981.
186. **Langston, W. J.,** The distribution of mercury in British estuarine sediments and its availability to deposit-feeding bivalves, *J. Mar. Biol. Assoc. U.K.*, 62, 667, 1982.
187. **Lobel, P. B. and Wright, D. A.,** Total body zinc concentration and allometric growth ratios in *Mytilus edulis* collected from different shore levels, *Mar. Biol.*, 66, 231, 1982.
188. **Fischer, H.,** Shell weight as an independent variable in relation to cadmium content of molluscs, *Mar. Ecol. Prog. Ser.*, 12, 59, 1983.

189. **Gault, N. F. S., Tolland, E. L. C., and Parker, J. G.,** Spatial and temporal trends in heavy metal concentrations in mussels from Northern Ireland coastal waters, *Mar. Biol.*, 77, 307, 1983.
190. **Fischer, H.,** Influence of temperature, salinity, and oxygen on the cadmium balance of mussels *Mytilus edulis, Mar. Ecol. Prog. Ser.*, 32, 265, 1986.
191. **Fischer, H.,** Cadmium in Kieler Förde mussels: an innovative monitoring programme, *Rapp. P.v. Reun. Cons. Int. Explor. Mer*, 186, 468, 1986.
192. **Orren, M. J., Eagle, G. A., Hennig, H. F.-K., and Green, A.,** Variations in trace metal content of the mussel *Choromytilus meridionalis* (Kr.) with season and sex, *Mar. Pollut. Bull.*, 11, 253, 1980.
193. **Lobel, P. B. and Wright, D. A.,** Gonadal and nongonadal zinc concentrations in mussels, *Mar. Pollut. Bull.*, 13, 320, 1982.
194. **Marina, M. and Enzo, O.,** Variability of zinc and manganese concentrations in relation to sex and season in the bivalve *Donax trunculus, Mar. Pollut. Bull.*, 14, 342, 1983.
195. **Roberts, D. F., Elliot, M. and Read, P. A.,** Cadmium contamination, accumulation and some effects of this metal in mussels from a polluted marine environment, *Mar. Environ. Res.*, 18, 165, 1986.
196. **Jackim, E., Morrison, G., and Steele, R.,** Effects of environmental factors on radiocadmium uptake by four species of marine bivalves, *Mar. Biol.*, 40, 303, 1977.
197. **Fowler, B. A., Wolfe, D. A., and Hettler, W. F.,** Mercury and iron uptake by cytosomes in mantle epithelial cells of quahog clams *(Mercenaria mercenaria)* exposed to mercury, *J. Fish. Res. Board Can.*, 32, 1767, 1975.
198. **Young, D. R. and Jan, T.-K.,** Metals in scallops, in *Southern California Coastal Water Research Project, Annual Report,* SCCWRP, Long Beach, CA, 1976, 117.
199. **Fowler, S. W. and Benayoun, G.,** Influence of evironmental factors on selenium flux in two marine invertebrates, *Mar. Biol.*, 37, 59, 1976.
200. **George, S. G., Pirie, B. J. S., and Frazier, J. M.,** Effects of cadmium exposure on metal-containing amoebocytes of the oyster *Ostrea edulis, Mar. Biol.*, 76, 63, 1983.
201. **Robinson, W. E., and Ryan, D. K.,** Metal interactions within the kidney, gill, and digestive gland of the hard clam, *Mercenaria mercenaria,* following laboratory exposure to cadmium, *Arch. Environ. Contam. Toxicol.*, 15, 23, 1986.
202. **Luoma, S. N. and Bryan, G. W.,** A statistical study of environmental factors controlling concentrations of heavy metals in the burrowing bivalve *Scrobicularia plana* and the polychaete *Nereis diversicolor, Estuarine Coastal Shelf Sci.*, 15, 95, 1982.
203. **Sturesson, U.,** Lead enrichment in shells of *Mytilus edulis, Ambio,* 5, 253, 1976.
204. **Sturesson, U.,** Cadmium enrichment in shells of *Mytilus edulis, Ambio,* 7, 122, 1978.
205. **Coombs, T. L. and Keller, P. J.,** *Mytilus* byssal threads as an environmental marker for metals, *Aquat. Toxicol.*, 1, 291, 1981.
206. **Koide, M., Lee, D. S., and Goldberg, E. D.,** Metal and transuranic records in mussel shells, byssal threads and tissues, *Estuarine Coastal Mar. Sci.*, 15, 679, 1982.
207. **Dermott, R. M. and Lum, K. R.,** Metal concentrations in the annual shell layers of the bivalve *Elliptio complanata, Environ. Pollut. Ser. B,* 12, 131, 1986.
208. **Farrington, J. W., Goldberg, E. D., Risebrough, R. W., Martin, J. H., and Bowen, V. T.,** U.S. "Mussel Watch" 1976—1978: an overview of the trace-metal, DDE, PCB, hydrocarbon, and artificial radionuclide data, *Environ. Sci. Technol.*, 17, 490, 1983.
209. **Goldberg, E. D., Koide, M., Hodge, V., Flegal, A. R., and Martin, J.,** U.S. Mussel Watch: 1977-1978 results on trace metals and radionuclides, *Estuarine Coastal Shelf Sci.*, 16, 69, 1983.
210. **Martin, M., Ichikawa, G., Goetzl, J., de los Reyes, M., and Stephenson, M. D.,** Relationships between physiological stress and trace toxic substances in the Bay mussel, *Mytilus edulis,* from San Francisco Bay, California, *Mar. Environ. Res.*, 11, 91, 1984.
211. **Martin, M. and Castle, W.,** Petrowatch: petroleum hydrocarbons, synthetic organic compounds, and heavy metals in mussels from the Monterey Bay area of central California, *Mar. Pollut. Bull.*, 15, 259, 1984.
212. **Martin, M.,** State mussel watch: toxics surveillance in California, *Mar. Pollut. Bull.*, 16, 140, 1985.
213. **Smith, D. R., Stephenson, M. D., and Flegal, A. R.,** Trace metals in mussels transplanted to San Francisco Bay, *Environ. Toxicol. Chem.*, 5, 129, 1986.
214. **Young, D. R. Heesen, T. C., and McDermott, D. J.,** An offshore biomonitoring system for chlorinated hydrocarbons, *Mar. Pollut. Bull.*, 7, 156, 1976.
215. **Curran, J. C., Holmes, P. J., and Yersin, J. E.,** Moored shellfish cages for pollution monitoring: design and operational experience, *Mar. Pollut. Bull.*, 17, 464, 1986.
216. **Ritz, D. A., Swain, R., and Elliot, N. G.,** Use of the mussel *Mytilus edulis planulatus* (Lamarck) in monitoring heavy metals in seawater, *Aust. J. Mar. Freshwater Res.*, 33, 491, 1982.
217. **Cain, D. J. and Luoma, S. N.,** Copper and silver accumulation in transplanted and resident clams *(Macoma balthica)* in South San Francisco Bay, *Mar. Environ. Res.*, 15, 115, 1985.

218. **Henze, M.,** Untersuchungen über das Blut der Asidien. I. Die Vanadiumbindung der Blutkörporchen, *Hoppe-Seyler's Z. Physiol. Chem.*, 72, 494, 1911.
219. **Bell, M. V., Pirie, B. J. S., McPhail, D. B., Goodman, B. A., Falk-Petersen, I.-B., and Sargent, J. R.,** Contents of vanadium and sulphur in the blood cells of *Ascidia mentula* and *Ascidiella aspersa, J. Mar. Biol. Assoc. U. K.*, 62, 709, 1982.
220. **Rowley, A. F.,** The blood cells of *Ciona intestinalis:* an electron probe X-ray microanalytical study, *J. Mar. Biol. Assoc. U.K.*, 62, 607, 1982.
221. **Pirie, B. J. S. and Bell, M. V.,** The localization of inorganic elements, particularly vanadium and sulphur in haemolymph from the ascidians *Ascidia mentula* (Müller) and *Ascidiella aspersa* (Müller), *J. Exp. Mar. Biol. Ecol.*, 74, 187, 1984.
222. **Papadopoulou, C. and Kanias, G. D.,** Tunicate species as marine pollution indicators, *Mar. Pollut. Bull.*, 8, 229, 1977.
223. **Martin, J. H. and Flegal, A. R.,** High copper concentrations in squid livers in association with elevated levels of silver, cadmium, and zinc, *Mar. Biol.*, 30, 51, 1975.
224. **Miramand, P. and Guary, J. C.,** High concentrations of some heavy metals in tissues of the Mediterranean octopus, *Bull. Environ. Contam. Toxicol.*, 24, 783, 1980.
225. **Smith, J. D., Plues, L., Heyraud, M., and Cherry, R. D.,** Concentrations of the elements Ag, Al, Ca, Cd, Cu, Fe, Mg, Mn, Pb and Zn, and the radionuclides ^{210}Pb and ^{210}Po in the digestive gland of the squid *Nototodarus gouldi, Mar. Environ. Res.*, 13, 55, 1984.
226. **Cross, F. A., Duke, T. W., and Willis, J. N.,** Biogeochemistry of trace elements in a coastal plain estuary: distribution of manganese, iron, and zinc in sediments, water and polychaetous worms, *Chesapeake Sci.*, 11, 221, 1970.
227. **Bryan, G. W.and Hummerstone, L. G.,** Adaptation of the polychaete *Nereis diversicolor* to estuarine sediments containing high concentrations of zinc and cadmium, *J. Mar. Biol. Assoc. U.K.*, 53, 839, 1973.
228. **Pesch, C. E. and Hoffman, G. L.,** Adaptation of the polychaete *Neanthes arenaceodentata* to copper, *Mar. Environ. Res.*, 6, 307, 1982.
229. **Packer, D. M., Ireland, M. P., and Wootton, R. J.,** Cadmium, copper, lead, zinc and manganese in the polychaete *Arenicola marina* from sediments around the coast of Wales, *Environ. Pollut. Ser. A*, 22, 309, 1980.
230. **Bryan, G. W. and Hummerstone, L. G.,** Adaptation of the polychaete *Nereis diversicolor* to manganese in estuarine sediments, *J. Mar. Biol. Assoc. U.K.*, 53, 859, 1973.
231. **Howard, L. S. and Brown, B. E.,** Natural variations in tissue concentration of copper, zinc, and iron in the polychaete *Nereis diversicolor, Mar. Biol.*, 78, 87, 1983.
232. **Luoma, S. N.,** The dynamics of biologically available mercury in a small estuary, *Estuarine Coastal Mar. Sci.*, 5, 643, 1977.
233. **Oshida, P. S. and Word, L. S.,** Bioaccumulation of chromium and its effects on reproduction in *Neanthes arenaceodentata* (Polychaeta), *Mar. Environ. Res.*, 7, 167, 1982.
234. **Popham, J. D. and D'Auria, J. M.,** A new sentinel organism for vanadium and titanium, *Mar. Pollut. Bull.*, 13, 25, 1982.
235. **Young, J. S. and Roesijadi, G.,** Reparatory adaptation to copper-induced injury and occurrence of a copper-binding protein in the polychaete, *Eudistylia vancouveri, Mar. Pollut. Bull.*, 14, 30, 1983.
236. **Howell, R.,** Levels of heavy metal pollutants in two species of marine nematodes, *Mar. Pollut. Bull.*, 13, 396, 1982.
237. **Anderlini, V.,** personal communication.

Chapter 7

TOXIC EFFECTS OF METALS AND THE INCIDENCE OF METAL POLLUTION IN MARINE ECOSYSTEMS

W. J. Langston

TABLE OF CONTENTS

I. INTRODUCTION

Metals occur naturally in seawater and many, such as Cu, Co, Fe, Mn, Ni, Se, V, and Zn, are used for essential purposes by marine organisms. In evolving mechanisms for the sequestration, transport, and utilization of these metals, the biota has developed a general capacity for metal tolerance. Once the assimilative capacity (threshold) of the system is overloaded, however, either by an excess of essential metal, or by the presence of unusually high levels of the rarer nonessential metals (such as Ag, Cd, Hg, and Pb), deleterious effects may occur.

The potential threat to marine organisms posed by metal pollution has been recognized for decades. The major concerns have focused on estuarine and nearshore waters where there are a number of examples of serious effects caused by industrial and domestic discharges. In the worst cases, the consequences have been so devastating that establishing cause-effect relationships has been relatively easy and subsequent "clean-up" procedures successfully implemented. For the majority of coastal and offshore environments, however, concentrations of metals are commonly below "effect levels" observed in field and laboratory tests and consequently the threat of chronic pollution, particularly at the ecosystem level, is still largely unresolved.

Nevertheless, in recent years, a variety of sensitive responses have been discovered in marine organisms which help to visualize the subtle effects of metal contamination. Although it is currently difficult to explain the ecological significance of many of these responses, further development and application of such indices is likely to become increasingly valuable in the assessment and prediction of damage caused by heavy metals and their compounds.

In the present chapter, we review some of the ways in which metals may exert their toxic effects, together with the methods used to establish threshold levels. Areas in the marine environment where potentially toxic metal concentrations have been found, and where biological impact is evident, are also considered.

II. TOXIC EFFECTS OF METALS

Since the most obvious manifestation of metal intoxication is death, early studies on the effects of metals were biased towards short term LC_{50}-type investigations. Such studies were useful for initial regulatory purposes in that they demonstrated the likely extent of direct contaminant-induced mortalities in the field. However, comparisons of dissolved metal concentrations from various sea areas (Table 1) with levels which are acutely toxic to marine species (reviewed for example by Mance[10]) suggest that, with the exception of a few heavily polluted sites, metals are unlikely to pose an instantaneous threat. For moderately contaminated areas, metal concentrations are considerably lower than acute thresholds and direct impact is likely to be restricted to areas immediately surrounding the site of release, such as industrial effluent pipes, mine adits, sewage outfalls, and dumpsites.

Dilution and removal processes are thus often seen as contributing to the self-cleansing process in the marine environment, but as a result, a much larger area is subjected to a moderate degree of contamination and recognition of chronic sublethal effects is clearly more relevant for practical assessments of damage. Before categorizing these effects, however, it is important to identify the parameters which influence the toxic action of metals and which may modify hazard evaluations.

A. FACTORS AFFECTING TOXICITY

Toxic effects occur when excretory, metabolic, storage, and detoxification mechanisms are no longer capable of matching uptake rates. This capacity may vary greatly between phyla, species, populations, even individuals and also may depend on the stage in the life

TABLE 1
Examples of the Range of Metal Concentrations in Various Sea Areas (µg/l)

	Offshore/coastal				Estuaries/embayments			
Metal	**Oceanic[1]**	**North Sea and U.K. Coastal[2-5]**	**Baltic Sea[5]**	**Mediterranean[6]**	**Bristol Channel, U.K.[7] (industry, sewage)**	**Poole Harbor, U.K.[a] (industry, sewage)**	**Derwent estuary[8] Tasmania (refinery, chlor-alkali)**	**Restronguet Creek,[9] U.K. S ‰ 18-34 (mining waste)**
As	1.35—2.5	1.52—2.4	0.52—7.4	—	—	0.54—1.38	1—6	3.3—65.4[a]
Cd	0.0002—0.025	0.005—0.48	0.03—0.07	<0.02—0.7	0.4—9.4	0.3—42	0.5—15	0.7—38
Cr	0.088—0.55	—	—	—	—	—	—	—
Cu	0.025—0.64	0.01—6.8	0.31—0.95	<0.04—5.8	0.6—5.4	0.2—28	10—27	3—176
Hg	0.004—0.012	0.003—0.08[a]	0.001—0.006	0.005—0.08	0.009—0.07[a]	0.002—0.13	<0.1—16	0.005—0.03[a]
Ni	0.18—0.70	0.01—4.8	0.41—1.76	—	0.2—3.0	—	—	1—18
Se	0.06—0.12	0.09—0.47[a]	—	—	0.2—1.8	0.3—29	—	0.1—0.3[a]
Pb	0.006—0.015	0.006—1.23	0.02—0.11	—	0.35—13	—	4—16	<2—4
Zn	0.0007—0.588	0.006—70	1.3—3.3	0.02—10.0	2.7—44	1—674	6—1500	22—20460

Note: Compiled from the sources indicated.

[a] Unpublished results.

history of the organism, thus explaining the range of sensitivities often observed in toxicity studies. In addition, many organisms may be better adapted to their natural environment than suggested by laboratory studies. *Metal tolerance,* either genetically based or resulting from previous exposure history can greatly influence apparent toxicity thresholds as demonstrated, for example, in the polychaete *Nereis diversicolor,* where individuals from highly metalliferous sediments (mining wastes) have become much more resistant to elevated levels of Cu and Zn than those from unpolluted sites.[11] The mechanisms which have evolved in organisms to counteract excess metals are many and varied and range from surface processes of exclusion/reduced permeability observed in *N. diversicolor* (Cu, Zn),[11] to the internal detoxification systems, including metallothioneins, lysosomes, and granules (Ag, Cd, Cu, Hg, Zn, Pb) described for representatives of many phyla.

Toxic effects are inexorably linked with bioaccumulation (since, in most cases, metals must first enter cells before exerting their influence) and many of the physical and chemical processes which control these events are identical. Metal burdens in organisms are usually a function of environmental concentrations; however, it is now widely recognized that *chemical speciation* of metals in seawater can greatly modify their bioavailability and hence toxicity. Generally, bioavailability is reduced by the presence of natural organic chelators and for metals, such as Cu and Cd, the free ion appears to be the most biologically active species. In addition, it is important to recognize that sediments represent a major source of contamination to many benthic organisms and that *partitioning* of metals among solid phases may, correspondingly, modify the bioavailability and toxicity of sediment-bound metals.[12]

As a general (though not universal) rule, high *temperature* and low *salinity* act synergistically with metals to increase mortality.[13] The higher toxicity of metals such as Cd, Cr, Pb, Hg, Ni and Zn at low salinity is usually attributable to more rapid accumulation rates and is clearly of greatest significance in estuaries.

It is usual for polluted environments to be contaminated with several, as opposed to single, metals and numerous laboratory studies, supported by residue data, point to *synergistic/antagonistic interactions.* In particular, there is consistent evidence of competition between a number of metals, such as Cd-Zn, Cu-Ag, Cu-Mn, Cd-Se, and Hg-Se.[9,14,15] In the latter case, the antagonistic effect of Se on Hg accumulation is often seen as fulfilling a detoxifying role, at least in some marine vertebrates, while in contrast, competition from Cu may result in deficiencies of essential Mn in algal cells.[14] For the majority of metal combinations, however, the consequences of their joint action under natural conditions is not understood.

Redox potentials may strongly influence speciation, availability, and thus toxicity of metals, particularly in sediments. Anoxic conditions are frequently accompanied by high metal levels (especially in pore water) at polluted sites and their effects may be synergistic as suggested, for example, by experiments with *Macoma balthica* subjected to combined oxygen and Cu stress.[16] In this study, it was suggested that the proportion of (bioavailable) Cu^{2+} ions was increased at the lower pH levels associated with low oxygen tensions, resulting in enhanced uptake and sublethal toxicity. Because of its impact on chemical speciation, it is perhaps surprising that the influence of pH on metal toxicity in marine organisms has not been studied more often. However, it is generally assumed that the buffering capacity of seawater is sufficient to resist major pH changes in most areas other than in the immediate vicinity of certain waste outfalls or mine drainage.

Bearing in mind the variety of factors which can modify toxicity, some of the ways in which metals exert their effects are now discussed.

B. MEASURING EFFECTS OF METALS

Techniques used to assess the effects of metals are varied and range across many levels of biological organization from cells to whole communities. It is worth noting, however, that effects rarely occur in isolation and impact is usually due to a combination or sequence of related events. Thus, for example, lower-order biochemical responses may be intimately linked with disturbances in essential life processes, such as growth and reproduction, which in turn may influence the health and survival of individuals, populations, and ultimately communities.

As a rule, the effects of metals take longer to develop for each increasing level of biological complexity and this has led to the concept of using stress indices (at the lower levels of organization) to predict higher-order effects.[17] However, although many of these indices have the advantage of extreme sensitivity and selectivity for metals, and undoubted potential as 'early warning systems' for metal pollution, their ecological relevance is not always evident: links between biochemical effects and physiological damage, for example, may appear obvious but are rarely demonstrated. In contrast, measurement of pollutant responses at the highest levels of organization maximizes ecological significance, but usually does so at the expense of sensitivity (thus population and community studies often fail to distinguish between natural and pollutant-induced changes in all but the most extreme cases). Consequently an integrated approach, using combinations of the techniques described below, is usually the only appropriate way of detecting the more subtle effects of metal contamination.

1. Biochemical Responses

There are a number of biochemical indices which may be used to reflect responses to specific contaminants, including metals, following their uptake by cells. Many are dealt with in greater detail in Chapter 8 of this volume, but two merit consideration here since they provide an insight into mechanisms of toxicity. Thus, both metallothionein (MT) induction and changes in lysosomal activity are frequently regarded as manifestations of the organism's effort to detoxify excess metals.

The production of MT (and other metal-binding proteins) appears to be particularly valuable as an index of stress in view of its specificity to metals (including Cd, Cu, Hg, and occasionally Ag, Zn), and a number of authors have reported that induction may be quantitatively related to metal contamination, both in the laboratory and in the field.[18,19] The detoxifying role of MT has been established in work with mussels, *Mytilus edulis,* which demonstrated that acquired tolerance to Hg was due to MT induction following pre-exposure to Hg (0.05 to 5 μg/l), Cu (5 μg/l), or Cd (50 μg/l).[20] It has been proposed that toxic effects occur as the binding capacity of MT becomes saturated, due to the interaction of excess free metal in the cell with the enzyme pool (''spillover'').[21] The precise significance of this displacement, in terms of decreased performance, is still open to question, however. Further development of methods to quantify and assess the consequences of MT induction is required in a wide range of species before this technique can be routinely used to monitor effects.

Impaired lysosomal function also reflects disturbances to the detoxification system of the cell and, as with induction and saturation of MT, may be seen as a forerunner, and possibly causitive feature, of various higher order effects. This potential has been demonstrated in studies with mussels which showed that reduced latency of lysosomal hexosaminidase, (a function of membrane destabilization in lysosomes, and inducible at Cu levels as low at 20 μg/l), is quantitatively related to physiological responses in the whole organism, such as clearance rates and growth.[17,22] Furthermore, studies with the hydroid, *Campanularia flexuosa,* showed that destabilization of lysosomes occurs at concentrations of Cu (1.2 to 1.9 μg/l), Cd (40 to 75 μg/l), and Hg (0.17 μg/l), almost ten times lower than levels causing reductions in growth, highlighting the value of such indices as early warning devices of deleterious effects.[17,23]

2. Whole Organism Responses

a. Growth

Of all the sublethal responses investigated to date, growth and condition are probably the most intensively studied. The effect of metals on growth has been assessed in a range of marine organisms and examples of threshold levels reported in experiments with algae, hydroids, mollusks, and fish are shown in Table 2.

Growth responses in algae, especially phytoplankton, are among the most sensitive indicators of heavy metal contamination and the use of ^{14}C incorporation as a measure of production has greatly improved the capability to determine impact. Also, following the discoveries that microalgal cultures contain large quantities of metal-chelating ligands and that, generally, toxicity is determined by free ions rather than complexed metals, recent studies have focused on the use of 'natural communities', either in the laboratory or in the field, to determine effects. The role of these chelating materials is still the subject of debate and may be linked with mobilization of essential elements, such as Fe and Mn. There is no doubt, however, that such complexes reduce the availability of several metals, especially Cu and Hg, thus fulfilling a detoxifying role.[27] As a result, metal concentrations reported to inhibit growth in microalgae vary greatly depending on the complexation capacity of the seawater and, as shown, for example, by the data of Sunda and Lewis (Table 2), the value of free ions (Cu^{2+}) causing growth inhibition may be considerably lower than levels of 'total' dissolved metal. It has even been suggested that Cu at natural levels might be inhibitory if uncomplexed, particularly near metal-rich upwellings. Although this claim has yet to be proved, there is clear evidence that Cu can be toxic at levels of 0.19 to 0.3 μg/l, only marginally above those found in oceanic waters.[14] Hg also reduces algal growth at low levels (~1 μg/l, Table 2) although such concentrations would be encountered only in highly contaminated estuaries and bays. Zinc, however, may reduce growth at 15 to 20 μg/l, concentrations not uncommon in some inshore areas.[26,30] Results for Cd are conflicting and variations in thresholds may be due to differences either in complexation or in tolerance of test populations. Arsenic (V) inhibits growth at concentrations as low as 5 μg/l and may possibly result in reduced productivity in As-contaminated estuaries and coastal waters.[31] While Pb reduces growth above 20 μg/l, no effects could be detected for Se, Ni, or Cr at levels around 50 μg/l and it seems unlikely that these elements would be toxic in most areas.

In one of the few studies using mixtures of metals, Thomas et al.[28] showed that the growth of diatom/dinoflagellate assemblages was inhibited at metal concentrations five times higher than levels typically found in moderately polluted estuaries: toxicity was mainly attributed to Cu (15 μg/l) and Hg (0.75 μg/l).

Based on results described here, and in similar studies, it is evident that metal contamination could reduce growth and photosynthesis at some polluted sites, although the extent to which metals influence the productivity of inshore waters has yet to be ascertained.

Experiments with macroalgae, *Laminaria saccharina,* have shown that growth may be affected by Hg, Cu, and Zn[32] and for the most sensitive life stages (sporelings) threshold concentrations are similar to those observed in microalgae (Table 2).

Among invertebrates, growth reductions have been demonstrated at concentrations sometimes encountered in contaminated areas (Table 2). For example, exposure to 14.3 μg/l Cu, 1.6 μg/l Hg, and 0.13 μg/l TBT (tributyltin) inhibited colonial size in hydroids *C. flexuosa.*[23] Although other responses may be more sensitive, growth parameters in hydroids have, nevertheless, proved useful in bioassays to reflect water quality, as influenced by sewage and industrial wastes. Thus, Karbe and co-authors[36] demonstrated that leaching of metals from Hamburg dredge spoils reduced growth in *Eirene viridula* colonies.

Mollusks are among the most frequently used organisms in pollution studies and effects of metals on growth in oysters, mussels, and clams are included in Table 2. Concentrations

TABLE 2
Sublethal Effects of Metals — Growth

Species	Metal concentration (μg/l) Ag	As	Cd	Cu	Hg	Ni	Pb	Zn	Response	Ref.
Phytoplankton										
Natural assemblage				0.3					Reduced ^{14}C fixation	14
Natural assemblage			1.0	1.0	1.0				Reduced ^{14}C fixation	24
Natural assemblage				10.0					Reduced ^{14}C fixation	25
Natural assemblage		23	112	6.4	<6.0	60[a]	20	20	Reduced growth	26
Monochrysis lutheri				21.6 (0.07 as Cu^{2+})					Reduced division	27
Natural assemblage					0.8				Reduced growth	28
Natural assemblage					1.0				Reduced productivity	29
Natural assemblage								15	Reduced photosynthesis	30
Natural assemblage		5							Reduced growth	31
Macroalgae										
Laminaria saccharina										
Sporeling				10	0.5 (0.5)[b]			100	Reduced growth	32
Sporophyte				50	50 (5.0)[b]			1000	Reduced growth	32
Hydroids										
Campanularia flexuosa			195	14.3	1.6			740	Reduced growth	23
Mollusks										
Mytilus edulis			10	3	0.3	>200[a]	>200[a]	10	Reduced shell growth	33
Mercenaria mercenaria	32			16	15	5700		195	Reduced growth[c]	34
Crassostrea virginica	25			33	12	1200			Reduced growth[c]	34
Fish										
Pleuronectes platessa			5	10					Reduced growth	25, 35

[a] No effect observed at this concentration.
[b] Methylmercury.
[c] For larvae: concentrations also represent LC_{50} values (8 to 12 d).

of dissolved metals reported to inhibit growth, and which occasionally prove toxic to larvae, may be environmentally significant for Cu, Hg, Zn, and Cd, while for Ni and possibly Pb, threshold values are seldom encountered in the field.

Various condition indices have also been used to assess the impact of pollutants in mollusks, and of these, scope for growth is advocated as a particularly valuable indicator of stress.[17] Thus, measurement of scope for growth (the energy available for growth and reproduction, calculated as the difference between energy consumed in food and that lost in respiration and excretion) showed that condition of transplanted mussels, *M. edulis,* declined along a pollution gradient in Narragansett Bay.[37] As with so many field studies, however, it was not possible to separate the effects of metals from other similarly distributed contaminants.

Marine fish are regarded generally as being less sensitive to the effects of metals than other aquatic organisms, possibly as a result of more highly developed detoxification systems. Nevertheless, inhibition of growth has been observed in plaice, *Pleuronectes platessa,* exposed to 5 μg/l Cd or 10 μg/l Cu.[25,35] In terms of environmental impact, however, it is likely that effects on reproductive success are most critical for fish stocks, as described in subsequent sections.

b. Morphology

Morphological responses to pollutants are usually mechanistically linked to growth disturbances and may be important indicators of metal exposure. For example, in studies with hydroids, *Eirene viridula,* Karbe et al.[36] have used alterations in hydranth morphology to evaluate the toxicity of As, Cd, Cu, Hg, Pb, Se, and Zn. Threshold values in these experiments were, with the possible exception of Cu, higher than concentrations present in most contaminated waters (Table 3).

Among the most important morphological and histopathological abnormalities caused by metals are those which occur in embryonic and larval stages and the implications for successful recruitment are obvious. Studies on early stages of mollusks, in particular oysters and mussels,[38,39] have shown that such effects might be expected at contaminated sites, at least for Cu, Hg, Cd, Zn, and possibly Ag, while abnormalities caused by As, Cr, Ni, Pb, and Se seem less likely (Table 3).

Some extraordinary metal-induced deformities have been discovered in stenoglossan gastropods. These include abnormalities in eyes and tentacles of *Urosalpinx cinerea* exposed to Hg (10 μg/l)[40] and mascularization of female dogwhelks, *Nucella lapillus,* exposed to TBT (1 ng/l).[41] The significance of this latter observation and evidence linking shell deformities in oysters with TBT is discussed later.

Recent suggestions that contaminants are responsible for the induction of fish lesions (skeletal and gill deformities, tumors, ulcers, and fin rot) have caused much concern. However, although a number of field studies have revealed correlations between the incidence of lesions and contaminant levels, it is frequently impossible to identify the causative agents (indeed there is considerable debate as to whether anthropogenic inputs are contributory factors at all in some areas). Evidence for the involvement of metals in the induction of gill and skeletal deformities is provided by Bengtsson and co-authors[42] who found a greater proportion of affected fish in populations from metal-polluted sites in the Baltic compared with 'clean' sites. In subsequent long-term experiments with sculpin, *Myoxocephalus quadricornis,* exposed to diluted smelter effluents,[43] effects due to metals (a mixture of As, Cd, Cu, Hg, Pb, Zn) were shown to be dose related, though significant abnormalities occurred even at the highest dilutions (Table 3). Such evidence is limited, however, and further verification is necessary before the incidence of deformity and disease in fish is widely used as an indicator of pollutant stress.

TABLE 3
Sublethal Effects of Metals — Morphology

Species	Metal concentration (μg/l)										Response	Ref.
	Ag	As	Cd	Cr	Cu	Hg	Ni	Pb	Se	Zn		
Hydroids												
Eirene viridula		300	100		10	3		300	3,000	1,500	Altered hydranth morphology[a]	36
Mollusks												
Crassostrea gigas	22	326	611	4538	5	7	349	476	>10,000	199	50% abnormal larvae	38
Mytilus edulis	14	>3,000	1200	4469	6	6	891	758	>10,000	175	50% abnormal larvae	38
Crassostrea virginica	24					11				206	50% abnormal larvae	39
Fish												
Myoxocephalus quadricornis		32	0.5		0.8	0.1		1.2		5.3	Increase in vertebral deformities[b]	43

[a] Lowest reported threshold concentration.

[b] Metals applied as a mixture.

c. Reproduction and Recruitment

In organisms which reproduce by both asexual and sexual means, the effect of metals may be to enhance or suppress one or other form. For example, sexual reproduction in the red seaweed, *Champia parvula,* as determined by the production of mature cystocarps, is reduced following exposure to As.[44] In contrast, hydroid colonies respond to metal intoxication by increasing the production of gonozooids (the sexual morphs), an adaptation which may help to disperse the colony away from polluted sites.[23] Threshold levels for gonozooid production in *Campanularia flexuosa* are among the lowest observed in sublethal tests with Cu and Zn (Table 4), and bioassays using this index have been used to assess water quality.[23]

For other organisms, inhibition of fertilization and embryogenesis are usually the most significant reproductive effects since they may ultimately determine the survival of the population. Thus, studies on the effects of metals on reproductive cycles in polychaetes (Table 4) have shown that reduced fecundity is likely to occur at concentrations considerably lower than levels reported to be acutely toxic in adults.[10] Based on these data, however, it seems unlikely that recruitment would be affected at all but the most severely contaminated sites. This agrees with observations from the field (described elsewhere in this chapter) that certain polychaetes are able to tolerate conditions, for example in highly metalliferous sediments, where more sensitive species are eliminated.[11] Resistance across the whole life cycle may, therefore, confer considerable adaptive advantages.

There are a number of observations of impaired fecundity in mollusks as a result of exposure to metals and their compounds and some of the most important discovered to date — those attributable to TBT — are discussed later. In experiments with mussels *Mytilus edulis,* Myint and Tyler[46] showed that oocyte development was suppressed by Cu and Zn at 50 and 200 μg/l, respectively. The sensitivity of eggs and sperm was also demonstrated in studies with the clam *Spisula solidissima* which showed that embryogenesis was affected to a far greater extent by prefertilization exposure of gametes to Ag (9.5 μg/l) than by postfertilization treatment.[47]

Germ cells and embryos of echinoderms are commonly used to assess the effects of pollutants on reproduction. Threshold concentrations associated with the arrest of fertilization and development in sea urchins range from 10 μg/l for Cu and Hg to $>$1000 μg/l for Pb, Cr, As, (Mn and Fe) as shown in Table 4. Gametes appear to be especially vulnerable and for sperm, effects are detectable at extremely low concentrations of Zn (5 μg/l). Echinoderm bioassays using fertilization, cleavage, and gastrulation as markers, have been successfully employed as indicators of water quality, demonstrating, for example, the extent of deleterious effects caused by metal refinery wastes.[51]

The incorporation of life cycle tests is an important feature of studies on crustaceans and has usually confirmed that sexual maturation and reproduction are the most sensitive stages. Threshold levels for some species (Table 4) suggest that effects might be expected at contaminated sites. Long term experiments with *Pontoporeia affinis,* for example, indicate that fecundity is reduced at around 5 μg/l Cd or Pb. Applied jointly, however, Pb appears to reduce the toxicity of Cd.[48]

Despite the relative tolerance displayed by adult fish, gametes and embryos have been identified as vulnerable to the effects of metals. Thus, fertilization and embryogenesis in spring-spawning herring are inhibited by Cu, Zn, and Cd in the range 5 to 10 μg/l, while hatching success in the eggs of *Fundulus heteroclitus* is reduced by similar levels of Hg.[52,53] Extremely low levels of free Cu^{2+} activity (Table 4) have been found to affect viability of eggs in spot, *Leiostoma xanthurus* and silverside *Menidia menidia*, prompting suggestions that natural Cu levels in sea water would, if uncomplexed, suppress hatching.[54] However, elevated metal concentrations in surface films of contaminated inshore areas, as described later, probably represent the greatest threat to the survival of free-floating fish eggs.

TABLE 4
Sublethal Effects of Metals — Reproduction and Development

Species	Metal concentration (μg/l) Ag	As	Cd	Cr	Cu	Hg	Ni	Pb	Zn	Response	Ref.
Macroalgae											
Champia parvula		60								Inhibition of sexual reproduction	44
Hydroids											
Campanularia flexuosa					0.05	0.01			500	Increase in gonozooid frequency[a]	23
Polychaetes											
Neanthes arenaceodentata			1000	50				3100	320	Reductions in reproduction	45
Capitella capitata			560	100				200	560	Reductions in reproduction	45
Ctenodrilus serratus			2500	50	100	50	500	1000	500	Reductions in reproduction	45
Bivalves											
Mytilus edulis					50				200	Development of oocytes supressed	46
Spisula solidissima (germ cells)	9.5									Impaired embryogenesis	47
Crustaceans											
Pontoporeia affinis			5.5					4.9		Fecundity reduced	48
Rhithropanopeus harrisii			50						25	Hatch time increased	49
Tigriopus japonicus			44		6.4					Generation time doubled	50
Echinoderms											
Sea urchin eggs (various spp.)		1500	600	1000	10	10	600	1000	30	Fertilization and development arrested[a]	51
Fish											
Spring-spawning herring			5		10				10	Reduced fertilization	52
Fundulus heteroclitus						10				Reduced hatch	53
Leiostomus xanthurus (eggs)					0.064[b]					Reduced hatch	54
Menidia menidia (eggs)					0.025[b]					Reduced hatch	54

[a] Lowest reported threshold concentration.
[b] Calculated as free ion (Cu^{2+}).

d. Behavior

Behavioral responses are often difficult to quantify in the field and except for Cu, Hg, and Cd, most information from laboratory studies suggests that threshold values are significantly above realistic environmental levels. Nevertheless, some responses, including food handling in hydroids[36] and (reduced) filtration in bivalves[17] are useful indicators of water quality and, in addition, they may be important precursors of other deleterious effects. For example, reduced filtering in bivalves during prolonged periods of metal exposure would clearly, in time, become detrimental to their condition. More importantly perhaps, complete valve closure can occur at relatively low metal levels, as observed in mussels exposed to 5 μg/l Cu.[33] Similar responses in sediment-dwelling species could lead to critical delays in burrowing and settlement. Thus in experiments with clams, *Macoma balthica,* McGreer[55] showed that inhibition in burrowing was linearly related to metal contamination in sediments and concluded that toxicity was probably due to Cd and Hg. At the highest sediment concentrations (1.4 μg/g Cd, 0.46 μg/g Hg), active avoidance was displayed by *Macoma*, demonstrating how recruitment could be affected at contaminated sites.

3. Community Responses

Through their effects on the previously described life processes, metals reduce the competitive ability of the most sensitive organisms and may ultimately lead to their exclusion from marine communities. Thus, one of the earliest observations of the effects of severe pollution was that opportunistic (tolerant) species became dominant at the expense of specialist species and often resulted in more simplified ecosystems.

Evidence for the impact of pollutants on plankton communities is derived largely from *in situ*, enclosure-type experiments using natural assemblages, and changes in dominance patterns of phytoplankton have been observed for several metals. As a general rule, sensitivity in algae appears to vary depending on the metal studied and it is not easy to predict resistant species. However, among diatoms, replacement of centric species by pennate forms is a consistent observation. Ibragim and Patin[24] showed, for example, that exposure of Mediterranean assemblages to Cd, Cr, Pb, or Hg resulted in a preponderance of *Nitzschia closterum* at most concentrations tested in the range 1 to 1000 μg/l. A similar transfer of dominance in favor of pennate diatoms has been observed by other workers at 5μg/l Cu,[56] 1 to 5 μg/l Hg,[29] and, perhaps most remarkably, at 5 μg/l As,[31] a concentration only five times greater than background.

Changes in the taxonomic diversity of phytoplankton communities are not necessarily the result of direct toxicity on sensitive species, however, but may be related to reduced grazing pressure.[29] Conversely, changes in phytoplankton diversity may have fundamental implications for higher trophic levels, since a change in dominant species, even if not accompanied by lowered productivity, could reduce the choice and abundance of preferred prey organisms. Sanders[57] has shown, for example, that algal population shifts from flagellates to smaller, tolerant diatoms, inducible at low levels of As (15 μg/l), may result in reduced copepod fecundity and survival, even though the copepods themselves are resistant to relatively high levels of As.

A degree of caution must be exercised when extrapolating the results of containment studies to natural ecosystems, nevertheless, they do provide valuable insights into the mechanisms of metal toxicity in plankton communities.

Despite difficulties in separating anthropogenic and naturally induced events, deleterious effects of metals in natural benthic marine communities have also been demonstrated. For example, using data from several Norwegian fjords, Rygg[58] was able to show that metal contamination in sediments (Cu, Pb, and Zn) reduces the diversity of benthic fauna. Based on the strength of (negative) correlations between species diversity and sediment metal concentrations, Cu was identified as the most likely agent responsible for the absence of

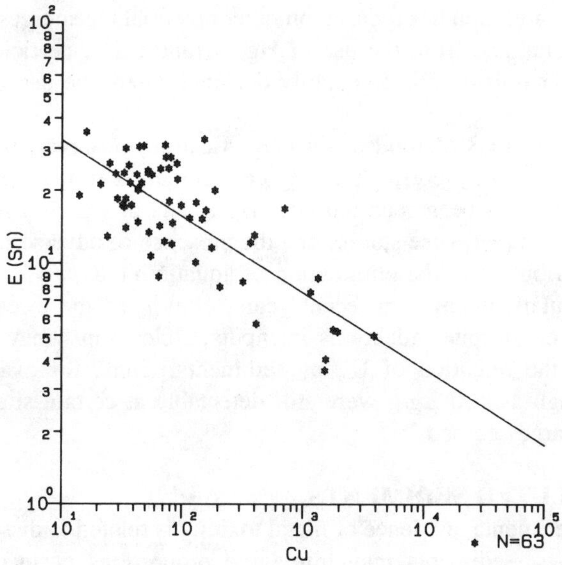

FIGURE 1. Correlation between faunal diversity, E (Sn), and sediment Cu concentrations (μg/g) for benthic populations in Norwegian Fjords. Regression equation is Log_{10} diversity = −0.3 $\log_{10}$ (Cu in sediment) + 4.2 (R = −0.76, $p \leq 0.001$). (From Rygg, B., *Marine Ecol. Progr. Ser.*, 25, 83, 1985. With permission.)

sensitive species (Figure 1). At sediment concentrations above 200 μg/g Cu (10 × background), diversity was reduced by 50%, while effects were evident even at values five times normal. The high proportion of carnivorous polychaetes surviving at polluted sites confirms earlier observations of their higher metal tolerance.

The foregoing discussion is by no means intended as a comprehensive review of metal toxicity, but nevertheless illustrates the ways in which deleterious effects may be detected. In the remainder of this chapter, we examine further evidence, from various areas of the marine environment, which indicates that metal contamination has resulted in biological impact.

III. DELETERIOUS EFFECTS OF METALS IN THE FIELD

A. MERCURY POLLUTION

It was the incidents involving human fatalities in Japan, from Hg poisoning, which initiated concern over metal pollution in aquatic environments. These cases have been well documented elsewhere and only a brief overview is presented here as an introduction to later discoveries. In the most important example, that of Minamata Bay, some 80 tonnes of Hg (used as a catalyst in the production of acetaldehyde and vinyl chloride) were discharged over a period from 1932 to 1968, with the result that sediments in some areas of the bay contained several hundred micrograms per gram Hg and concentrations in water as high as 3.6 μg/l Hg were recorded.[59] By the early 1950s, there were mortalities in fish and invertebrates from the bay and cats, dogs, rats, and waterfowl inhabiting the waterfront displayed symptoms of poisoning. Although the first human case was recorded in 1956, it was not until the end of the decade that mobilization of methyl mercury (MeHg) through the food chain was identified as the cause of the problem and by 1975, there were more than 800 verified cases of 'Minamata disease' of which more than 100 proved fatal, primarily among fishermen and their families.

Since Minamata, attention has focused on other potential Hg sources (notably discharges from chlor-alkali plants and from the use of Hg as fungicides, pesticides, and antifouling preparations) and contamination is frequently detected where the use of Hg by industry is high (Table 5).

Apart from observations of food chain magnification, however, there are surprisingly few substantiated reports of the effects of Hg contamination in marine biota. The absence of intertidal organisms has been reported near Hg inputs or where levels in sediments are high,[8,60] although in none of these studies can the presence of adverse factors other than Hg be excluded as contributing to the elimination of aquatic biota.

Widespread regulations have, in recent years, eliminated the threat of Minamata-type catastrophes; however, despite reductions in inputs, "clean-up" may take many years in some areas due to the retention of Hg by sediments. Thus, for example, sediment Hg concentrations as high as 100 μg/g were still detectable at certain sites in Minamata Bay 10 years after discharges ceased.[59]

B. METAL-POLLUTED SEDIMENTS

Most of the experimental evidence of metal toxicity is related to dissolved forms though, in estuaries and bays, sediments may trap large proportions of incoming contaminants (tragically apparent at Minamata) so that sediment metal concentrations often exceed those in overlying water by several orders of magnitude. Examples of values from polluted sites are shown in Table 5. Uptake by organisms of even a minute fraction of sediment-bound metal could have considerable toxicological significance especially where conditions favor bioavailability.[12] Elevated metal concentrations in pore water may also contribute significantly to sediment toxicity.

Evidence for deleterious effects is usually established by the absence of sensitive species or by the induction of resistance mechanisms and adaptation in tolerant forms. There are few studies, however, where effects can be related to a single contaminant and in order to confirm that pollution (as opposed to natural variation) is the cause of change, biological observations are usually backed up by chemical measurements and bioassay response. Applying this combined approach at sites in Puget Sound, Long and Chapman[63] were able to show that, where combined contaminant levels (metals plus organics) were high in sediments and subsequently induced toxicity in bioassays (amphipod lethality; oligochaete respiration; larval development in fish, oysters, and polychaetes; cytotoxicity and mutagenicity in fish cells), lowered diversity in infauna could be expected. Reduced numbers of echinoderms and arthropods and a predominance of polychaetes and mollusks, were consistent features of benthic communities at polluted stations. Although agreement between toxicological observations and sediment pollution gradients (up to 581, 2109, and 1190 μg/g for Cu, Pb, and Zn) were generally good, biological responses could not always be satisfactorily predicted from chemical data, highlighting the need to consider site-specific factors which affect bioavailability.

C. SEWAGE DISPOSAL AND SLUDGE DUMPING

The discharge and dumping of sewage and industrial spoils may release or redistribute significant quantities of heavy metals in the marine environment. In the U.K. for example, 10 million tons of sewage (25% of the U.K. total) are discharged annually to coastal waters.[10] Studies near outfalls in California[61] have shown that, locally, there may be strong enrichment of metals in sediments (up to 200 times background, see Table 5). However, although chemically, it may be easy to distinguish the presence of contamination — and for metals such as Ag, Cd, and Cu, enhanced bioaccumulation is a frequent observation — evidence of deleterious effects in biota is limited. Impact depends largely on the dispersive character of the area with problems occurring mainly at accumulative sites. McGreer[64] has shown,

TABLE 5
Examples of Metal Concentrations (μg/g) in Contaminated Sediments[a] (Compiled from Sources Indicated)

Metal	Baltic Sea[5] (various sources)	Bristol Channel/ Severn estuary, U.K.[9] (industry, sewage)	Mersey estuary U.K.[60] (sewage, industry including chlor-alkali)	Los Angeles outfall, California[61] (sewage)	Derwent estuary Tasmania[8] (refinery, chlor-alkali)	Restronguet Creek, U.K.[9] (mining)	Port Pirie, Australia[62] (smelter)
As		8	71			2520 (13)	151 (1.0)
Cd	8.1 (<0.01)	1.1	3.9	66 (0.3)	862	1.2 (0.3)	267 (0.5)
Cu	283 (1.0)	54	144	940 (8.3)	>400	2,540 (19)	122 (3.0)
Hg	9 (0.01)	0.48	6.2	5.4 (0.04)	1,130	0.22 (0.12)	8
Ni	920 (1.0)	33	44	130 (9.7)	42	32 (28)	19.4 (12)
Pb	400 (2)	88	205	580 (6.1)	>1,000	400 (2)	5,270 (2)
Zn	2,090 (6)	255	255	2,900 (43)	>10,000	2,090 (6)	16,667 (11)

[a] Maximum concentrations shown together with local background values (in parentheses), where given.

for example, that sediment-bound metals, originating from sewage discharge in the Fraser River estuary, Canada, were influencing the distribution of clams, *Macoma balthica*. Although metal concentrations (0.89 μg/g for Hg, 234 μg/g for Cu, 264 μg/g for Zn) were not exceptional, or even toxic to adults, their joint action was considered sufficient to prevent settlement and survival of juveniles near to the outfall.

In a rare demonstration in the field of the potential value of biochemical indices, Brown et al.[21] discovered an increase in Cd, Cu, and Zn binding to MT, accompanied by 'spillover' of metals into the enzyme pool, in mussels collected in the vicinity of a sewage input. Interestingly, these observations correlated with decreasing mussel population densities near to the outfall, although a direct link between the two events was not established.

D. IMPACT FROM METAL MINING AND SMELTING

Based on levels of contamination, the most significant anthropogenic releases of metals to the marine environment originate from the disposal of mine tailings and the discharge of wastes from smelters and refineries. Consequently some of these localities provide conclusive evidence of the deleterious effects of metals.

Southwest England has for centuries been the scene of widespread mining for metals, such as Sn, Cu, Zn, As, Pb, and Ag and a number of estuaries are heavily contaminated. Metal concentrations in water and sediments of Restronguet Creek, for example, exceed 'normal' values by orders of magnitude (Tables 1 and 5). Biological surveys have revealed that while some species, such as the polychaete (*Nereis diversicolor*), crabs (*Carcinus maenas*), and oysters (*Ostrea edulis*), have adapted to conditions in the creek, many others are absent from sites where metal contamination is highest.[11] Bivalves including mussels *Mytilus edulis,* cockles *Cerastoderma edule,* and clams, *Scrobicularia plana* are conspicuously absent when compared with nearby uncontaminated estuaries. Bryan and co-authors[11] considered Cu and Zn to be responsible for the impoverished fauna of Restronguet Creek, largely as a result of reduced recruitment and survival of juvenile forms.

Large reductions in clam production and a decrease in species diversity of benthic invertebrate communities were also evident in Goa estuaries heavily impacted with mine wastes,[65] though high turbidity, blanketing of bottom fauna, and lowered oxygen tensions probably contributed significantly to these effects. Similarly, in the Southeast Asia tin mining strip, secondary factors, notably increased suspended solids resulting from tailings dispersal, may exert the greatest impact. Corals in particular are highly intolerant of turbid waters.[66]

At the other end of the extraction process, wastes from smelters and refineries are potentially most hazardous since metals are usually in a more chemically active form. Studies by Ward and co-authors[62] at the site of the world's largest Pb smelter at Port Pirie in Australia (where some 600 km^2 have been contaminated with Cd, Pb, Zn, As, Cu, Mn, and Sb), have shown that metals are involved in the elimination of 20 species of fish and crustaceans from sea-grass communities. Sediments, the most appropriate measure of contamination, were considered to play a significant role in the toxicity of metals. Although effects were greatest near the effluent source, where sediment values 200 to 1000 times background were recorded (Table 5), species richness was reduced by sediment concentrations (0.5 to 0.7 μg/g Cd, 10 to 14 μg/g Pb, 62 to 92 μg/g Zn) only five times background. The authors concluded that changes in community structure were probably due to an array of sublethal effects rather than acute toxicity, and that, although Zn was probably most toxic, biological impact (observed over an area of 100 km^2) was due to the joint action of several metals.

In situ cage-culture techniques have sometimes been used to evaluate the impact of smelting and mining wastes. For example, in studies with phytoplankton assemblages transplanted along contamination gradients in Norwegian fjords (Zn, Cu, Pb, Cd, and Hg concentrations up to 610, 4, 2, 3 and 0.4 μg/l, respectively) Eide and co-authors[67] were able to demonstrate that elimination of sensitive species and growth inhibition in tolerant forms

TABLE 6
Effects of Tributyltin (TBT) on Marine and Estuarine Organisms

Species	Concentration of TBT in water (ng/1 as Sn)	Effect	Ref.
Nucella lapillus, dogwhelk	1—10	Imposex, impaired reproduction	41
Crassostrea gigas, oyster[a]	8	Shell thickening	70
	20	Reduced growth, viability	71
Mytilus edulis, mussel[a]	40	Reduced viability	72
Venerupis decussata, clam[a]	40	Reduced growth, viability	73
Gammarus oceanicus, amphipod[a]	120	Reduced growth	74
Homarus americanus, lobster[a]	400	Reduced viability	75
Pavlova lutheri, *Skeletonema costatum*, *Dunaliella tertiolecta* (microalgae)	40—400	Reduced growth	76

[a] Larvae.

were a function of reduced water quality. Evidence for the involvement of smelter discharges in the increased incidence of fish deformities has been discussed previously.

Residues from the production of titanium dioxide have been dumped in the sea or released into estuaries for many years. Possible effects could arise from high acidity, ferrous sulfate, and some trace metals (Mn, Ti, Cr, V, Cd). However, although tests show the potential for harmful effects, these are not easily demonstrated in the field. Complexation of metals with rapidly precipitating ferric oxide may help to reduce their toxicity.

E. ANTIFOULING AGENTS — METALS AND TBT

Because of their toxic properties, metals, particularly Cu, have been incorporated into antifouling paints since the beginning of this century. Following observations of Cu tolerance in some fouling organisms, a number of other biocides, including organomercury, arsenic, and lead compounds, have been used to enhance the toxic properties of Cu. The use of these 'boosters' was phased out around 1970 because of fears concerning their environmental toxicity, though there is little published information on their impact in the field.

Since 1970, TBT compounds have been used increasingly in antifouling preparations (for use on boat hulls and fish-farming gear), both in conjunction with Cu and as the sole biocidal agent. Although TBT eventually degrades in the environment, effects on nontarget organisms have recently been recognized at lower levels than was ever envisaged. Shell abnormalities and reduced growth and recruitment in oysters, *Crassostrea gigas,* sampled near marinas, were the first indications of the "TBT problem"[68] and subsequently effects have been demonstrated in a number of marine and estuarine species. TBT concentrations in inshore waters typically range from a few nanograms per liter in boat-free areas to levels in excess of 600 ng/l (as Sn) near marinas,[69] often exceeding toxic thresholds observed for aquatic organisms, especially mollusks (Table 6). The most sensitive species identified so far is the dogwhelk, *Nucella lapillus,* which exhibits 'imposex' (the imposition of male characteristics on females) at concentrations as low as 1 ng/l TBT (as Sn), and complete reproductive failure at 10 ng/l. *N. lapillus* is currently in decline along much of the south coast of England and there appear to be few populations in England where the effects of TBT are not readily observed.[41]

Following the discovery of the highly toxic nature of TBT-based paints, regulations have recently been placed on their usage on small boats in France (1982) and the U.K. (1987). In view of the reservoir of TBT in estuarine sediments and its continued application on large vessels, however, it may be some time before contamination is eliminated from many inshore localities.

The association of TBT with sediments may have important consequences for infauna, especially bivalves, which have become increasingly scarce at some highly contaminated sites in the U.K.[69] However, although laboratory experiments show that spiked sediments (1 μg/g TBT) are toxic to clams, *S. plana*,[80] further evidence is required to confirm whether or not TBT is the cause of the decline in field populations.

F. ORGANOLEAD

In addition to TBT, a number of other organometallic compounds have proved to be extremely toxic to marine life as a result of their release into the sea. For example alkyl Pb compounds, originating from the manufacture of petrol additives, were almost certainly responsible for bird kills in the Mersey estuary, U. K., during 1979 to 1982. Consumption of clams and worms which had accumulated trimethyl Pb from water was identified as the critical pathway to birds.[77] Fortunately, as a result of improved effluent standards, Pb burdens in invertebrate fauna in the Mersey have declined substantially in recent years[60] and to date no further bird kills have been reported.

G. SEA-SURFACE MICROLAYER

Metal concentrations in the surface microlayer of the sea can often exceed subsurface concentrations by orders of magnitude at inshore sites and may, therefore, threaten the survival of organisms which congregate in the upper layers.[78] Recent studies,[78,79] in Puget Sound, for example, have shown that the abundance of neustonic organisms and also numbers, chromosomal integrity, and viability of fish eggs, were reduced in surface film samples from urban areas (Σ Cd, Pb, Zn, Ag, and Cu up to 1420 μg/l) compared with samples from control sites (combined metals 58 μg/l). The most significant metal enrichments at urban sites were for Cu and Pb, the latter as a result of petrol combustion. Enrichments for Ag were attributed to sewage inputs. However, although metal-induced effects would be expected at these high concentrations, it seems likely that other contaminants which are similarly magnified in the surface film, including TBT, PCBs, and hydrocarbons, will also play a considerable role in determining overall toxicity.

IV. SUMMARY

Following the development of a wide range of criteria to assess the impact of metals on marine organisms, effects are now recognized at much lower levels than in earlier LC_{50} studies. Growth, reproduction, and recruitment are generally the processes most susceptible to metal stress and sublethal damage may be detected at concentrations more than an order of magnitude lower than in acute tests. Effects on embryonic and larval development probably have the greatest ecological significance in terms of preserving the health of marine communities. The trend towards increasingly more sensitive indices may eventually enable us to visualize effects at even lower levels; biochemical studies, related to the induction and saturation of detoxification mechanisms, seem particularly promising in this respect, providing their ecological relevance can be established.

On present evidence, Cu and Hg are potentially the most hazardous metals present in the marine environment and their harmful effects have been demonstrated in many studies. Overall, Cu is perhaps most significant and, arguably, may be toxic at levels only marginally above background. Although Hg is extremely toxic, concentrations in the field rarely exceed even the lowest experimentally derived thresholds and, despite the dangers in extrapolating data from the laboratory to nature, it would seem that recent regulations on Hg emissions have considerably reduced the direct threat of Hg pollution. Nevertheless, the consequences of food chain biomagnification of MeHg residues still remain a justifiable cause for concern in a number of contaminated areas.

Recently, the impact of organic forms of several other metals has assumed increasing importance. TBT, for example, is one of the most toxic compounds released into the marine environment, directly affecting recruitment and survival of aquatic organisms, especially mollusks. In the case of organolead, effects may, as with MeHg, be most severe in top predators.

For metals such as Ag, Cd, Zn, Pb, and Cr, effects might be expected only in highly contaminated inshore areas, particularly estuaries and embayments, and then usually as a result of their joint action; the individual impact of these metals has rarely been demonstrated in the field. Because of their low toxicities, Se and Ni present few problems for aquatic organisms. Arsenic appears to be relatively harmless to marine fauna but may, surprisingly, affect phytoplankton at concentrations only five times greater than background.

However, although the temptation to generalize about pollutant effects may be great, extreme caution is warranted in view of the large number of environmental and physiological factors which influence the toxicity of metals, notably the wide range of tolerance displayed by different organisms. In addition, indirect effects caused by the elimination of sensitive species could have far greater significance for marine communities than indicated in toxicity studies with single species. Consequently, the incidents of metal pollution described in this chapter merely serve to highlight the types of change which can occur at contaminated sites and do not necessarily signify universally applicable responses. Clearly, reasonable assessment and prediction of environmental impact is best made on a regional basis, following a thorough consideration of local conditions.

ACKNOWLEDGMENTS

Parts of this work were supported by the Department of the Environment under contract PECD 7/7/226.

REFERENCES

1. **Burton, J. D. and Statham, P. J.,** Occurrence distribution and chemical speciation of some minor dissolved constituents in ocean waters, in *Environmental Chemistry,* Vol. 2, Bowen, H. J. M., Ed., Royal Society of Chemistry, London, 1982, 234.
2. **Topping, G., Bewers, J. M., and Jones, P. G. W.,** A review of the past and present measurements of selected trace metals in sea water in the Oslo Commission and ICNAF/NAFO areas, ICES Cooperative Research Report, 97, 1980, 43 pp.
3. **Mart, L. and Nurnberg, H. W.,** Cd, Pb, Cu, Ni and Co distribution in the German Bight, *Mar. Chem.,* 18, 197, 1986.
4. **Duinker, J. C. and Nolting, R. F.,** Dissolved copper, zinc and cadmium in the Southern Bight of the North Sea, *Mar. Pollut. Bull.,* 13, 93, 1982.
5. **Brugman, L.,** Heavy metals in the Baltic Sea, *Mar. Pollut. Bull.,* 12, 214, 1981.
6. **Huynh-Ngoc, L. and Fukai, R.,** Levels of trace metals in open Mediterranean surface waters — summary report, in *IV journées d'études sur les pollutions marines en Méditerranée,* Workshop on pollution of the Mediterranean, Monaco, CIESM, 1979, 171.
7. **Abdullah, M. I. and Royle, L. G.,** A study of dissolved and particulate trace elements in the Bristol Channel, *J. Mar. Biol. Assoc. U.K.,* 54, 581, 1974.
8. **Bloom, H. and Ayling, G. M.,** Heavy metals in the Derwent estuary, *Environ. Geol.,* 2, 3, 1977.
9. **Bryan, G. W., Langston, W. J., Hummerstone, L. G., and Burt, G. R.,** A guide to the assessment of heavy-metal contamination in estuaries using biological indicators, *J. Mar. Biol. Assoc. U.K.,* occasional publication No. 4, 1985.
10. **Mance, G.,** *Pollution Threat of Heavy Metals in Aquatic Environments,* Elsevier, London, 1987.
11. **Bryan, G. W., Gibbs, P. E., Hummerstone, L. G., and Burt, G. R.,** Copper, zinc and organotin as long-term factors governing the distribution of organisms in the Fal estuary in Southwest England, *Estuaries,* 10, 208, 1987.

12. **Langston, W. J. and Bryan, G. W.,** The relationships betwwen metal speciation in the environment and bioaccumulation in aquatic organisms, in *Complexation of Trace Metals in Natural Waters,* Kramer, C. J. M. and Duinker, J. C., Eds., Nijoff/Junk, The Hague, 375.
13. **McLusky, D. S., Bryant, V., and Campbell, R.,** The effects of temperature and salinity on the toxicity of heavy metals to marine and estuarine invertebrates, *Oceanogr. Mar. Biol. Annu. Rev.,* 24, 481, 1986.
14. **Sunda, W. G., Barber, R. T., and Huntsman, S. A.,** Phytoplankton growth in nutrient rich sea water: importance of copper-manganese cellular interactions, *J. Mar. Res.,* 39, 567, 1981.
15. **Pelletier, E.,** Mercury-selenium interactions in aquatic organisms: a review, *Mar. Environ. Res.,* 18, 111, 1985.
16. **Neuhoff, H. G.,** Synergistic physiological effects of low copper and various oxygen concentrations on *Macoma balthica, Mar. Biol.,* 77, 39, 1983.
17. **Bayne, B. L., Brown, D. A., Burns, K., Dixon, D. R., Ivanovici, A., Livingstone, D. R., Lowe, D. M., Moore, M. N., Stebbing, A. R. D., and Widdows, J.,** *The Effects of Stress and Pollution on Marine Animals,* Praeger, 1985.
18. **Langston, W. J. and Zhou, M.,** Evaluation of the significance of metal-binding proteins in the gastropod *Littorina littorea, Mar. Biol.,* 92, 505, 1986.
19. **Roesijadi, G., Drum, A. S., Thomas, J. T., and Fellingham, G. W.,** Enhanced mercury tolerance in marine mussels and relationship to low molecular weight, mercury-binding proteins, *Mar. Pollut. Bull.,* 13, 250, 1982.
20. **Roesijadi, G. and Fellingham, G. W.,** Influence of Cu, Cd and Zn pre-exposure on Hg toxicity in the mussel *Mytilus edulis, Can. J. Fish. Aquat. Sci.,* 44, 680, 1987.
21. **Brown, D. A., Gosset, R. W., Hershelman, P., Schaefer, H. A., Jenkins, K. D., and Perkins, E. M.,** Bioaccumulation and detoxification of contaminants in marine organisms from Southern California coastal waters, in *Waste Disposal in the Oceans,* Soule, D. F. and Walsh, D., Eds., Westview Press, Boulder, CO, 1983, 171.
22. **Moore, M. N., Widdows, J., Cleary, J. J., Pipe, R. K., Salkeld, P. N., Donkin, P., Farrar, S. V., Evans, S. V., and Thomson, P. E.,** Responses of the mussel *Mytilus edulis* to copper and Phenanthrene: interactive effects, *Mar. Environ. Res.,* 14, 167, 1984.
23. **Stebbing, A. R. D. and Brown, B. E.,** Marine ecotoxicological tests with coelenterates, in *Ecotoxicological Testing for the Marine Environment,* Vol. 1, Persoone, G., Jaspers, E., and Claus, C., Eds., State University Ghent and Institute of Marine Scientific Research, Bredene, Belgium, 1984, 307.
24. **Ibragim, A. M. and Patin, S. A.,** Effect of mercury, lead, cadmium and copper on primary production and phytoplankton in some coastal regions of the Mediterranean and Red Seas, *Oceanology Moscow,* 15, 589, 1976.
25. **Saward, D., Stirling, A., and Topping, G.,** Experimental studies on the effects of copper on a marine food chain, *Mar. Biol.,* 29, 351, 1975.
26. **Hollibaugh, J. T., Seibert, D. L. R., and Thomas, W. H.,** A comparison of the acute toxicities of ten heavy metals to phytoplankton from Saanich Inlet, B. C., Canada, *Estuarine Coastal Mar. Sci.,* 10, 93, 1980.
27. **Sunda, W. G. and Lewis, J. M.,** Effect of complexation by natural organic ligands on the toxicity of copper to a unicellular alga *Monochrysis lutheri, Limnol. Oceanogr.,* 23, 870, 1978.
28. **Thomas, W. H., Hollibaugh, J. T., Seibert, D. L. R., and Wallace, G. T.,** Toxicity of a mixture of ten metals to phytoplankton, *Mar. Ecol. Prog. Ser.,* 2, 213, 1980.
29. **Thomas, W. H., Seibert, D. L. R. and Takahashi, M.,** Controlled ecosystem pollution experiment; effect of mercury on enclosed water columns. III. Phytoplankton dynamics and production, *Mar. Sci. Commun.,* 3, 331, 1977.
30. **Davies, A. G. and Sleep, J.,** Photosynthesis in some British Coastal waters may be inhibited by zinc pollution, *Nature (London),* 277, 292, 1979.
31. **Sanders, J. G. and Vermersch, P. S.,** Response of marine phytoplankton to low levels of arsenate, *J. Plankton Res.,* 4, 881, 1982.
32. **Thompson, R. S. and Burrows, E. M.,** The toxicity of copper, zinc and mercury to the brown macroalga *Laminaria saccharina,* in *Ecotoxicological Testing for the Marine Environment,* Vol. 2, Persoone, G., Jaspers, E., and Claus, C., Eds., State University Ghent and Institute Marine Scientific Research, Bredene, Belgium, 1984, 259.
33. **Strömgren, T.,** Effect of heavy metals (Zn, Hg, Cu, Cd, Pb, Ni) on the length growth of *Mytilus edulis, Mar. Biol.,* 72, 69, 1982.
34. **Calabrese, A., MacInnes, J. R., Nelson, D. A., and Miller, J. E.,** Survival and growth of bivalve larvae under heavy metal stress, *Mar. Biol.,* 41, 179, 1977.
35. **Westernhagen, H., von, Dethlefsen, V., and Rosenthal, H.,** Correlation between cadmium concentration in the water and tissue residue levels in dab, *Limanda limanda* L., and plaice, *Pleuronectes platessa* L., *J. Mar. Biol. Assoc. U.K.,* 60, 45, 1980.

36. **Karbe, L., Borchardt, T., Dannenberg, R., and Meyer, E.,** Ten years of experience using marine and freshwater hydroid bioassays, in *Ecotoxicological Testing for the Marine Environment,* Vol. 2, Persoone, G., Jaspers, E., and Claus, C., Eds., State University Ghent and Institute Marine Scientific Research, Bredene, Belgium, 1984, 99.
37. **Widdows, J., Phelps, D. K., and Galloway, W.,** Measurement of physiological condition of mussels transplanted along a pollution gradient in Narragansett Bay, *Mar. Environ. Res.,* 4, 181, 1981.
38. **Martin, M., Osborn, K. E., Billig, P., and Glickstein, N.,** Toxicities of ten metals to *Crassostrea gigas* and *Mytilus edulis* embryos and *Cancer magister* larvae, *Mar. Pollut. Bull,* 12, 305, 1981.
39. **MacInnes, J. R. and Calabrese, A.,** Response of embryos of the American oyster, *Crassostrea virginica,* to heavy metals at different temperatures, in *Physiology and Behaviour of Marine Organisms,* McLusky, D. S. and Berry, A. J., Eds., Pergamon Press, Oxford, 1978, 195.
40. **Reinhart, K. and Myers, T. D.,** Eye and tentacle abnormalities in embryos of the Atlantic oyster drill *Urosalpinx cinerea, Chesapeake Sci.,* 16, 286, 1975.
41. **Bryan, G. W., Gibbs, P. E. Hummerstone, L. G., and Burt, G. R.,** The decline of the gastropod *Nucella lapillus* around South-West England; evidence for the effect of tributyltin from antifouling paints, *J. Mar. Biol. Assoc. U.K.,* 66, 611, 1986.
42. **Bengtsson, B. E., Bengtsson, A., and Himberg, M.,** Fish deformities and pollution in some Swedish waters, *Ambio,* 14, 32, 1985.
43. **Bengtsson, B. E. and Larsson, A.,** Vertebral deformities and physiological effects in fourhourn sculpin *(Myoxocephalus quadricornis)* after long-term exposure to a simulated heavy metal-containing effluent, *Aquat. Toxicol.,* 9, 215, 1986.
44. **Thursby, G. B. and Steele, R. L.,** Toxicity of arsenite and arsenate to the marine macroalga *Champia parvula* (Rhodophyta), *Environ. Toxicol. Chem.,* 3, 391, 1984.
45. **Reish, D. J.,** The effects of heavy metals on polychaetous annelids, *Rev. Int. Oceanogr. Med.,* 49, 99, 1978.
46. **Myint, V. M. and Tyler, P. A.,** Effects of temperature, nutritive and metal stressors on the reproductive biology of *Mytilus edulis, Mar. Biol.,* 67, 209, 1982.
47. **Eyster, L. S. and Morse, M. P.,** Development of the surf clam *(Spisula solidissima)* following exposure of gametes, embryos, and larvae to silver, *Arch. Environ. Contam. Toxicol.,* 13, 641, 1984.
48. **Sundelin, B.,** Single and combined effects of lead and cadmium on *Pontoporeia affinis* (Crustacea, Amphipoda) in laboratory soft-bottom microcosms, in *Ecotoxicological Testing for the Marine Environment,* Vol. 2, Persoone, G., Jaspers, E., and Claus, C., Eds., State University Ghent and Institute Marine Scientific Research, Bredene, Belgium, 1984, 237.
49. **Benijts-Claus, C. and Benyts, F.,** The effect of low lead and zinc concentrations on the larval development of the mud-crab *Rhithropanopeus harrisii* Gould, in *Sublethal Effects of Toxic Chemicals on Aquatic Animals,* Koeman, J. M. and Strik, T. W. A., Eds., Proc. Swed. Neth. Symp., Wageningen, Elsevier, Oxford, 1975, 43.
50. **D'Agostino, A. and Finney, C.,** The effect of copper and cadmium on the development of *Tigriopus japonicus,* in *Pollution and Physiology of Marine Organisms,* Vernberg, F. J. and Vernberg, W. B., Eds., Academic Press, New York, 1974, 445.
51. **Kobayashi, N.,** Marine ecotoxicological testing with echinoderms, in *Ecotoxicological Testing for the Marine Environment,* Vol. 1, Persoone, G., Jaspers, E., and Claus C., Eds., State University Ghent and Institute Mar. Scient. Res., Bredene, Belgium, 1984, 341.
52. **Ojaveer, E., Annist, J., Jankowski, H., Palm, T., and Raid, T.,** On effect of copper, cadmium and zinc on the embryonic development of Baltic spring spawning herring, *Finn. Mar. Res.,* 247, 135, 1980.
53. **Sharp, J. R. and Neff, J. M.,** Effects of the duration of exposure to mercuric chloride on the embryogenesis of the estuarine teleost *Fundulus heteroclitus, Mar. Environ. Res.,* 3, 195, 1980.
54. **Engel, D. W. and Sunda, W. G.,** Toxicity of cupric ion to eggs of the spot *Leiostomus xanthurus* and the Atlantic silverside *Menidia menidia, Mar. Biol.,* 50, 121, 1979.
55. **McGreer, E. R.,** Sublethal effects of heavy metal contaminated sediments on the bivalve *Macoma balthica* (L), *Mar. Pollut. Bull,* 10, 259, 1979.
56. **Thomas, W. H. and Seibert, D. L. R.,** Effects of copper on the dominance and the diversity of algae: controlled ecosystem pollution experiment, *Bull. Mar. Sci.,* 27, 23, 1977.
57. **Sanders, J. G.,** Direct and indirect effects of arsenic on the survival and fecundity of estuarine zooplankton, *Can. J. Fish. Aquat. Sci.,* 43, 694, 1986.
58. **Rygg, B.,** Effect of sediment copper on benthic fauna, *Mar. Ecol. Prog. Ser.,* 25, 83, 1985.
59. **Tsubaki, T. and Irukayama, K.,** *Minamata Disease,* Kodansha, Tokyo and Elsevier, Amsterdam, 1977.
60. **Langston, W. J.,** Metals in sediments and benthic organisms in the Mersey estuary, *Estuarine Coastal Shelf Sci.,* 23, 239, 1986.
61. **Hershelman, G. P., Schafer, H. A., Jan, T.-K., and Young, D. R.,** Metals in marine sediments near a large California municipal outfall, *Mar. Poll. Bull.,* 12, 131, 1981.

62. **Ward, T. J., Warren, L. J., and Tiller, K. G.,** The distribution and effects of metals in the marine environment near a lead-zinc smelter, South Australia, in *Environmental Impacts of Smelters,* Nriagu, J. O., Ed., John Wiley & Sons, New York, 1984, chap 1.
63. **Long, E. R. and Chapman, P. M.,** A sediment quality triad: measurements of sediment contamination, toxicity and faunal community composition in Puget Sound, *Mar. Pollut. Bull.*, 16, 405, 1985.
64. **McGreer, E. R.,** Factors affecting the distribution of the bivalve, *Macoma balthica* (L) on a mudflat receiving sewage effluent, Fraser River estuary, British Columbia, *Mar Environ. Res.*, 7, 131, 1982.
65. **Parulekar, A. M., Ansari, Z. A., and Ingole, B. S.,** Effect of mining activities on the clam fisheries and bottom fauna of Goa estuaries, *Proc. Indian Acad. Sci. (Anim. Sci.),* 95, 325, 1986.
66. **Brown, B. E. and Howard, L. S.,** Assessing the effects of "stress" on reef corals, *Adv. Mar. Biol.*, 22, 1, 1985.
67. **Eide, I., Jensen, A., and Melsom, S.,** Application of *in situ* cage cultures of phytoplankton for monitoring heavy metal pollution in two Norwegian Fjords, *J. Exp. Mar. Biol. Ecol.*, 37, 271, 1979.
68. **Alzieu, C. and Heral, M.,** Ecotoxicological effects of organotin compounds on oyster culture, in *Ecotoxicological Testing for the Marine Environment,* Vol. 2, Persoone, G., Jaspers, E. and Claus, C., Eds., State University Ghent and Institute Marine Scientific Research, Bredene, Belgium, 1984, 187.
69. **Langston, W. J., Burt, G. R., and Zhou, M.,** Tin and organotin in water, sediments and benthic organisms of Poole Harbour, *Mar. Pollut. Bull.*, 18, 634, 1987.
70. **Thain, J. E., Waldock, M. J., and Waite, M. E.,** Toxicity and degradation studies of tribulyltin (TBT) and dibutyltin (DBT) in the aquatic environment, in *Oceans 87 Conference Record, Vol. 4,* Organotin Symposium, New York Institute of Electrical and Electronics Engineers, New York, 1987, 1398.
71. **His, E. and Robert, R.,** Development des veligeres de *Crassostrea gigas* dans le basin d'Arcachon. Etudes sur les mortalites larvaires, *Rev. Trav. Inst. Peches Marit.*, 47, 63, 1985.
72. **Beaumont, A. R. and Budd, M. D.,** High mortality of the larvae of the common mussel at low concentrations of tribulytin, *Mar. Pollut. Bull.*, 15, 402, 1984.
73. **Thain, J. E. and Waldock, M. J.,** The impact of tributyl tin (TBT) antifouling paints on molluscan fisheries, *Water Sci. Technol.*, 18, 193, 1986.
74. **Laughlin, R. B., Norlund, K., and Linden, O.,** Long-term effects of tribultyltin components on the Baltic Amphipod, *Gammarus oceanicus, Mar. Environ. Res.*, 12, 243, 1984.
75. **Laughlin, R. B. and French, W. J.,** Comparative study of the acute toxicity of a homologous series of trialkyltins to larval shore crabs *Hemigrapsus nudus* and lobster, *Hommarus americanus, Bull. Environ. Contam. Toxicol.*, 25, 802, 1980.
76. **Beaumont, A. R. and Newman, M. D.,** Low levels of tributyltin reduce growth of marine macroalgae, *Mar. Pollut. Bull.*, 17, 457, 1986.
77. **Wilson, K. W., Head, P. C., and Jones, P. D.,** Mersey Estuary (U.K.) bird mortalities — causes, consequences and correctives, *Water Sci. Technol.*, 18, 171, 1986.
78. **Hardy, J., Kiesser, S., Antrim, L., Stubin, A., Kocan, R., and Strand, J.,** The sea-surface micro layer of Puget Sound. I. Toxic effects on fish eggs and larvae, *Mar. Environ. Res.*, 23, 227, 1987.
79. **Hardy, J. T., Crecelius, E. A., Antrim, V. L., Broadhurst, C. W., Apts, J. M., Curtisen, J. M., and Fortman, T. J.,** The sea-surface microlayer of Puget Sound. II. Concentrations of contaminants and relation to toxicity, *Mar. Environ. Res.*, 23, 251, 1987.
80. **Langston, W. J.,** unpublished data.

Chapter 8

BIOCHEMICAL AND CYTOLOGICAL ASSESSMENTS OF METAL TOXICITY IN MARINE ANIMALS

Stephen G. George

TABLE OF CONTENTS

I. INTRODUCTION

A. ESSENTIAL AND POLLUTANT METALS

This chapter deals with the cellular controls, biochemical and cytological toxicities of pollutant and essential trace metals, and assessments of toxicities in marine animals. On present evidence, 20 of the 90 naturally occurring elements appear to be essential to life, and a further 6 are desirable in that animals can survive deficiencies in them but are not completely healthy. These are classified in Table 1.

Many of the essential elements are metals, whose functions in oxygen transport (Cu,Fe), enzyme co-factors or metalloenzymes (Cu,Co,Mn,Mo,Se,Zn), and redox (electron transport) activities are well understood. The normally low tissue concentrations found for the rest of the naturally occuring elements generally reflect environmental exposure. Many of these are now causing concern due to the contamination of the marine environment by man's expanding activities, and the demonstration of toxic effects upon the normal function of cells, tissues, or organs caused by excessive uptake. The heavy metals of most concern to human health are cadmium, lead, and mercury which exhibit a cumulative toxicity, while copper, chromium, and organotin are also important environmentally because of their properties as biocides.

B. METAL TOXICITIES

Several trace metals (e.g., Cu, Zn) are chemically highly reactive (hence their utilization as catalysts) and while they are essential and show a "beneficial" dose response curve at low concentrations, with increasing concentration, they may become inhibitory and ultimately toxic. There are several reasons for this effect; the catalytic activities of metalloenzymes are a feature of their specific three-dimensional configuration and, therefore, if a different metal is bound, the geometry of the active site and hence enzyme activity, will be altered. Metal ions bind to organic ligands with different affinities, following the Irving-Williams series. For example, the relative stability constants for cysteine –SH co-ordination are $Cu^{++} > Ni^{++} = Zn^{++} > Co^{++} > Fe^{++} > Mn^{++}$; thus Cu will displace Zn, Fe, etc. When free metal ion concentrations rise, these metal substitution reactions will become significant and inhibition of enzyme activity or destabilization of structural components of cellular molecules may occur. Toxicity may also be indirect, the displaced metal itself exerting toxic effects by binding to other cellular components. These substitution and displacement reactions are particularly important for metal ions which show closely related chemical behavior, such as Ag^{+}, Cu^{+}, Hg^{+}, Cd^{++}, and Zn^{++}. Thus the toxicity of nonessential "pollutant" metals e.g., Ag, Cd, Hg is due to Cu/Zn competition as well as to "spillover" onto other binding ligands. For other metals, toxic effects are quite specific. For example, in the case of Cr^{3+} which appears to function only in the glucose tolerance factor, toxicity is due to covalent binding to proteins and nucleic acids.[1] Lead on the other hand binds to nucleic acids and is substituted for Ca in bone, and the toxic effects of organometallic compounds, such as methylHg and tributylSn are similarly specific (see Section V).

C. HOMEOSTASIS

The intracellular concentrations of essential metal ions are generally maintained at optimal levels by homeostatic mechanisms. In animal cells, the concentrations of the major ions (Table 1) are controlled by energy-dependent pumping mechanisms, and those of the essential trace elements are "buffered" by storage or "detoxification" systems.[2,3] At least one of these systems is inducible and when "switched on" appears to confer tolerance to elevated environmental metal levels. Theoretical "dose/effect" curves for an essential and a nonessential metal which is toxic are plotted in Figure 1. Displacement of the inhibition curves to the right occurs when tolerance is induced, either by genetic selection or by

TABLE 1
Classification of Essential Elements

Element group	Subgroup	Number	Members	Tissue conc.
Major	Bulk elements	5	C,H,N,O,S	g/kg
	Major ions	6	Na,K,Ca,Mg,P,Cl	g/kg
Trace	Essential	9	Co,Cu,I,Fe,Mn,Mo,Se,Si, and Zn	μg—mg/kg
	Desirable	6	As,Cr,F,Ni,Sn,V	μg/kg

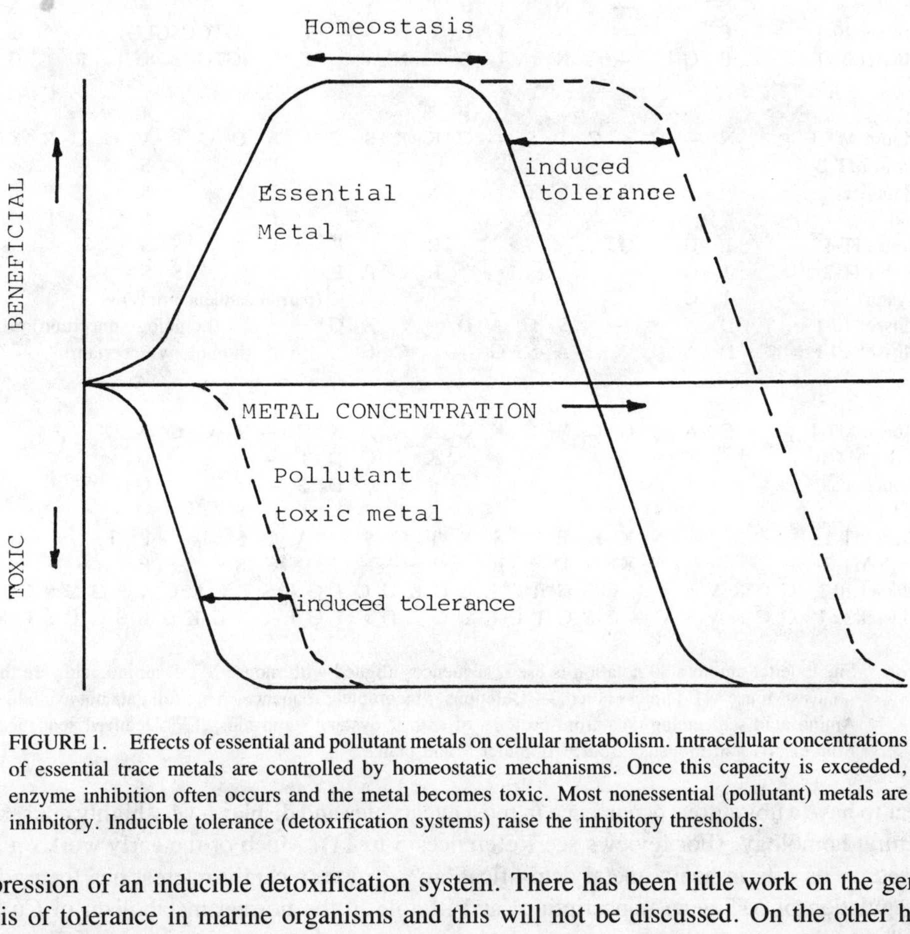

FIGURE 1. Effects of essential and pollutant metals on cellular metabolism. Intracellular concentrations of essential trace metals are controlled by homeostatic mechanisms. Once this capacity is exceeded, enzyme inhibition often occurs and the metal becomes toxic. Most nonessential (pollutant) metals are inhibitory. Inducible tolerance (detoxification systems) raise the inhibitory thresholds.

expression of an inducible detoxification system. There has been little work on the genetic basis of tolerance in marine organisms and this will not be discussed. On the other hand, there are now considerable data on accumulation and detoxification of metals in marine organisms. Analysis of the status of these systems and the metals associated with them often provides the first method of assessing the sublethal impact of elevated metal concentrations on marine organisms.

II. METAL DETOXIFICATION SYSTEMS

A. HEAVY METAL-BINDING PROTEINS

The most studied and well characterized are the metallothioneins (MTs), a group of low molecular weight (6 to 7 kDa) proteins unique in their high cysteine content (22 to 33 mol%) and paucity of aromatic amino acid residues. MTs specifically bind 4 to 12 atoms per mole of heavy metal ions, such as Zn, Cu, Cd, Hg, and Ag by clusters of thiolate bonds. They

TABLE 2
Amino Acid Sequences of Metallothioneins from Marine Animals Compared with Sequence of Mouse MT-I

	1									10										20		
Mouse MT-I	AcM	D	P	N	C	S	C	S	T	G	G	S	C	T	C	T	S	S	—	C	A	C	S
Trout MT-B				—		E			K	T				N		G	G				K		
Flounder				—		E			K	T		T		N		G	G				T		K
Plaice				—		E			K	T		T		N		G	G				T		K
Crab MT-I	P	G		—		—		—	—	N	D	K		V		K	E	G	—		G		K
Crab MT-II	N	D		—		—		—	—	N	D	K		N		K	E	G	—		E		K
Oyster	S			—		N		I	E	T		T		A		S	D	?	—		P	—	A
Mussel 10-1	P	A		—		N		I	E	T	N	V		I		(DTGCSGEG)					R		G
Mussel 20-1	P	G		—		N		I	E	T	N	V		I		(GTGCSGKG)					R		G

									30										40				
Mouse MT-I	N	—	C	K	C	T	S	--	C	K	K	S	C	C	S	C	C	P	V	G	C	S	K
Trout MT-B		—		A											P				S	D			
Flounder				S			T			N					P				S			P	
Plaice				S			T			N					P				S			P	
Crab MT-1	E	G		Q						R	C		P		E	K		S	S			—	
Crab MT-2	T	G								R	C	P	P		E	Q		S	S			—	
Oyster	T	G				G	P					. . . (partial sequence only)											
Mussel 10-1	D	A				S	G	A	D		C	K	G		. . . (homology uncertain)								
Mussel 20-1	D	A				A	S	G	—		C	G		. . . (homology uncertain)									

						50											60		
Mouse MT-I	C	A	Q	G	C	V	C	K	G	A	A	D	K	C	—	T	C	C	A	
Trout MT-B			S							K	T	C	D	T	—	S			Q	
Flounder			S							K	T	C	D	T	—				Q	
Plaice			S							K	T	C	D	T	—	S			Q	
Crab MT-1			N	K	E	E		S	K	T	C	S		A	C	S			P	T
Crab MT-2			N	K	E	D		R						P	C	S			P	
Mussel 10-1	G C K V V C K C S G R C E C G K G C T G P S T C K C A P G C S C K																			
Mussel 20-1	G C K V V C K C S G T C K C G C D C T G P T N C K C E S G C S C K																			

Note: Single letter amino acid notation is used, sequences aligned with mouse MT-I, amino acids are the same as Mouse MT-I unless noted. — Deletions, *incomplete sequence, . . . full data not available. Amino acid sequencing data from proteins of crab,[14] oyster,[155] and mussel.[156] Deduced sequences from cDNA's of mouse,[18] trout,[15] flounder,[12] and plaice.[13]

appear to have a ubiquitous occurrence in most eucaryotes and display a very highly conserved structural homology. (For reviews see References 5 to 11). Much of the early work on MT focused on its role in heavy metal detoxification;[6,7,9] more recently interest has focused on the regulation of MT gene transcription[5] and its role in the normal metabolism of Cu and Zn.[7,10,11]

Metallothioneins and MT-like proteins have been reported in many marine animals from sea urchins to mammals. Amino acid sequence data are only available for a few marine species. The amino acid sequences of flounder[12] and plaice[13] thioneins are remarkably homologous to the mammalian class 1 metallothioneins; those of the invertebrate thioneins, crab,[14] oyster,[155] and mussel,[156] show greater divergence but still display a high degree of homology in residues 1 to 30 of the protein (Table 2). In common with the fungal[17] thionein, that of the sea urchin[16] shows little homology and is classified as a type 2 metallothionein.[6]

Low molecular weight metal-binding proteins have been purified from many other marine animals, and are sufficiently well characterized (metal-binding properties, amino acid composition, etc.) to warrant their identification as MTs (* in Table 3), although in a number of cases, particularly in the invertebrates, some of these properties appear to be somewhat

TABLE 3
Occurence of MTs and MT-like Proteins in Marine Animals

Mammals: *Halichoerus* (fur seal); *Seriola* (grey seal); *Monodon* (Narwhal); *Physeter* (sperm whale); *Zalophus* (sea lion)

Birds: *Fulmaris* (fulmar)

Fish: *Anguilla* (eel)*; *Enophyrys* (sculpin); *Fundulus* (killifish)*; *Leptocottus* (sculpin); *Morone sp.* (white perch and striped bass); *Mugil* (mullet), *Oncorhynchus* (coho salmon); *Parophys* (sole); *Pleuronectes* (plaice)*; *Platichthys* (flounder)*; *Scyliorhinus* (dogfish); *Sebastodes* (rockfish)

Crustaceans: *Acetes* (shrimp), *Palaemon* (prawn); *Callinectes**, *Cancer**, *Carcinus*, *Scylla** (crabs); *Homarus* (lobster)

Mollusks: *Cryptochiton* (Chiton); *Crassostrea, Ostrea* (oysters)*; *Mytilus* (mussels)*; *Murex* (whelk); *Pecten* (scallop)$; *Protothaca* (clam); *Littorina* (periwinkle); *Patella* (limpet)$

Others: Polychaetes: *Eudystilia* (fan worm); *Neanthes* (bristle worm); Sea urchin, *Strongylocentrotus**

Note: Genera containing low molecular weight metal-binding proteins which display characteristic properties of MTs (amino acid composition, etc.,) are denoted *; those with proteins not fully characterized and possessing some different properties $; in the remainder, characterization of proteins is generally limited to the presence of Cd-, Cu-, or Zn-binding protein of apparent molecular weight of 8 to 15 kDa on gel permeation chromatography.

different ($ in Table 3). Such differences may be due to real differences in the structure of the protein (as found with the fully characterized sea urchin thionein), or inadequate purification and the presence of contaminants or degradation products. Degradation by proteases is a major problem in isolation of the protein from digestive glands. It can be reduced by addition of protease inhibitors, heat treatment, and the use of ion-exchange chromatography as an initial stage.[19] Other problems, encountered when significant Cu concentrations are present (especially in crustaceans which contain hemocyanin) are Cu-displacement reactions and oxidation of Cu-MT. Oxidation can be reduced by purification in an inert atmosphere and addition of a reducing agent (e.g., 2-mercaptoethanol or dithiothreitol).[19-22]

Several other low molecular weight cysteine-rich metal-binding proteins that are not MTs have now been isolated from a number of species including several marine organisms (Cd-binding proteins reviewed by Stone and Overnell[23]). These are cited in Table 4. Some of these are similar in molecular weight to MTs but contain aromatic amino acids and have low cysteine contents. The proteins of the whelk,[24] oysters[25,26] and mussels[32] appear to be induced by Cd while the crab[21,27] and lobster[28,29] proteins are present in animals which have not been exposed to elevated metal levels.

B. METAL SEQUESTRATION IN LYSOSOMES

Cu-MT is extremely susceptible to oxidation and when significant quantities are present it polymerizes and accumulates in lysosomes.[11] A similar condition has been observed in the white perch, *Morone americana*, from Chesapeake Bay which naturally accumulates vast amounts of Cu both in MT[33] and in tertiary lysosomes.[34] Although this lysosomal accumulation of Cu from MT occurs in vertebrates, lysosomal accumulation of Cu and other metals is very much more pronounced in invertebrates where it represents a major mechanism for metal detoxification. The relationship between MT and these lysosomes has been investigated in the mussel, *Mytilus edulis*,[35-37] and a working hypothesis has been proposed to explain differences in subcellular distribution of Cu, Cd, and Zn, based upon the susceptibility of their MTs to lysosomal degradation and the lysosomal turnover time in different tissues (Figure 2).

TABLE 4
Low Molecular Weight non-MT Metal-Binding Proteins in Marine Animals

		Metal(s) bound	Mol wt. (kDa)	Characteristics	Ref.
Enoplus	Nematode	Cu, Cd	28		30
			63(450)		
Nereis	Polychaete	Cd	10(20)	1% Cys, 13% Asp, 13% Glu, 10% aromatics	153
Buccinum	Whelk	Cd	8	5% Cys, 20% CHO	24
		Cd	13	4% Cys, 9% CHO	
Crassostrea	Oyster	Cd	10,20	18% Cys	21
Ostrea	Oyster	Cd	6	Heat stable	26
		Cu, Zn	0.2—2	Several peptides	31
Pecten	Scallop	Cd	55	9% Cys	20
Carcinus	Crab	Zn	4		21
Callinectes	Crab	Zn			27
Homarus	Lobster	Cd	25—30		28
		Zn	40—45		29
Palaemon	Prawn	Cu, Cd	5	Impure	22
		Zn	4	Impure	

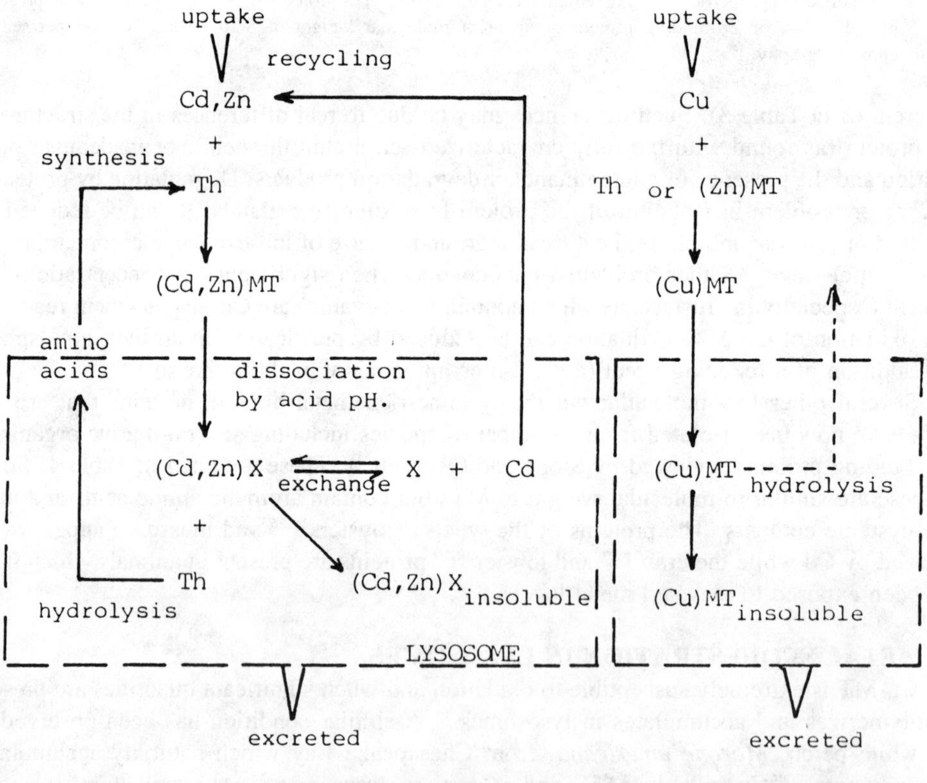

FIGURE 2. Control of subcellular metal distribution by lysosomes: an example using different tissues of the mussel, *Mytilus edulis*. In the digestive gland, lysosomes are excreted with a $t^1/_2$ of about 4 d, Cd- and Zn-Metallothionein are mainly cytosolic since in the lysosomes the metals are rapidly released and recycled back to the cytosol. Cu is only slowly released from MT in the lysosomes (and also polymerizes) and, therefore, is cleared on lysosome exocytosis. In the kidney the lysosomes have a $t^1/_2$ of about 45 d and, therefore, metals become trapped by peroxidation and accumulate in the tertiary lysosomes.

The matrix of tertiary lysosomes represents the insoluble indigestible remnants of cellular macromolecules and is an insoluble lipoprotein pigment, lipofuschin. Lipofuschin granules are formed by the peroxidation of polyunsaturated fatty acids and the formation of intramolecular cross-linkages with lipids and proteins.[36,39] This peroxidation is catalyzed by Cu^{++} or Fe^{++}. Thus when Cu- or Fe-metalloproteins are taken up by lysosomes and the metals are released, either by proteolysis or the acid pH, lipid peroxidation increases and the metals become trapped within the cross-linked matrix of the lipofuschin granule. Thus tertiary lysosomes almost invariably contain Fe derived from degradation products of heme pigments and the Fe-storage protein, ferritin (known as hemosiderin). In some animals which utilize the Cu-containing respiratory protein, hemocyanin (i.e. some mollusks and crustaceans) high concentrations of Cu (and S) become trapped in tertiary lysosomes in an analogous manner.[40,41] Uptake and immobilization of metalloprotein-bound metals in lysosomes can be explained by this process. However, a variety of other metals have also been found in lysosomes, including Al, Am, As, Co, Pb, Pu, Sn.[42-45] Their presence was explained by studies with isolated tertiary lysosomes of mussels in which it was shown that the latter can act as nonspecific chelators of other metals.[36] Therefore, following peroxidation and cross-linkage, any chelated metals will also become trapped and thus detoxified by insolubilization. Several colloidal/particulate metals (e.g., Pb, Th, Fe) are taken into invertebrate epithelial cells (particularly the gills and mantle) by endocytosis and after fusion of these endocytic vesicles with lysosomes, these metals also become immobilized.[43,46,47]

Immobilization of metals by sequestration in lysosomes represents a major nonspecific mechanism for detoxification of many metals in animals exposed to excessively high levels. In many invertebrates where there appears to be little regulation of metal accumulation, much of the high metal burdens observed are attributable to metals sequestered in these membrane-limited vesicles (tertiary lysosomes) rather than in the cytoplasm.[43,44,46,48] The vesicles are usually found in epithelia of tissues, such as the hepatopancreas and kidney which have a lumen in contact with the exterior, and microscopic studies have demonstrated that they may be excreted by exocytosis.[46,48,49] From studies utilizing radioisotopes of metals, estimates of the turnover times of these tertiary lysosomes can be made.[38] The turnover time of tertiary lysosomes in the hepatopancreas of crustaceans and mollusks appears to be quite rapid ($t^1/_2$ about 3 d[37]) while in the kidney they are much longer (e.g., $t^1/_2$ of 60 d in mussel kidney for Zn.[48] The determination of the concentrations of heavy metals in lysosomes of these invertebrate tissues can, therefore, provide an assessment of the past exposure history of the animal to a number of metals, probably as an integration over a longer time scale than that provided by MT determinations. Two approaches to this determination are available (1) homogenization of the tissue, fractionation by centrifugation, and metal analysis by AAS or ICPMS;[39] or (2) fixation, embedding, and sectioning of the tissue followed by metal determination using light microscope histochemistry,[50] electron microscope X-ray microanalysis, or laser mass spectroscopy (LAMMA).[51]

In mammals (e.g., in the cases of Cu-poisoned sheep, Wilson's disease, or transfusion-induced siderosis in humans) and in fish (e.g., *Morone americana*, noted earlier) such metal-containing tertiary lysosomes accumulate in the liver and are not readily excreted.

C. METAL SEQUESTRATION IN CALCIUM GRANULES

Intracellular granules (concretions), 0.5 to 2 μm in diameter, predominately made up of Ca and Mg phosphates are particularly common in crustaceans and mollusks, although they have been found in organisms ranging from amoeba to man.[43,52] The granules are found within membrane-limited vesicles of the endoplasmic reticulum/golgi system, usually in tissues of the alimentary canal and kidney with a lumen opening to the exterior. They are virtually insoluble in saline, this insolubility being increased by the presence of pyrophosphate,[53] urates, and oxalates.[54] Chemical analyses have shown that this type of concretion also accumulates other metals, including Al, Ba, Co, Mn, Pb, and Zn.[3,52,55] Therefore, the granules are important in accumulation and detoxification of several heavy metals, especially Zn, in barnacles,[56] scallops,[57,58] clams,[58,59] and winkles.[60]

While these phosphate granules can be purified and analyzed easily to give some qualitative indication of heavy metal exposure of the animals, the kinetics of uptake and excretion of metals in phosphate granules are extremely complex. Studies with vascularly injected Co, Mn, and Zn in gastropod mollusks have demonstrated that the metals are sequestered by the granules from the blood quite rapidly.[60,61] However, the excretion times vary quite markedly between species. For example on exposure to high Mn concentrations, preexisting granules are excreted from the hepatopancreas of gastropod mollusks relatively quickly (about 3 to 7 d[60,61]), while in other animals, excretion may be episodic. In crabs, hepatopancreas granules may be lost at molt[62] and in scallops, they are lost on spawning. In both clam and scallop kidneys, isotopic and ultrastructural data indicate that the concretions have a much longer half life of at least 3 to 6 months.[63]

D. SEQUESTRATION OF METALS IN SPECIALIZED CELLS OF INVERTEBRATES

1. Hemocytes

Many of the hemocytes in invertebrates are similar to mammalian macrophages. They are amoeboid and move in and out of the body tissues where they phagocytose foreign materials, dead cells, and denatured proteins. Since metals are associated with many of these materials, they are, therefore, taken up and accumulate in the hemocyte lysosomes. In bivalve mollusks exposed to Fe[65] and Pb[45,66] or injected with particulate ThO_2,[66] pinocytosis of the metals by hemocytes has been observed. It is believed that the amoebocytes transport metals from the sites of uptake to storage tissues, or to the kidney for eventual excretion[64,65] and that this is a nonspecific function of macrophages. In gastropods, it has been postulated that the hemocytes transport Ca during shell repair.[55]

In ascidians and oysters, there are granular hemocytes which accumulate very large deposits of specific metals. In *Ascidia mentula*, certain blood cells, termed vanadocytes, accumulate up to 1 *M* vanadium in a solution of H_2SO_4 within membrane-limited vesicles.[67] The function of these cells is unknown although a role in tunic synthesis has been postulated. In 1886, Cu-containing blood cells were described in oysters.[68] In recent studies, it has been found that a hemocyte containing membrane-limited deposits of both Cu and Zn is present in all oysters[69-71], while the European oyster, *Ostrea edulis*, also contains metal-specific cells, a basophilic cell containing up to 1 *M* Zn in association with P and an acidophilic cell containing up to 300 m*M* Cu in association with S.[71] The exact nature of the metal complexes is not yet known although several peptide and amino acid complexes have been identified.[31] Clearly these hemocytes are an extremely effective means of detoxifying very large amounts of Cu and Zn, since even in oysters containing Cu and Zn concentrations up to 10% of their body weight[72] the concentrations of these metals in other cells are not elevated.[71] In a study of European oysters, it has been shown that Cu and Zn are almost entirely sequestered in these cells rather than by MT. In response to Cd exposure, however, there was induction of MT synthesis which was accompanied by a reduction in numbers of Cu/Zn amoebocytes.[73,74] The exact role and function of these hemocytes is not known. It has been suggested that they may have an important function in release of Cu at sites of cellular injury and trauma,[74] and it may be postulated that oyster hemocytes respond to stress by release of metals rather like mast cells release histamine in higher organisms. A role in metal clearance has been postulated from the observed loss of the cells at epithelial surfaces by diapedesis.[75,76] From the observed accumulations of metals[72] and the long half lives ($t^1/_2$ for Zn about 9 months[77]), it may be concluded that this is not the major role.

2. Connective Tissue/Pore Cells

Ultrastructural and histological studies of several mollusks, particularly gastropods, have shown the presence of connective tissue cells containing granular deposits of calcium carbonate, typically 0.5 to 2 μm in diameter.[51,55] These deposits are formed in sacs of the golgi apparatus and often occupy most of the cell volume. Since these sacs are in contact with the extracellular fluid via pores in the plasma membrane, the granules may, therefore, be regarded as extracellular. *In vitro* they are soluble in saline.[52] *In vivo* they are readily solubilized during dietary and osmotic imbalance, shell repair, acidosis, and anoxia[79] and, therefore, enable maintenance of Casaturation of the body fluids[80] providing a readily available storage system for calcium. While they are usually almost pure $CaCO_3$, in the mussel (*Mytilus edulis*) chronically exposed to Pb, similar deposits of $PbCO_3$ have been observed.[3,81] These may be indicative of a competitive effect of Pb on Ca metabolism (seen in deposition of Pb in bone of mammals) and the ease of carbonate precipitation in body fluids supersaturated with CO_2.

III. TOXIC EFFECTS OF HEAVY METALS AND THEIR ASSESSMENT

Some marine animals naturally accumulate abnormally high levels of heavy metals even from uncontaminated environments and, therefore, may be unsuitable for use as indicator organisms. The best known example is the oyster discussed above. Scallops naturally accumulate up to 500 ppm dry wt Cd in the digestive gland,[63] (some being bound to MT but the majority to higher molecular weight proteins[20]), and up to 19 mg/g Zn and 15 mg/g dry wt Mn in the kidneys (as concretions of Ca/Mn/Zn phosphate[57]). The accumulation of up to 2 mg/g Cu in the livers of *Morone americana* from Chesapeake Bay both in MT[33] and tertiary lysosomes[34] appears to be a feature of this particular species since the related *Morone saxatilis* from the same bay contains normal hepatic Cu levels. Unexplicably high Cd concentrations have been reported in hepatopancreas of crabs (*Cancer pagurus*) from the north Scottish islands of Orkney and Shetland (up to 55 ppm wet wt, bound to MT), and the oceanic deep water crab (*Geryon affinis*) from Rockall (88 ppm)[82] possibly as a result of geochemical anomalies. Other oceanic animals contain unexplicably high Cd levels, including several species of birds and the oceanic sea-skating insect, *Halobates*.[83,84]

A. ANALYSIS OF METAL-BINDING COMPONENTS

Determination of total body or tissue metal concentrations can be used as a method of assessing past exposure to metals which accumulate throughout the life span and are not excreted, e.g., organomercury. For metals which are accumulated and detoxified in multiple compartments, each of which has different kinetics of accumulation and excretion, analysis of the distribution in the different pools can provide a more detailed indication of past exposure history. Thus in mussels, ZnMT has a shorter half life than Zn in tertiary lysosomes (about 60 d[38,47]) and theoretically, integrations of Zn exposure over two time scales are possible (this has yet to be tested practically). In clams and scallops, analysis of the kidney metal concretions should provide an integration of 4 months or more metal exposure. Preliminary data of environmental variability have been reported.[58,59]

1. Metallothionein as an Indicator of Metal Exposure

Low concentrations of MT are present in tissues under normal conditions. Exposure to sublethal doses of metals, such as Cd, Cu, Hg, and Zn induces thionein synthesis and levels can rise by 20 to 30-fold.[5-10] Thus MT levels in the liver and kidney may provide an index of environmental exposure to these metals.[85-87] Since metals do not bind to MT with equal strength (in decreasing order the affinity is Hg>Cu>Cd>Zn), displacement reactions can

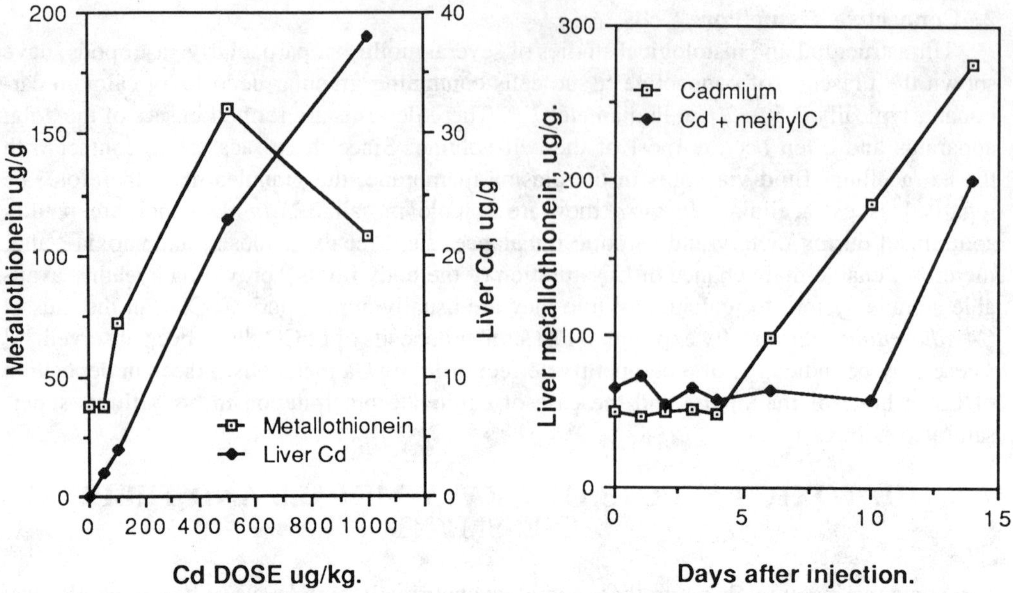

FIGURE 3. Plaice (*Pleuronectes platessa*) liver MT. (A) Dose response. Animals analysed 6-d post-intraperitoneal injection $CdCl_2$ in saline. (B) Time course after intraperitoneal injection 1 mg/kg $CdCl_2$. Filled symbols, animals receiving co-injection of polyaromatic hydrocarbon (3-methylcholanthrene).

occur and the metal content of MT varies both with these relative binding affinities and the metal concentrations present.

Before measures of MT levels can be used as an indicator of past metal exposure in field studies, the characteristics of the response in the test organism must be known. Thus the dose/response, longevity, and magnitude of effect (induction and degradation times), as well as information on seasonal variations, effects of reproduction, and other stressors are required. The kinetics of induction and degradation appear to be inexplicably longer in marine animals than in terrestrial mammals, even allowing for temperature effects. In plaice liver, induction of MT by Cd and Zn is apparent about 7 d after exposure[88,89] (Figure 3) and Zn-thionein is degraded with a half life of about 27 d.[90] This induction period may be altered by other stressors e.g., polyaromatic hydrocarbons[88] (Figure 3). There have been few dose-response studies in marine animals; some data for plaice are given in Figure 3. In mammals, two- to fourfold increases in MT levels are induced by various stress conditions that raise steroid hormone levels (hot, cold, etc.) and by tissue injury which produces an inflammatory response, e.g., after injecting agents such as turpentine, endotoxins, etc.[11] In wild fish, the interanimal variation is probably too large for such changes to be detected,[88-91] and indeed it has been reported that dexamethasone, turpentine, and endotoxin had no effect on plaice liver MT levels 7 d after injection.[90] Results of studies in plaice and rainbow trout[89,91] have shown that liver MT levels in female fish are two to threefold higher during exogenous vitellogenesis. Thus with current data, it is probably reasonable to conclude that MT levels in fish which are more than four times greater than control values are truly indicative of metal exposure, and that the effect will still be apparent about 30 d after exposure. MT levels can be measured in a number of ways (1) Cd- or Hg-saturation assays,[92,93] (2) by polarography,[93,94] (3) by gel filtration of cytosol and conventional metal analysis or (4) by immunoassay.[85,95] Each has shortcomings. The Cd-saturation assay is the most convenient but does not detect Cu- or Hg-thioneins, polarography is costly to set up, gel permeation chromatography/AAS is laborious and can only be used when other low molecular weight metal-binding proteins are absent. Production of good antisera to MT is difficult but immunoassays developed for perch[157] and trout[158] MTs have recently been described. While these antisera show very little cross reactivity with mollusk and mammalian MTs, for fish they display wide cross-species reactivity which will enable their wide application.

Studies on invertebrates have largely focused on determining the presence of MT and MT-like proteins. In crustaceans, any relation of MT levels with environmental exposure is seriously complicated by the presence of fluctuating Cu, Zn, hemocyanin, and MT levels during the molt cycle.[40,97,98] While MT-like metal-binding proteins can be induced in bivalve mollusks, other metal detoxification systems are present which may have greater metal capacities e.g., oyster hemocytes,[70,71,73,74] or concretions in clam and scallop kidneys.[57,58] In mussels, Zn is preferentially accumulated in kidney tertiary lysosomes.[47,99] Since the metals in these lysosomes are insoluble and do not affect metabolism of other heavy metals, the use of mussel as Cd, Cu, or Hg indicators is justified. Cd- and Hg-thioneins have been detected in mussels (by increases in MT-bound metal) after 7-d exposure.[3,100] *De novo* synthesis of MT has been demonstrated (by ^{35}S incorporation) in mussel gills after 48-h Cu exposure.[101] Degradation and dose-response studies have yet to be reported. In 36-h post-fertilization gastrulae of sea urchins, induction of MT mRNA has been demonstrated after 4-h $ZnSO_4$ treatment.[102]

2. Assessment of Toxicity by Changes in Intracellular Metal Distributions

A *sine qua non* for detoxification of metals by specialized binding proteins, such as MT or as insoluble organic granules (lipofuschins in tertiary lysosomes) or inorganic phosphate granules, is that binding to other cellular components and inhibition of essential cellular functions are prevented. The sequestration capacity of these detoxification systems are limited by factors, such as the amounts of thionein present, the metals already bound to the existing MT pool, the rate of *de novo* synthesis upon metal induction, the rate of transport of metals into granules or lysosomes, etc. Under conditions of high metal uptake when the sequestration capacity of the detoxification systems is exceeded, toxicity will occur. Since the initial report of a correlation between pathological effects of Cd and ''spillover'' of Cd from the MT fraction into the high molecular weight soluble proteins of mouse,[103] there have been numerous attempts to utilize such a metal distribution as a diagnostic measure of metal stress in marine organisms.[44,86,96,104-107] The distribution of metals among particulate fractions (i.e., lysosomes), high and low molecular weight proteins and low molecular weight ligands has been investigated both in marine organisms sampled from the field or exposed in the laboratory to heavy metals.[104-107] The clearest results have been obtained from laboratory experiments where animals were exposed to controlled concentrations of metal in the water. In crab larvae exposed to Cd, growth and development were reduced when the MT-containing fraction appeared to be saturated.[105,106] In the polychaete, *Neanthes arenaceodentata*, exposed to different concentrations of Cu, CuMT concentrations increased proportionately with Cu dosage, and a reduction in fecundity and reproductive success corresponded with an appearance of Cu in the very low molecular weight fraction of the cytosol.[107,108] Studies of this nature provide a promising potential method of diagnosing the onset of toxicity in a number of marine invertebrates.

B. EFFECTS OF METALS ON BLOOD AND TISSUE ENZYME ACTIVITIES IN FISH

Damage to specific organs of mammals due to disease, chemical toxicity, or neoplasia, has long been diagnosed by the appearance of tissue-specific enzyme activities in the blood and attempts have been made to establish similar relationships in fish.[110,111] The lack of the necessary knowledge of the organ and isoenzyme distribution and effects of other stressors in marine animals, however, precludes making full use of these parameters at the moment.

In salmonids, for which most data are available, glutamate-pyruvate transaminase (GPT), whose activity is high in liver and kidney, may provide a sensitive indicator of pathology[111] and plasma levels appear to be elevated after Pb exposure.[112] Probably the enzyme of greatest diagnostic value is D-amino levulinic acid dehydratase (ALA-D) which is the key enzyme of heme synthesis. This enzyme, present in high levels in fish erythrocytes, is strongly and specifically inhibited by Pb, the activity being inversely related to the Pb exposure time and blood Pb concentration.[112,113]

Greater attention has been paid to changes in tissue enzyme activities after metal exposure of fish, but the results appear to be quite variable and incomplete (reviewed in reference 111). Inhibition of gill Na^+K^+-ATPase activity by concentrations of Cu, Cd, Hg, and Zn much lower and in shorter exposure times than those required for significant impairment of osmoregulation appears to be a reproducible efect in several teleost fish, although the effect may vary according to fresh or sea water adaptation.[111,114-116]

C. METAL-INDUCED CHANGES IN LEVELS OF BIOCHEMICALS IN THE BLOOD AND TISSUES OF FISH

The biological significance of pollutant-induced changes in serum components, such as cholesterol, cortisol, glucose, lactate, pyruvate, protein, and C-reactive protein in fish has been reviewed by Wedermeyer and Yasutake.[117] However, many other factors, such as catching stress, season, sex, temperature, reproductive cycles, etc. also alter many of these parameters and must be accounted for before they can be utilized diagnostically.[111] In several species, including catfish, flounder, mullet, and rainbow trout, there appears to be a generalized stress response to several organic pollutants and Cd which causes a significant elevation of plasma cortisol followed by an alpha-hyperglycemia and depletion in muscle glycogen levels.[117-121] In view of the complex nature of these responses, many of which are due to perturbations in the hypothalamo-pituitary axis, it is probable that they will be of little diagnostic use in assessing specific metal toxicities. Many essential processes are dependent upon maintenance of hepatic glutathione levels and long term studies have demonstrated that these levels are increased after exposure to Cd, Hg, or Pb in mullet and croaker.[122-124]

IV. CYTOCHEMICAL ASSESSMENTS OF HEAVY METAL TOXICITY

Many pollutants, including heavy metals, interact with cell membranes and alter their function. These have been extensively studied in mammals and include changes in erythrocyte membrane permeability produced by Au, Ag, Cu, Hg, and Pb, and inhibition of glucose uptake by Cu and Hg, etc. in rat diaphragm.[125] Interactions with membranes of subcellular organelles are less well studied and include effects of Cu and Hg on mitochondrial membranes.[126,127] In the electric organ of *Torpedo marmorata*, swelling of mitochondria and an elevation in the numbers of lipofuschin-containing tertiary lysosomes were observed after Cu-exposure.[128] The activity of lysosomal hydrolases is normally latent. They are only active within the lysosomes unless released by damage to the membrane, e.g., by *in vitro* treatment with detergents, freezing and thawing, osmotic shock,[109] Hg^{++} or organo-mercurials.[125] If the membrane is damaged *in vivo*, release of free hydrolases into the cell cytoplasm results in autolytic cell damage.[129] A reduction in latency, presumably due to a destabilization of the lysosomal membrane, can be measured histochemically as the ''labilization period'', i.e., the length of time of preincubation of a tissue section at pH 4.5 that is required to give maximum staining intensity for the lysosomal hydrolases, arylsulfatase, β-glucuronidase or N-acetyl-β-hexosaminidase.[130] An increase in this labilization period for lysosomal hydrolases in mollusk (*Mytilus* and *Littorina*) tissues has been utilized as an indicator of pollutant-

TABLE 5
Some Examples of Metals Localized in Lysosomes of Marine Animals by X-ray Microanalysis

Metal	Exposure	Animal genus	Tissue	Ref.
As	Lab	*Ictalurus*	Liver	140
Cu	Lab	*Fundulus*	Liver	141
Cu	Field	*Morone*	Liver	34
Cu	Field	*Balanus*	Body parenchyma	136
Pb, Cu	Field	*Asellus*	Gut	150
Cu, Zn	Field	*Penaeus*	Gut	40
Cu	Field	*Littorina*	Gut, kidney	78
Cu	Lab	*Eudistylia*	Gills	152
Cu, Fe	Field	*Nereis*	Gut	154
Cd, Zn, Pb	Lab	*Mytilus*	Kidney, gut	35,45,47
Fe	Lab, Field	*Mytilus*	Gill, gut, kidney	64

induced lysosomal membrane damage,[37,130-132] and appears to correlate with another generalized index of stress measured physiologically, "scope for growth".[132] While most studies have focused on polyaromatic hydrocarbons as the causative agents of decreased lysosomal latency, the latency of *Mytilus* digestive gland lysosomes is reduced after Cu exposure.[37,131]

Accumulations of metals in tertiary lysosomes (noted earlier) can be observed both at the light and electron microscope level. Light microscope histochemistry has been used to demonstrate lysosomal accumulation of Cu by rubeanic acid staining in mussel digestive gland[37], oyster amoebocytes[75,78], and fish liver;[34] of Zn by dithiocarbamate staining in mussel kidney[40] and oyster amoebocytes;[75,78] of Fe by Perl's reaction in mussel tissues;[65] and nonspecific staining for sulfide-precipitable heavy metals (Cu, Zn, Hg, and possibly Cd) by silver substitution and development (Timm's reaction) in mussels and oysters.[133] Electron microscope X-ray microanalysis has been extensively utilized for definitive subcellular localization of heavy metals in organisms, but great care has to be taken to minimize redistribution artifacts during specimen preparation.[35,49,50,70,71] Some examples of heavy metal localization in marine animals using the electron microscope are given in Table 5. A review of histopathological effects of heavy metals on marine organisms is outside the scope of the present work. For marine animals, there is not the wealth of data that exists for freshwater animals.

V. TOXIC EFFECTS OF ORGANOMETALLIC COMPOUNDS

A. METHYL MERCURY

The early life stages in the life history of fishes are generally the most sensitive to heavy metals and the embryotoxic effects of organomercurials have been studied in a number of species. Very low concentrations of both inorganic or organic Hg (15 to 30 μg/1, = ppb) are sufficient to induce teratogenic effects, including skeletal defects (spinal curvature), cardiovascular abnormalities (abnormal development of heart and blood vessels), and craniofacial defects (cyclopia, etc.), in *Brachydania*, *Fundulus*, and *Oryzias*.[137-140] For adult fish 96-h LC_{50}s are 10 to 100 times greater than this teratogenic dose.[140] The incidence of spinal curvature in *Fundulus heteroclitus* has been proposed as a method of assessing population or ecosystem effects of methylHg. This symptom is not specific and may be produced by a number of toxicants.[140,141] A manifestation of early subacute methylHg poisoning in mammals is generally neurological, the peripheral nervous system being affected. In adult fish from the methylHg-polluted Minimata Bay, impaired movements and brain lesions were recorded.[143] Both inorganic and methylHg cause some loss of osmoregulatory capacity in fish through inhibition of branchial Na uptake and renal Na transport.[142]

B. TRIBUTYL TIN

Tributyl tin oxide (TBTO), widely used as a biocide in industrial cooling water, a slimicide in paper manufacture, and an antifoulant, is extremely toxic to marine animals at extremely low dissolved concentrations in sea water. Effects on shellfish, particularly oysters and whelks, were first noticed around marinas where TBTO-based antifoulants were used on small boats. Their use has led to a dramatic decline in populations of the dogwhelk, *Nucella lapillus* around the U.K. Exposure of juveniles to a few nanograms TBTO per liter before maturity is sufficient to induce imposex (the growth of a penis and vas deferens in females) and cause sterility.[144,145,155] The mechanism is unclear but the action is thought to be due to effects on sex hormone balance. Measurement of imposex is regarded as the best method of assessing environmental TBTO contamination. In oysters, shell growth is inhibited at TBTO levels of <0.02 μg/l, and there is abnormal shell thickening and chambering at about 0.08 μg/l.[147] While the toxicity to adult fish is somewhat less, the 96-h LC_{50} being around 30 μg/l, damage to the gill epithelium occurs at lower concentrations and 1 μg/l has been proposed as the "safe" level. A study with rainbow trout yolk sac fry reported acute toxicity in fish exposed to 5 μg/l tributyl tin for 10 d and reduction in growth rates in fish exposed to 0.2 μg/l for 110 d.[149]

REFERENCES

1. **George, S. G.,** Cell biochemistry and transmembrane transport of some metals, in *Metals and Their Compounds in the Environment,* Vol.1, Merian, E., Ed., VCH Verlagsgesselschaft, Weinheim, 1988, chap. 11.
2. **Frausto da Silva, J. J. R. and Williams, R. J. P.,** The uptake of elements by biological systems, in *Structure and Bonding,* Vol. 2, Dunitz et al., Eds., Springer Verlag, Berlin, 1976, 69.
3. **George, S. G.,** Subcellular accumulation and detoxication of metals in aquatic animals, in *Physiological Mechanisms of Marine Pollutant Toxicity,* Vernberg, W. B., Calabrese, A., Thurberg, F. P., and Vernberg, J. F., Eds., Academic Press, New York, 1982, 3.
4. **Williams, R. J. P.,** Physico-chemical aspects of inorganic element transfer through membranes, *Philos. Trans. R. Soc. London Ser. B,* 294, 57, 1981.
5. **Hamer, D. H.,** Metallothionein, *Annu. Rev. Biochem.,* 55, 913, 1987.
6. **Kagi, J. H. R. and Kojima, Y.,** Metallothionein II, Experientia Supplementum, Vol. 52, Birkhauser Verlag, Basel, 1987.
7. **Karin, M.,** Metallothioneins: proteins in search of a function, *Cell,* 41, 9, 1985.
8. **Kojima, Y. and Kagi, J. H. R.,** Metallothionein, *Trends Biochem. Sci.,* 3, 403, 1978.
9. **Webb, M.,** *The Chemistry, Biochemistry and Biology of Cadmium,* Elsevier/North Holland, Amsterdam, 1979.
10. **Brady, F. O.,** The physiological function of metallothionein, *Trends Biochem. Sci.,* 7, 173, 1982.
11. **Bremner, I.,** Involvement of metallothionein in the hepatic metabolism of copper, *J. Nutr.,* 117, 19, 1987.
12. **Chan, K. M., Davidson, W. S., Hew, C. L., and Fletcher, G. L.,** Winter flounder metallothionein cDNA, *Eur. J. Biochem.,* submitted.
13. **Leaver, M. J. and George, S. G.,** Characterisation of a metallothionein cDNA clone from the plaice, *Pleuronectes platessa* and comparison with other fish metallothioneins, *Fish, Physid. Biochem.,* submitted.
14. **Lerch, K., Ammer, D., and Olafson, R. W.,** Crab metallothionein. Primary structures of metallothioneins 1 and 2, *J. Biol. Chem.,* 257, 2420, 1982.
15. **Bonham, K., Zafarullah, M., and Gedamu, L.,** The rainbow trout metallothioneins: molecular cloning and characterisation of two distinct cDNA sequences, *DNA,* 6, 519, 1987.
16. **Nemer, M., Wilkinson, D. G., Travaglini, E. C., Sternberg, E. J., and Butt, T. R.,** Sea urchin metallothionein sequence: key to an evolutionary diversity, *Proc. Natl. Acad. Sci. U.S.A.,* 82, 4992, 1985.
17. **Lerch, K.,** Amino acid sequence of copper-metallothionein from *Neurospora crassa,* in *Metallothionein,* Kagi, J. H. R., and Nordberg, M., Eds., Birkhauser Verlag, Basel, 1979, 173.
18. **Durnam, D. M., Perrin, F., Gannon, F., and Palmiter, R. D.,** Isolation and characterisation of the mouse metallothionein-I gene, *Proc. Natl. Acad. Sci. U.S.A.,* 77, 6511, 1980.
19. **Winge, D. R. and Brouwer, M.,** Discussion summary. Techniques and problems in metal-binding protein chemistry and implications for proteins in nonmammalian organisms, *Env. Health Perspect.,* 65, 211, 1986.

20. **Stone, H. C., Wilson, S. B., and Overnell, J.,** Cadmium-binding proteins in the scallop, *Pecten maximus, Env. Health Perspect.*, 65, 189, 1986.
21. **Wong, V. M. T. and Rainbow, P. S.,** Two metallothioneins in the shore crab *Carcinus maenas, Comp. Biochem. Physiol.*, 83A, 149, 1986.
22. **White, S. L. and Rainbow, P. S.,** A preliminary study of Cu-, Cd- and Zn-binding components in the hepatopancreas of *Palaemon elegans* (Crustacea: Decapoda), *Comp. Biochem. Physiol.*, 83C, 111, 1986.
23. **Stone, H. and Overnell, J.,** Non-metallothionein cadmium binding proteins, *Comp. Biochem. Physiol.*, 80C, 9, 1985.
24. **Dohi, Y., Ohba, K., and Yoneyama, Y.,** Purification and molecular properties of two cadmium-binding glycoproteins from the hepatopancreas of a whelk, *Buccinum tenuissimum, Biochim. Biophys. Acta*, 745, 50, 1983.
25. **Fowler, B. A., Engel, D. W., and Brouwer, M.,** Purification and characterisation studies of cadmium-binding proteins from the American oyster, *Crassostrea virginica, Environ. Health Perspect.*, 65, 63, 1986.
26. **Nordberg, M., Nuottaniemi, I., Cherian, G., Nordberg, G. F., Kjellstrom, T., and Garvey, J. S.,** Characterisation studies on the cadmium-binding proteins from two New Zealand oysters, *Environ. Health Perspect.*, 65, 57, 1986.
27. **Engel, D. and Brouwer, M.,** Cadmium-binding proteins in the blue crab, *Callinectes sapidus:* laboratory field comparison, *Mar. Environ. Res.*, 14, 139, 1984.
28. **Ray, S. and White, M.,** Metallothionein-like protein in lobsters *(Homarus americanus), Chemosphere*, 10, 1205, 1981.
29. **Engel, D. W., and Brouwer, M.,** Cadmium and copper metallothioneins in the american lobster, *Homarus americanus, Environ. Health Perspect.*, 65, 87, 1986.
30. **Howell, R. and Smith, L.,** Binding of heavy metals by the marine hematode, *Enoplus brevis* Bastian, 1865, *Nematologica*, 29, 39, 1983.
31. **Liu Fayi and George, S. G.,** Purification of very low molecular weight Cu-complexes from the European oyster, in *Marine Pollution and Physiology: Recent Advances*, Vernberg, F. J., Thurberg, F. P., Calabrese, A., and Vernberg, W., Eds., University of South Carolina, Columbia, S.C., 1985, 145.
32. **Frazier, J. M., George, S. G., Overnell, J., Coombs, T. L., and Kagi, J. H. R.,** Characterisation of two molecular weight classes of cadmium-binding proteins from the mussel *Mytilus edulis* (L.), *Comp. Biochem. Physiol.*, 80C, 275, 1985.
33. **Anderson, R., Frazier, J., and Huang, P. C.,** Transition metal-binding proteins from three Chesapeake Bay fish species, *Environ. Health Perspect.*, 65, 149, 1986.
34. **Bunton, T. E., Baksi, S. M., George, S. G., and Frazier, J. M.,** Abnormal hepatic copper storage in a teleost fish *(Morone americanus), Vet. Pathol.*, 24, 515, 1987.
35. **George, S. G.,** Heavy metal detoxication in the mussel *Mytilus edulis* — composition of Cd-containing kidney granules (tertiary lysosomes), *Comp. Biochem. Physiol.*, 76C, 53, 1983.
36. **George, S. G.,** Heavy metal detoxication in *Mytilus* kidney — an *in vivo* study of Cd- and Zn-binding to isolated tertiary lysosomes, *Comp. Biochem. Physiol.*, 76C, 59, 1983.
37. **Viarengo, A., Zanicchi, G., Moore, M. N., and Orenesu, M.,** Accumulation and detoxication of copper by the mussel *Mytilus galloprovincialis* Lam.: a study of the subcellular distribution in the digestive gland, *Aquat. Toxicol.*, 1, 147, 1981.
38. **George, S. G. and Viarengo, A.,** A model for heavy metal homeostasis and detoxication, in *Marine Pollution and Physiology: Recent Advances*, Vernberg, F. J., Thurberg, F. P., Calabrese, A., and Vernberg, W., Eds., University of South Carolina, Columbia, 1985, 125.
39. **George, S. G., Pirie, B. J. S., and Coombs, T. L.,** Characterisation of metal-containing granules from the kidney of the mussel, *Mytilus edulis, Biochem. Biophys. Acta.*, 716, 61, 1982.
40. **Al-Mohanna, S. Y. and Nott, J. A.,** The accumulation of metals in the hepatopancreas of the shrimp *Penaeus semisulcatus* de Haan (Crustacea: decapoda) during the moult cycle, in *Proc. Conf. Marine Environment and Pollution*, Academic Press, Oxford, 1985, 195.
41. **Icely, J. D. and Nott, J. A.,** Accumulation of copper within the ''Hepatopancreatic'' caeca of *Corophium volutator* (Crustacea: Amphipoda), *Mar. Biol.*, 57, 193, 1980.
42. **Sternlieb, I. and Goldfischer, S.,** Heavy metals and lysosomes, in *Lysosomes in Biology and Pathology*, Vol. 5, Dingle, J. T. and Dean, R. T., Eds., American Elsevier, New York, 1976, 185.
43. **Coombs, T. L. and George, S. G.,** Mechanisms of immobilisation and detoxication of metals by marine organisms, in *Proc. 12th Eur. Symp. Marine Biology*, McClusky, D. S. and Berry, A. J., Eds., Pergamon Press, Oxford, 1979, 179.
44. **Brown, D. A.,** Increases of Cd and the Cd:Zn ratio in the high molecular weight protein pool from apparently normal liver and tumour-bearing flounders *(Parophys vetulus), Mar. Biol.* 44, 203, 1977.
45. **Schulz-Baldes, M.,** Lead transport in the common mussel, *Mytilus edulis*, in *Proc. 12th Eur. Symp. Marine Biology*, McClusky, D. S. and Berry, A. J., Eds., Pergamon Press, Oxford, 1979, 211.
46. **Bevelander, G. and Nakahara, H.,** Correlation of lysosomal activity and ingestion by the mantle epithelium, *Biol. Bull. Woods Hole, Mass.*, 131, 76, 1966.

47. **George, S. G. and Pirie, B. J. S.,** Metabolism of zinc in the mussel, *Mytilus edulis,* L.: a combined ultrastructural and biochemical study, *J. Mar. Biol. Assoc. U.K.* 60, 575, 1980.
48. **Lowe, D. M. and Moore, M. N.,** The cytochemical distribution of zinc (ZnII) and iron (FeIII) in the common mussel *Mytilus edulis* and their relationship with lysosomes, *J. Mar. Biol. Assoc. U.K.*, 59, 851, 1979.
49. **George, S. G., Pirie, B. J. S., and Coombs, T. L.,** Use of electron microscope X-ray analysis in the determination of detoxication mechanisms for heavy metals in shellfish, in *Analytical Techniques in Environmental Chemistry,* Albaiges, J., Ed., Pergamon Press, Oxford, 1980, 477.
50. **Mason, A. Z. and Nott, J. A.,** A rapid routine technique for the X-ray microanalysis of microincinerated cryosections: an SEM study of inorganic deposits in tissues of the gastropod *Littorina littorea* (L.), *J. Histochem. Cytochem.*, 28, 1980.
51. **Mason, A. Z. and Nott, J. A.,** The role of intracellular biomineralized granules in the regulation and detoxication of metals in gastropods with special reference to the prosobranch *Littorina littorea, Aquat. Toxicol.*, 1, 239, 1981.
52. **Simkiss, K.,** Biomineralisation and detoxication, *Calcif. Tissue Res.*, 24, 199, 1977.
53. **Howard, B., Mitchell, P. C. H., Ritchie, A., Simkiss, K., and Taylor, M.,** The composition of intracellular granules from the metal-accumulating cells of the common garden snail *(Helix aspersa), Biochem. J.*, 194, 507, 1981.
54. **Bouillon, J.,** Ultrastructure des cellules renales des mollusques. I. Gasteropode pulmones terrestres, *Ann. Sci. Nat. Zool. Biol. Anim.*, 2, 719, 1960.
55. **Simkiss, K. and Mason, A. Z.,** Metal ions: metabolic and toxic effects, in *The Mollusca,* Vol. 2, Wilbur, K., Ed., Academic Press, New York, 1983, chap. 4.
56. **Walker, G., Rainbow, P. S., Foster, P., and Holland, D. L.,** Zinc phosphate granules in tissue surrounding the midgut of the barnacle *Balanus balanoides, Mar. Biol.* 33, 161, 1975.
57. **George, S. G., Pirie, B. J. S., and Coombs, T. L.,** Isolation and elemental analysis of metal-rich granules from the kidney of the scallop, *Pecten maximus* (L.), *J. Exp. Mar. Biol. Ecol.*, 42, 143, 1980.
58. **Doyle, L. J., Blake, N. J., Woo, C. C., and Yevich, P.,** Recent biogenic phosphorite: concretions in mollusc kidneys, *Science,* 199, 1431, 1978.
59. **Carmichael, N. G., Squibb, K. S., and Fowler, B. A.,** Metals in the molluscan kidney: a comparison of two closely related species *(Argopectan),* using X-ray microanalysis and atomic absorption spectrophotometry, *J. Fish. Res. Board Can.*, 36, 1149, 1979.
60. **Simkiss, K.,** Calcium pyrophosphate and cellular pollution, *Trends Biochem. Sci.*, 3, 111, 1981.
61. **Mason, A. Z. and Simkiss, K.,** Sites of mineral deposition in metal-accumulating cells, *Exp. Cell Res.*, 139, 383, 1982.
62. **Becker, G. L., Chen, C. H., Greenawalt, J. W., and Lehninger, A. L.,** Calcium phosphate granules in the hepatopancreas of the blue crab, *Callinectes sapidus, J. Cell Biol.* 61, 316, 1974.
63. **Bryan, G. W.,** The occurrence and seasonal variation of trace metals in the scallops, *Pecten maximus* (L.), and *Chlamys opercularis* (L.), *J. Mar. Biol. Assoc. U.K.*, 53, 145, 1973.
64. **George, S. G., Pirie, B. J. S., and Coombs, T. L.,** The kinetics of accumulation and excretion of ferric hydroxide by common mussel *Mytilus edulis* (L.) and its distribution in the tissues, *J. Exp. Mar. Biol. Ecol.*, 42, 143, 1976.
65. **Coombs, T. L.,** Uptake and storage mechanisms of heavy metals in marine organisms, *Proc. Anal. Div. Chem. Soc.*, 14, 219, 1977.
66. **Nakahara, H. and Bevelander, G.,** An electron microscope study of ingestion of thorotrast by amoebocytes of *Pinctada radiata, Tex. Rep. Biol. Med.*, 27, 101, 1969.
67. **Macara, I. G.,** Vanadium — an element in search of a role, *Trends Biochem. Sci.*, 2, 92, 1980.
68. **Lankester, E. R.,** On green oysters, *Q. J. Microsc. Sci.*, 26, 71, 1886.
69. **Ruddell, C. L. and Rains, D. W.,** The relationship between zinc and copper and the basophils of two in oysters, *Crassostrea gigas* and *Crassostrea virginica, Comp. Biochem. Physiol.*, 51A, 585, 1975.
70. **Pirie, B. J. S., George, S. G., Lytton, D. L., and Thomson, J. D.,** Metal containing blood cells of oysters: ultrastructure, histochemistry and x-ray microanalysis, *J. Mar. Biol. Assoc. U.K.*, 64, 115, 1984.
71. **Thomson, J. D., Pirie, B. J. S., and George, S. G.,** Cellular metal distribution in the pacific oyster, *Crassostrea gigas* (Thun.) determined by quantitative X-ray microprobe analysis, *J. Exp. Mar. Biol. Ecol.*, 85, 37, 1985.
72. **Thrower, S., and Eustace, I. J.,** Heavy metal accumulation in oysters grown in Tasmanian waters, *Food Technol. Aust,* Nov., 546, 1973.
73. **George, S. G., Pirie, B. J. S., and Frazier, J. M.,** Effects of cadmium exposure on metal-containing amoebocytes of the oyster, *Ostrea edulis, Mar. Biol.*, 76, 63, 1983.
74. **Frazier, J. M. and George, S. G.,** Cadmium kinetics in oysters — a comparative study of *Crassostrea gigas* and *Ostrea edulis, Mar. Biol.*, 76, 55, 1983.
75. **Ruddell, C. L.,** Elucidation of the structure and function of the granular oyster amoebocytes through histochemical studies of normal and traumatised oyster tissues, *Histochemie,* 26, 95, 1971.

76. **Brown, A. C. and Brown, R. J.,** The fate of thorium dioxide injected into the pedal sinus of *Bullia* (Gastropoda: prosobranchiata), *J. Exp. Biol.,* 42, 509, 1965.
77. **Wolfe, D. A.,** Levels of stable Zn and ^{65}Zn in *Crassostrea virginica* from North Carolina, *J. Fish. Res. Board Can.* 7, 47, 1970.
78. **Mason, A. Z., Simkiss, K., and Ryan, K.,** The ultrastructural localisation of metals in specimens of *Littorina littorea* collected from clean and polluted sites, *J. Mar. Biol. Assoc. U.K.,* 64, 699, 1984.
79. **Wieser, W.,** Responses of *Helix pomatia* to anoxia: changes in solute activity and other properties of the haemolymph, *J. Comp. Physiol.,* 141, 503, 1981.
80. **Burton, R. F.,** Calcium balance in *Helix aspersa,* in *Perspectives in Experimental Biology,* Spencer Davis, P., Ed., Pergamon Press, Oxford, 1976, 7.
81. **Marshall, A. T. and Talbot, V.,** Accumulation of cadmium and lead in the gills of *Mytilus edulis:* X-ray microanalysis and chemical analysis, *Chem. Biol. Interact.,* 27, 111, 1979.
82. **Overnell, J. and Trewhella, E.,** Evidence for the natural occurence of (cadmium-copper)-metallothionein in the crab *Cancer pagurus, Comp. Biochem. Physiol.,* 64C, 69, 1979.
83. **Osborn, D.,** A naturally occurring cadmium and zinc binding protein from the liver and kidney of *Fulmaris glacialis,* a pelagic North Atlantic seabird, *Biochem. Pharmacol.,* 27, 822, 1978.
84. **Bull, K. R., Murton, R. K., Osborn, D., Ward, P., and Cheng, L.,** High levels of cadmium in Atlantic seabirds and sea-skaters, *Nature (London),* 269, 507, 1977.
85. **Mehra, R. K. and Bremner, I.,** Development of a radioimmunoassay for rat liver metallothionein. I. Application to the analysis of rat plasma and kidneys, *Biochem. J.,* 213, 459, 1983.
86. **Hamilton, S. J. and Mehrle, P. M.,** Metallothionein in fish: review of its importance in assessing stress from metal contaminants, *Trans. Am. Fish. Soc.* 115, 596, 1986.
87. **Klaverkamp, J. F., MacDonald, W. A., Duncan, D. A., and Wagemann, R.,** Metallothionein and acclimation to heavy metals in fish: a review, in *Contaminant Effects on Fisheries,* Cairns, V. W., Hodson, P. V., and Nriagu, J. O., Eds., John Wiley & Sons, New York, 1984, 100.
88. **George, S. G. and Young, P.,** The time course of effects of cadmium and 3-methylcholanthrene on enzymes of xenobiotic metabolism and metallothionein levels in plaice, *Pleuronectes platessa, Comp. Biochem. Physiol,* 83C, 37, 1986.
89. **Overnell, J., McIntosh, R., and Fletcher, T.,** The levels of liver metallothionein and zinc in plaice, *Pleuronectes platessa* L., during the breeding season, and the effect of oestradiol injection, *J. Fish Biol.,* 30, 539, 1987.
90. **Overnell, J., McIntosh, R., Fletcher, T.,** The enhanced induction of metallothionein by zinc, its half life in the marine fish *Pleuronectes platessa,* and the influence of stress factors on metallothionein levels, *Experientia,* 43, 178, 1987.
91. **Olsson, P-E., Haux, C., and Forlin, L.,** Variations in hepatic metallothionein, zinc and copper levels during an annual reproductive cycle in rainbow trout, *Salmo gairdneri, Fish Physiol. Biochem.,* 3, 39, 1987.
92. **Eaton, D. L. and Toal, B. F.,** Evaluation of the Cd/Hemoglobin affinity assay for the rapid determination of metallothionein in biological tissues, *Toxicol. Appl. Pharmacol.,* 66, 134, 1982.
93. **Onosaka, S. and Cherian, M. G.,** Comparison of metallothionein determination by polarographic and cadmium saturation methods, *Toxicol. Appl. Pharmacol.,* 63, 270, 1982.
94. **Olafson, R. W. and Sim, R. G.,** An electrochemical approach to quantitation and characterisation of metallothioneins, *Anal. Biochem.,* 100, 343, 1979.
95. **Garvey, J. S., Vander Mallie, R. J., and Chang, C. C.,** Radioimmunoassay of metallothioneins, *Methods Enzymol.,* 84, 1212, 1982.
96. **Brown, D. A. and Parsons, T. R.,** Relationship between cytoplasmic distribution of mercury and toxic effects on zooplankton and chum salmon *(Oncorhynchus keta)* exposed to mercury in a controlled ecosystem, *J. Fish Res. Board Can.,* 35, 880, 1978.
97. **Engel, D.,** Metal regulation and molting in the blue crab, *Callinectes sapidus.* Copper, zinc and metallothionein, *Biol. Bull. Woods Hole, Mass.,* 172, 68, 1987.
98. **Engel, D. and Brouwer, M.,** Metal regulation and molting in the blue crab, *Callinectes sapidus,* metallothionein function in metal metabolism, *Biol. Bull. (Woods Hole, Mass.)* 173, 239, 1987.
99. **Lobel, P. B.,** Role of the kidney in determining the whole soft tissue zinc concentration of individual mussels *(Mytilus edulis), Mar. Biol.,* 92, 355, 1986.
100. **Roesijadi, G.,** Uptake and incorporation of mercury into mercury-binding proteins in gills of *Mytilus edulis* as a function of time, *Mar. Biol.,* 66, 151, 1982.
101. **Viarengo, A., Pertica, M., Mancinelli, G., Zanicchi, G., and Orunesu, M.,** Rapid induction of copper-binding proteins in the gills of metal exposed mussels, *Comp. Biochem. Physiol.,* 67C, 215, 1980.
102. **Nemer, M., Travaglini, E. C., Rondinelli, E., and D'alonzo, J.,** Developmental regulation, induction and embryonic tissue specificities of sea urchin metallothionein gene expression, *Dev. Biol.,* 102, 471, 1984.

103. **Winge, D., Krasno, J., and Colucci, A. V.,** Cadmium accumulation in rat liver; correlation between bound metal and pathology, in *Trace Element Metabolism in Animals,* Vol. 2., Hoekstra, W. G., Suttie, J. W., Ganther, H. E., and Mertz, W., Eds., University Park Press, Baltimore, 1973, 500.
104. **Jenkins, K. D., Brown, D. A., Oshida, P. S., and Perkins, E. M.,** Cytosolic metal distribution as an indicator of toxicity in sea urchins from the southern Carlifornian bight, *Mar. Pollut. Bull.,* 13, 413, 1982.
105. **Sanders, B. M., Jenkins, K. D., Sunda, W. G., and Costlow, J. D.,** Free cupric ion activity in seawater: effects on metallothionein and growth in crab larvae, *Science (Washington, D. C.),* 222, 53, 1983.
106. **Sanders, B. M. and Jenkins, K. D.,** Relationships between free cupric ion concentrations in sea water and copper metabolism and growth in crab larvae, *Biol. Bull. Woods Hole, Mass.,* 167, 704, 1984.
107. **Jenkins, K. D. and Sanders, B. M.,** Relationships between free cadmium ion activity in seawater, cadmium accumulation and subcellular distribution, and growth in polychaetes, *Environ. Health Perspect.,* 65, 205, 1986.
108. **Jenkins, K. D.,** unpublished results.
109. **de Duve, C. and Wattiaux, R.,** Functions of lysosomes, *Annu. Rev. Physiol.,* 28, 435, 1965.
110. **Gingerich, W. H.,** in *Aquatic Toxicology,* Vol. 1., Weber, L. J., Ed., Raven Press, New York, 1982, 55.
111. **Neff, J. M.,** Use of Biochemical Measurements to Detect Pollution Mediated Damage in Fish, ASTM Special Publication 854, American Society for Testing and Materials, Washington, D.C., 1985, 155.
112. **Hodson, P. V., Blunt, B. R., and Spry, D. J.,** Chronic toxicity of water-borne and dietary lead levels to rainbow trout *(Salmo gairdneri)* in lake Ontario water, *Water Res.,* 12, 809, 1978.
113. **Hodson, P. V., Blunt, B. R., Spry, D. J., and Austin, K.,** Evaluation of erythrocyte δ-amino levulinic acid dehydratease activity as a short term indicator of harmful exposure to lead, *J. Fish Res. Board Can.,* 34, 501, 1977.
114. **Bouquegneau, J. M.,** ATPase activity in mercury intoxicated eels, *Experientia,* 33, 941, 1977.
115. **Lorz, H. W., Williams, R. H., and Fitish, C. A.,** Effects of Several Metals on Smolting of Coho Salmon, EPA-600/3-78-090, *U.S. Environmental Protection Agency,* Washington, D.C., 1978.
116. **Lorz, H. W. and McPherson, B. P.,** Effects of copper and zinc in fresh water on the adaptation to sea water and ATPase activity, and the effects of copper on migratory disposition of coho salmon *(Oncorhynchus kisutch), J. Fish. Res. Board Can.,* 33, 2023, 1976.
117. **Wedermeyer, G. and Yasutake, W. T.,** Clinical methods for the assessment of the effects of environmental stress on fish health, Paper 89, *Technical Papers of the U.S. Fish and Wildlife Service,* Washington, D.C., 1977, 18.
118. **Larsson, A. and Haux, C.,** Altered carbohydrate levels in fish exposed to sublethal levels of cadmium, *J. Environ. Biol.,* 3, 263, 1982.
119. **Srivastava, D. K.,** Comparative effects of copper, cadmium and mercury on tissue glycogen of the catfish, *Heteropneustes fossilis* (Block), *Toxicol. Lett.,* 11, 135, 1982.
120. **Thomas, P. and Neff, J. M.,** Plasma corticosteroid and glucose responses to pollutants in striped mullet: different effects of naphthalene, benzo(a)pyrene and cadmium exposure, in *Marine Pollution and Physiology: Recent Advances,* Vernberg, F. J., Thurberg, F. P., Calabrese, A., and Vernberg, W., Eds., University of South Carolina, Columbia, 1985, 63.
121. **Thomas, P., Bally, M., and Neff, J. M.,** Ascorbic acid status of mullet, *Mugil cephalus* Linn., exposed to cadmium, *J. Fish Biol.,* 20, 183, 1982.
122. **Thomas, P., Wofford, W., and Neff, J. M.,** Effect of cadmium on glutathione content of mullet *(Mugil cephalus),* in *Physiological Mechanisms of Marine Pollutant Toxicity,* Vernberg, W. B., Calabrese, A., Thurberg, F. P., and Vernberg, J. F., Eds., Academic Press, New York, 1982, 109.
123. **Wofford, H. W. and Thomas, P.,** Interactions of glutathione and other acid soluble thiols with cadmium in the striped mullet *(Mugil cephalus* L.), *Mar. Environ. Res.,* 14, 119, 1984.
124. **Thomas, P. and Juedes, M. J.,** Altered glutathione status in Atlantic croaker *(Micropogonias undulatus)* tissues exposed to lead, *Mar. Environ. Res.,* 17, 192, 1985.
125. **Verity, M. A. and Reith, A.,** Effect of mercurial compounds on structure-linked latency of lysosomal hydrolases, *Biochem. J.,* 105, 685, 1967.
126. **Scott, R. L. and Gamble, J. L., Jr.,** Effect of mercurial compounds on potassium binding by mitochondria, *J. Biol. Chem.,* 236, 570, 1961.
127. **Lehninger, A. L.,** Water uptake and extrusion by mitochondria in relation to oxidative phosphorylation, *Physiol. Rev.,* 42, 467, 1962.
128. **Aloj Totaro, E., Pisanti, F. A., Glees, P., and Continillo, A.,** The effect of copper pollution on mitochondrial degeneration, *Mar. Environ. Res.,* 18, 245, 1986.
129. **Koenig, H.,** Lysosomes in the nervous system, in *Lysosomes in Biology and Pathology,* Vol. 2, Dingle, J. T., and Fell, H. B., Eds., North Holland/American Elsevier, Amsterdam, 1969, 111.
130. **Moore, M. N.,** Cytochemical demonstration of latency of lysosomal hydrolase in digestive cells of the common mussel, *Mytilus edulis,* and changes induced by thermal stress, *Cell Tissue Res.,* 175, 279, 1976.

131. **Harrison, F. L. and Berger, R.,** Effects of copper on the latency of lysosomal hexosaminidase in the digestive cells of *Mytilus edulis, Mar. Biol.*, 68, 109, 1982.
132. **Bayne, B. L., Livingstone, D. R., Moore, M. N., and Widdows, J.,** A cytochemical and a biochemical index of stress in *Mytilus edulis, Mar. Pollut. Bull.*, 7, 221, 1976.
133. **Lytton, D. G., Eastgate, A. R., and Ashbolt, N. J.,** Digital mapping of metal-bearing amoebocytes in tissue sections of the pacific oyster by scanning microphotometry, *J. Microsc.*, 138, 1, 1985.
134. **Sorenson, E. M. B. and Smith, N. K. R.,** Hemosiderin granules: cytotoxic response to arsenic exposure in channel catfish, *Bull. Environ. Contam. Toxicol.*, 27, 645, 1981.
135. **Weis, P., Bogden, J. D., and Enslee, E. C.,** Hg- and Cu-induced hepatocellular changes in the mummichog, *Fundulus heteroclitus, Environ. Health Perspect.*, 65, 167, 1986.
136. **Walker, G.,** ''Copper'' granules in the barnacle *Balanus balanoides, Mar. Biol.* 39, 343, 1977.
137. **Heisinger, F. J. and Green, W.,** Mercuric chloride uptake by eggs of the rice fish and resulting teratogenic effects, *Bull. Environ. Contam. Toxicol.* 14, 665, 1975.
138. **Weis, J. S. and Weis, P.,** Effects of heavy metals on development of the killifish, *Fundulus heteroclitus, J. Fish Biol.*, 11, 49, 1977.
139. **Kihlstrom, J. E. and Hulth, L.,** The effect of phenylmercuric acetate upon the frequency of hatching in the zebrafish, *Bull. Environ. Contam. Toxicol.*, 7, 111, 1972.
140. **Sharp, J. R. and Neff, J. M.,** Age-dependent differences of *Fundulus heteroclitus* embryos following chronic exposure to mercury, in *Marine Pollution and Physiology: Recent Advances,* Vernberg, F. J., Thurberg, F. P., Calabrese, A., and Vernberg, W., Eds., University of South Carolina, Columbia, 1985, 281.
141. **Bengtsson, B. E.,** Biological variables, especially skeletal deformaties in fish, for monitoring marine pollution, *Philos. Trans. R. Soc. London Ser. B.* 286, 457, 1979.
142. **Schmidt-Neilson, B., Sheline, J., Miller, D. S., and Deldonno, M.,** Effect of methyl mercury upon osmoregulation, cellular volume, and ion regulation in winter flounder, *Pseudopleuronectes americanus,* in *Physiological Responses of Marine Biota to Pollutants,* Vernberg, F. J., Calabrese, A., Thurberg, F. P., and Vernberg, W. B., Eds., Academic Press, New York, 1977, 105.
143. **Takeuchi, T., Shiraishi, Y., Hirata, Y., and Sasaki, M.,** Pathological studies on the fish caught at the coast where Minimata disease occurs, *J. Kumamoto Med. Soc.*, 3 (Suppl. 34), 542, 1960.
144. **Bryan, G. W., Gibbs, P. E., Hummerstone, L. G., and Burt, G. R.,** The decline of the gastropod *Nucella lapillus* around south-west England. Evidence for the effect of tributyltin from antifouling paints, *J. Mar. Biol. Assoc. U.K.*, 66, 611, 1986.
145. **Bryan, G. W., Gibbs, P. E., Burt, G. R., and Hummerstone, L. G.,** The effects of tributyltin (TBT) accumulation on adult dog whelks, *Nucella lapillus*: long term field and laboratory experiments, *J. Mar. Biol. Assoc. U.K.*, 67, 525, 1987.
146. **Thain, J. E. and Waldock, M. J.,** The growth of bivalve spat exposed to organotin leachates from antifouling paints, *I.C.E.S.*, 28, 6, 1985.
147. **Waldock, M. J. and Thain, J. E.,** Shell thickening in *Crassostrea gigas:* organotin antifouling or sediment induced?, *Mar. Pollut. Bull.*, 14, 411, 1983.
148. **Chliamovitch, Y. P. and Kuhn, C.,** Behavioural, haematological and histological studies on acute toxicity of bis-(tri-n-butyltin) oxide on *Salmo gairdneri* Richardson and *Tilapia rendalli* Boulenger, *J. Fish Biol.*, 10, 575, 1977.
149. **Seinen, W., Helder, T., Vernij, H., Penninks, A., and Leeuwangh, P.,** Short term toxicity of tri-n-butyltinchloride in rainbow trout *(Salmo gairdneri* Richardson) yolk sac fry, *Sci. Total Environ.*, 19, 155, 1981.
150. **Brown, B. E.,** Lead detoxication by a copper-tolerant isopod, *Nature,* 276, 388, 1978.
151. **George, S. G. and Pirie, B. J. S.,** The occurrence of cadmium in subcellular particles in the kidney of the marine mussel, *Mytilus edulis* exposed to cadmium. The use of electron microprobe analysis, *Biochim. Biophys. Acta,* 580, 234, 1979.
152. **Young, J. R., Adee, R. R., Piscopo, I., and Buschbom, R. L.,** Effects of copper on the sabellid polychaete, *Eudistylia vancouveri.* II. Copper accumulation and tissue injury in the branchial crown, *Arch. Environ. Contam. Toxicol.*, 10, 87, 1981.
153. **Nejmeddine, A., Dhainaut-Courtois, N., Baert, J.-L., Sautiere, P., Fournet, B., and Boulenger, P.,** Purification and characterization of a cadmium-binding protein for *Nereis diversicolor* (Annelida, Polychaeta), *Comp. Biochem. Physiol.*, 89C, 321, 1988.
154. **George, S. G.,** unpublished data, 1988.
155. **Langston, W. J.,** Toxic affects of metals and the incidence of metal pollution in marine ecosystems, this volume, chap. 7.
156. **Roesijadi, G., Kielland, S., and Klerks, P.,** Oyster metallothioneins: purification and properties, *Mar. Environ. Res.*, in press.

157. **Mackay, E. A., Overnell, J., Dunbar, B., Hunziker, P. E., Kagi, J. H. R., and Fothergill, J. R.,** Complete amino acid sequences of five dimeric and four monomeric forms of metallothionein from the edible mussel, *Mytilus edulis,* in prep.
158. **Hogstrand, C., Olsson, P.-E., and Haux, C.,** A radioimmunoassay for metallothionein in fish, *Mar. Environ. Res.,* in press.
159. **Norey, C. G., Lees, W. E., Darke, B. M., Stark, J. M., Baker, T. S., Cryer, A., and Kay, J.,** Immunological distinction between piscine and mamalian metallothioneins, *Comp. Biochem. Physiol.,* in press.

Chapter 9

METAL LEVELS IN MARINE VERTEBRATES

David R. Thompson

TABLE OF CONTENTS

I. INTRODUCTION

Following the documentation of anthropogenic mercury contamination of aquatic and terrestrial food webs in Japan, Sweden, and Iraq,[8,16,60,99] together with cadmium poisoning, also in Japan,[59] the distribution of these and other heavy metals within the marine environment has been increasingly studied in recent years. The potential risks to human health have prompted many investigations with respect to the more toxic metals cadmium, lead, and mercury. Of these metals, lead and mercury, in that order, are thought to show the highest values for the ratio anthropogenic/natural input[109] and are likely, therefore, to pose the most serious pollution threat. Furthermore, Bryan[22] states that mercury, and more specifically, methylmercury, is the only metal for which evidence exists for bioamplification up marine food chains.

The heavy metals can be divided into nonessential elements (lead, mercury, and probably cadmium) and essential elements with relatively well-defined roles and functions (copper, iron, selenium, and zinc). For other heavy metals, few data on levels in marine vertebrates exist, although a short summary of levels of arsenic, chromium, manganese, and nickel is given in Section VIII.

Account should be taken of the limitations of data reported on metal levels in marine vertebrates. Seasonal variation in metal concentrations in seabirds with respect to molt cycles, for example, can be considerable (see chapter by Walsh in this volume). Seabirds sampled at the start of molt will contain relatively high concentrations of mercury in liver tissues, for example, but as the 'body pool' of this metal diminishes with the formation of new feathers, birds sampled at a later stage of the molt cycle will contain relatively less mercury in liver tissues. This variation has been clearly demonstrated in the black-eared kite *Milvus migrans lineatus*, as reported by Honda et al.[88] and is likely to be common to all seabirds exposed to mercury. Similar variations in metal levels may exist in other marine groups and should be considered in any interpretation of data. The sources of many samples used for metal analysis pose other limitations. Biases may be introduced by the use of seabirds 'found dead' or by the use of beached or stranded marine mammals, although the collection of fresh samples of this latter group is both difficult and restricted by the scarcity of many species. Furthermore, with the enhancement and refinement of analytical techniques over the last 10 years, it is clear that the results from early investigations into metal levels in marine vertebrates are less reliable than those from recent studies.

In this review, the marine vertebrates are for convenience, treated in the following subgroups: marine fish, both Chondrichthyes and Osteichthyes; seabirds, excluding sea ducks and divers (loons); marine mammals (which have been further subdivided into the Pinnipedia, the Cetacea, and the Sirenia with other marine mammal groups). The metal concentration data which have been presented in the various tables are not intended to be a complete record of all work on each respective group for a particular metal; the data included in the tables are a representative selection of work, used here to illustrate particular trends or differences, other work being referred to in the text. Unless otherwise stated, the data presented in the various tables refer to samples made up of both male and female individuals, and age/length data are included only when accumulation effects of a particular metal are likely to be a factor.

II. CADMIUM

A. INTRODUCTION

In vertebrates, cadmium is usually located in a low molecular weight metalloprotein, metallothionein, but is believed to have little or no biological function.

B. CADMIUM IN MARINE FISH

Cadmium levels in fish are low, often below instrumental detection limits with values rarely exceeding 0.2 μg/g wet weight in muscle tissue.[3,10,36,49,70,71,74,75,85,91,112,118,132-135,151-153,168,176,179,186-188] Values tend to be higher in liver and kidney tissues,[15,70,74,133,135,151,168,186-188] Mackay et al.,[118] for example, reporting cadmium concentrations in liver tissues of black marlin *Makaira indica* ranging from 0.2 to 83.0 μg/g wet weight with a mean value of 9.2 μg/g wet weight.

Changes in cadmium concentration with increasing fish age/size have not been studied to any great extent; several studies report no correlation with increasing size/weight/length of fish,[85,91,118] while in others, cadmium has been shown to increase with increasing fish weight/length[7,36] which would seem appropriate for this metal, given its tendency to increase in concentration with increasing age/size in other groups of marine vertebrates. Relatively

TABLE 1
Summary of the References to Heavy Metals in Marine Fish

Class	
Chondrichthyes	28 29 40 49 55 57 71 74 76 77 108 115 116 127 132—135 154 168 175 179 183 184 187
Osteichthyes	3 7 10 11 14 15 28 31 34 36 40 48 49 57 58 70 73—79 82 83 85 91 101 105 108 110 112 113 118 122 127 132—135 151—154 158 160 163 164 166 173 175 186—188
Location	
European coastal	3 7 48 70 82 83 85 91 112 113 132—135 151 153 154 168 176 179 186 188
Australian coastal	10 29 31 40 49 71 115 116 118 122 152 175 183
West Atlantic Ocean	11 14 34 57 58 74—77 187
East Atlantic Ocean	73
Pacific Ocean	15 36 55 73 78 79 105 110 127 158 160 163 164 166 173
Indian Ocean	73 101 108 184
Metal	
Cadmium	3 7 10 15 36 49 70 71 74 75 82 83 85 91 112 118 132—135 151—154 168 176 179 186—188
Copper	3 10 15 34 49 70 71 74 75 83 91 112 118 132—135 151—154 168 176 179 186—188
Iron	34 83 152 179
Lead	3 7 10 15 40 70 71 74 75 82 83 112 118 132—135 152—154 168 176 179 186
Mercury	10 11 28 29 31 34 40 48 55 57 58 70 73—79 83 101 105 108 112 113 115 116 118 127 132—135 151 153 154 158 163 164 166 173 175 183 184 186 187
Selenium	10 28 58 71 101 113 116 118 122 164
Zinc	3 7 10 15 34 36 49 71 74 75 82 83 91 112 118 132—135 151—154 168 176 179 186—188
Others	10 14 15 34 49 70 71 74 75 83 91 110 112 118 132—135 152—154 160 179 187 188

high cadmium levels have been attributed to high incidence of crustaceans in the diet[82] and possible pollution effects.[83] Positive correlations between cadmium and zinc concentrations in marine fish have been shown to exist.[91,118] A summary of references to cadmium levels in marine fish is presented in Table 1.

C. CADMIUM IN SEABIRDS

Cadmium levels in seabird tissues tend to decrease in the order kidney > liver > muscle with very low or undetectable levels in feathers and eggs.[2,13,21,25,90,92-94,130,138,140,141,145,149,155,157,167,177] The extremely high cadmium concentrations in eggs of sooty terns *Sterna fuscata* from Hawaii reported by Stoneburner and Harrison[170] (mean concentration 75.04 μg/g wet weight) would appear to be anomalous and are in total contrast to cadmium levels in eggs of the same species, also from Hawaii, reported by Ohlendorf and Harrison[146] which failed to exceed the limit of detection (0.1 μg/g wet weight). Similarly, the relatively high cadmium concentrations (up to 27 μg/g wet weight) in feathers of Hawaiian seabirds reported by Cheng et al.[30] are difficult to explain. Representative data from northern and southern hemisphere studies are presented in Table 2. The variations in kidney cadmium levels in seabirds are likely to reflect both dietary differences and, although not studied extensively, age accumulation effects. Cadmium concentration in the kidney has been shown to correlate positively with age in the great skua *Catharacta skua skua*[61] and also increases with age in laughing gulls *Larus atricilla*,[155] royal terns *Sterna maxima*, and sandwich terns *Sterna sandvicensis*.[120]

Those seabirds which feed predominantly on pelagic cephalopods, in particular, and large euphausiid crustaceans which accumulate cadmium to relatively high levels, tend to

TABLE 2
Cadmium Concentrations (μg/g Wet Weight) in Liver and Kidney Tissue of Seabirds from the Antarctic (A), Gough Island (G) and Spitsbergen (S)

Species	Number sampled	Liver Mean	Liver Range	Kidney Mean	Kidney Range	Locality	Ref.
Adelie penguin *Pygoscelis adeliae*	10	4	1—8	51	24—93	A	90
Chinstrap penguin *Pygoscelis antarctica*	13	2	1—4	19	9—39	A	141
Rockhopper penguin *Eudyptes crestatus*	12	14	4—26	72	32—122	G	130
Macaroni penguin *Eudyptes chrysolophus*	9	9	4—34	49	18—166	A	141
Wandering albatross *Diomedea exulans*	2	32	24—41	137	127—148	G	130
Yellow-nosed albatross *Diomedea chlororhynchos*	9	9	3—17	25	15—46	G	130
Sooty albatross *Phoebetria fusca*	8	26	22—33	76	58—92	G	130
Southern fulmar *Fulmarus glacialoides*	6	5	2—10	38	16—75	A	141
Northern fulmar *Fulmarus glacialis*	10	17	6—32	55	22—114	S	141
Atlantic petrel *Pterodroma incerta*	13	19	9—40	61	42—102	G	130
Kerguelen petrel *Pterodroma brevirostris*	14	15	10—21	45	22—68	G	130
Soft-plumaged petrel *Pterodroma mollis*	18	15	8—41	48	32—90	G	130
Broad-billed prion *Pachyptila vittata*	31	16	9—26	33	19—72	G	130
Great shearwater *Puffinus gravis*	12	15	6—27	74	38—99	G	130
Little shearwater *Puffinus assimilis*	13	14	9—21	43	23—71	G	130
Grey-backed storm petrel *Garrodia nereis*	8	12	8—18	23	18—36	G	130
White-faced storm petrel *Pelagodroma marina*	7	8	6—12	33	25—55	G	130
White-bellied storm petrel *Fregetta grallaria*	8	11	9—15	21	18—26	G	130
Common diving petrel *Pelecanoides urinatrix*	17	7	3—14	32	17—74	G	130
Tristan skua *Catharacta skua hamiltoni*	13	3	1—5	26	13—45	G	130
Brown skua *Catharacta skua lonnbergi*	8	5	4—7	33	25—42	A	141
South polar skua *Catharacta maccormicki*	8	5	<1—7	25	4—36	A	141
Glaucous gull *Larus hyperboreus*	11	4	<1—9	23	4—58	S	141
Little auk *Alle alle*	9	4	2—6	21	6—34	S	141
Brunnich's guillemot *Uria lomvia*	9	4	2—11	16	7—38	S	141

TABLE 3
Cadmium Concentrations (μg/g Wet [w] or Dry [d] Weight) in Liver (L), Kidney (K), and Muscle (M) Tissue of Pinnipeds

Species	No. sampled	Tissue	Mean	Range	Locality	Age (years)	Ref.
California sea lion	10	L	15.1	5.7—90.0[d]	California	10—14	123
Zalophus californianus	10	K	115.0	85.0—569.0[d]	(females)	10—14	
Steller sea lion	19	L	0.7	0.1—1.3[w]	Hokkaido, Japan	<1—8.8	80
Eumetopias jubata	21	K	4.2	0.8—10.0[w]		<1—8.8	
	15	M	<0.1	<0.1—0.2[w]		<1—8.8	
Northern fur seal	9	L	1.8	0.6—4.6[w]	Washington coast	1—20	1
Callorhinus ursinus	9	K	5.8	0.2—15.6[w]	U.S.	1—20	
Harbor seal	31	L	—	<0.1—0.2[w]	German North Sea	<1—8	45
Phoca vitulina	16	K	—	<0.1—0.4[w]	coast	<1—8	
Ringed seal	29	L	7.3	2.7—14.9[w]	West Greenland	—	98
Phoca hispida	29	K	37.4	9.0—146.2[w]		—	
	29	M	<0.1	<0.1—0.4[w]		—	
Harp seal	57	L	12.0[w]	—	Gulf of St. Lawrence	6+	161
Phoca groenlandica	56	K	38.8[w]	—	(females)	6+	
	56	M	0.1[w]	—		6+	
Ribbon seal	2	L	2.6	2.1—3.1[w]	Okhotsk Sea	3—5	81
Histriophoca fasciata	16	M	—	<0.1—0.3[w]		1—16	
Gray seal	66	L	0.8	<0.1—8.5[w]	East coast Scotland	1—9	119
Halichoerus grypus	70	K	2.1	0.1—15.1[w]		1—9	
Leopard seal	15	L	5.1	0.8—29.0[w]	Antarctic	—	180
Hydrurga leptonyx	15	M	<0.1	—		—	
Weddell seal	2	L	1.1	1.0—1.3[w]	Antarctic	Adult	189
Leptonychotes weddellii	2	K	6.4	2.9—9.9[w]		Adult	
	2	M	0.2	<0.1—0.3[w]		Adult	
Crabeater seal	5	L	38.8[d]	—	Antarctic	—	167
Lobodon carcinophagus	5	K	102.1[d]	—		—	
	5	M	0.4[d]	—		—	
Ross seal	20	L	103.9	33.0—422.0[d]	Antarctic	—	117
Ommatophoca rossi							
Elephant seal	1	M	0.4[w]	—	Antarctic	—	43
Mirounga leonina							

exhibit relatively elevated cadmium concentrations.[90,130,141,149] There is, however, some evidence to suggest that cadmium is regulated to some extent in seabirds. Muirhead and Furness[130] noted that the distributions of cadmium levels in a range of seabird species from Gough Island did not differ significantly from Gaussian. This trend was also found for the essential metals copper and zinc, although the intraspecific variation of cadmium was higher than for these latter metals. From the data in Tables 2 to 4 it can be seen that cadmium concentrations in seabird kidney tissues range from 4 to 166 μg/g wet weight. The range of concentrations in seal kidney tissues is 0.1 to 146.2 μg/g wet weight while in whales and dolphins, a range of <0.1 to 205.4 μg/g wet weight is seen. Hence, in seabirds, the range of cadmium values in kidney tissues tends to be of an order of magnitude less than in both seals and whales and dolphins. Therefore, although cadmium shows age-related concentration increases, seabirds would appear to show some weak regulation of cadmium levels. Evidence of kidney damage was noted by Nicholson and Osborn[139] in northeast Atlantic seabirds, this being attributed largely to naturally occurring high cadmium concentrations and existing despite the presence of metallothionein. The latter is thought to offer some protection against cadmium toxicity, especially in kidney tissue, cadmium binding strongly to this protein. Zinc is also bound by metallothioneins. Hence, zinc and cadmium concentrations have been shown to be positively correlated in kidney tissue of many seabirds.[94,129,130,138,141,157]

TABLE 4
Cadmium Concentrations (μg/g Wet Weight) in Liver (L), Kidney (K), and Muscle (M) Tissue of Cetaceans

Species	No. sampled	Tissue	Mean	Range	Locality	Length[a]	Ref.
Beluga	1	L	0.9	—	Baltic Sea	2.7	84
Delphinapterus leucas	1	K	1.9	—		2.7	
	1	M	<0.1	—		2.7	
Narwhal	37	L	32.0	1.3—130.8	Pond Inlet	3.7	182
Monodon monoceros	54	K	63.5	1.0—205.4	Canada	3.8	
	58	M	0.2	<0.1—1.1		4.3	
Striped dolphin	57	L	6.3	<0.1—11.1	East coast Japan	—	89
Stenella coeruleoalba	54	K	24.8	<0.1—69.6		—	
	58	M	0.1	<0.1—0.3		—	
Short-finned pilot whale	1	L	69.4	—	North Queensland	—	41
Globicephala macrorhynchus	1	K	121.5	—		—	
	1	M	0.4	—		—	
Harbor porpoise	17	L	0.2	<0.1—0.9	East coast	—	50
Phocoena phocoena	17	K	1.1	0.2—2.9	Scotland (males)	—	
Goose-beaked whale	2	L	50.5	37.0—64.0	Bermuda	—	104
Ziphius cavirostris	2	K	47.5	33.0—62.0		—	
	4	M	0.2	—		—	
Bottlenose whale	1	L	5.6	—	North Sea	5.7	84
Hyperoodon ampullatus	1	M	<0.1	—		5.7	
Bowhead whale	1	L	1.5	—	Alaska	10.0	26
Balaena mysticetus	2	K	1.4	<0.1—2.8		9.4	
	2	M	<0.1	—		9.4	

[a] Mean length in meters.

D. CADMIUM IN MARINE MAMMALS

1. Pinnipeds

Seals show the same general pattern of cadmium distribution as seabirds in that levels tend to decrease in the order kidney > liver > muscle.[1,23,27,43,45,80,81,98,117,119,123,159,161,167,180,189] Cadmium concentrations in a range of species are given in Table 3. Comparisons between and within species are hindered by presentation of data by different authors either on wet or dry weight bases, often without appropriate conversion factors, and further complicated by age-accumulation effects for this element.

Cadmium concentration has been shown to increase with age (size) in several species.[45,81,123,159,161] Diet would appear to play an important role in determining cadmium levels in seals; those species feeding on cephalopods and crustaceans, for example, Ross seals *Ommatophoca rossi*, showing high levels[117] while predominantly fish-eating species, for example, harbor seals *Phoca vitulina* and gray seals *Halichoerus grypus* show lower cadmium levels.[27,45,46,84,86,159] Cadmium levels may correlate with liver selenium levels[123,161] suggesting a degree of protection by selenium to cadmium toxicity,[121] although the role of metallothionein, especially in kidney tissue, is important.[111]

2. Cetaceans

In common with seabirds and seals, the whales and dolphins show the same general body distribution pattern of cadmium in that organ concentrations decrease in the order kidney > liver > muscle.[26,41,50,51,84,89,104,169,182] Representative data are presented in Table 4. Comparisons are, again, hindered by age-accumulation effects, although there is general agreement between the data presented in Table 4 and those from other studies.[51,169] Cadmium concentrations have been shown to be positively correlated with body length[50,89] and similarly to body weight and age.[89]

A positive correlation between zinc and cadmium levels has been demonstrated in the liver and kidney tissues of striped dolphin *Stenella coeruleoalba*.[89] Dietary sources are likely to be important since high cadmium levels are known to be found in cephalopods[124] while fish tend to exhibit lower cadmium concentrations. The low cadmium levels in harbor porpoise *Phocoena phocoena*,[50,84] for example, may reflect low cadmium levels in fish which predominate in the diet of this cetacean.

3. Sirenians and Other Groups

The dugong *Dugong dugon* shows the same pattern of cadmium distribution as found in other marine mammals and seabirds[42] with the kidney exhibiting the highest concentrations (ranging from <0.04 to 59 μg/g wet weight). The two studies of cadmium in dugong muscle are in general agreement[42,128] with levels in the range <0.02 to 0.12 μg/g wet weight. A positive correlation between cadmium concentration and age and with zinc in both liver and kidney tissue has been demonstrated by Denton et al.[42] The correlation between metals may simply be a reflection of the tendency for both to increase with age. Cadmium levels in polar bear *Ursus maritimus* liver have been shown to correlate positively with age by Norstrom et al.,[144] but the levels are generally low (<1 μg/g wet weight) reflecting the tendency for polar bears to eat seal skin and blubber which contain relatively low levels of cadmium. Cadmium concentrations in polar bears were found to be higher in the west Arctic compared to the east Arctic.[144] Cadmium concentrations in otters *Lutra lutra* from the Orkney Islands have been found to be generally low with liver levels ranging from "not detected" to 0.39 μg/g wet weight. Kidney concentrations were slightly higher, falling between 0.08 and 0.56 μg/g wet weight.[126]

E. CONCLUSIONS

Highest levels of cadmium reported in marine vertebrates have been found in kidney tissue of Ross seals (422 μg/g dry weight),[117] narwhals *Monodon monoceros* (205 μg/g wet weight),[182] and macaroni penguins *Eudyptes chrysolophus* (166 μg/g wet weight).[141] Generally, cadmium tends to accumulate to a greater extent and to relatively high concentrations in kidney tissue of higher marine vertebrates. This pattern is less well defined in marine fish, although this may reflect the comparatively low levels found in fish and the difficulty in measuring accurately small increases or fluctuations. Being nonessential and so not metabolically regulated but accumulated and stored in a nontoxic situation in most marine vertebrates, cadmium shows age-related concentration increases, especially in marine mammals,[45,50,80,89,123,161,159] this often in association with similar and correlated increases in zinc concentrations.

There is some evidence to suggest that cadmium levels can be better regulated by seabirds than by marine mammals,[130] and that age accumulation is more pronounced in mammals. It would seem likely that, along with mercury and, to a lesser extent lead, cadmium is potentially a pollution threat, although clear-cut examples of anthropogenic influence causing environmental damage by cadmium are few. Long-lived seabirds and marine mammals feeding on cephalopods appear to accumulate cadmium to the greatest extent[117,130,141] and may be most vulnerable to cadmium pollution.

III. COPPER

A. INTRODUCTION

Copper is an essential element in vertebrates, being associated with numerous metalloenzymes and metalloproteins.

B. COPPER IN MARINE FISH

Copper concentrations in fish tissues show little variation with location or species and would appear to be under close physiological regulation. Generally, muscle copper levels have means of around 0.3 to 0.8 μg/g wet weight with few values less than 0.2 μg/g or greater than 2.5 μg/g.[3,10,34,70,83,118,132-135,151-153,187] Copper concentrations tend to be higher in liver and kidney tissues relative to muscle.[70,118,151,168,179,187] From studies of copper concentrations in cartilaginous fish,[71,168,179,187] there would appear to be little difference in copper levels between sharks and their allies and the bony fish. Copper concentration shows little or no tendency to increase or decrease with increasing fish length which is not unexpected for an essential element.[34,71,91,118] A summary of references to copper levels in marine fish is presented in Table 1.

C. COPPER IN SEABIRDS

Copper concentrations in seabird tissues exhibit low variation on both interspecies and geographical bases, again suggesting close metabolic regulation of this metal. Mean copper concentration in liver tissue tend to be around 6 μg/g wet weight with few values greater than 10 μg/g. In general, kidney copper concentrations tend to be lower than liver concentrations with levels in muscle lower still.

Representative copper concentration data from studies of a range of species from the southern and northern hemispheres are presented in Table 5. Other studies have yielded similar results to those presented. Data for birds from the Antarctic[2,90,141,167] are in close agreement with those in Table 5, although the high muscle copper concentrations in three species of petrel presented by Anderlini et al.[2] are difficult to explain. Studies of metal levels in crested terns *Sterna bergii*[92] and fairy prions *Pachyptila turtur*[21] from Australia, brown pelicans *Pelecanus occidentalis* from the U.S.[13] and kelp gulls *Larus dominicanus* and silver gulls *Larus novaehollandiae* from estuarine regions of New Zealand[177] have produced comparable copper concentrations to those above. In a study of Antarctic and northern hemisphere birds,[141] those birds from Spitsbergen showed significantly higher copper levels relative to Antarctic species. However, the difference was not great.

D. COPPER IN MARINE MAMMALS

1. Pinnipeds

Copper concentrations in a selection of seal species are presented in Table 6. From these and other data, copper levels tend to decrease from liver > kidney > muscle.[45,46,51,84,98,123,161,167,180,189] Within a given species, there is variation in tissue copper concentration. However, this variation is not thought to represent accumulation of copper with increasing size/age of animal.[45,117,161] Martin et al.[123] found significantly higher copper concentrations in liver and kidney tissue of female California sea lions *Zalophus californianus* with premature pups when compared to females with normal term pups. In this case, the balance of other elements, particularly mercury, selenium, and bromine was reported to be important. Geographical variation of copper levels as expressed by this group is unclear. Harbor seals from areas of the North and Wadden Seas[45,46,84] might be expected to have elevated copper levels in view of the relatively polluted nature of such locations. This does not seem to be the case (Table 6). Furthermore, within a study of copper levels in the harp seal *Phoca groenlandica* from several locations, no clear geographical trend could be deduced.[161] Diet has been cited as playing an important role in determining the copper burden of seals, in that high levels of copper observed in Ross seals are thought to be natural and reflect the relatively high levels of this metal in squid, the main prey of this species.[117]

2. Cetaceans

Copper levels in various whale species are presented in Table 7. As with the data for seal tissue copper concentrations, the levels in cetaceans generally decrease in the order liver > kidney > muscle. The values reported in Table 7 are in close agreement with each other, despite large geographical variation in the origins of species analyzed. Close regulation of copper levels is not unexpected given that it is an essential element. Wagemann et al.[182]

TABLE 5
Copper Concentrations (μg/g Wet Weight) in Liver and Kidney Tissue of Seabirds from the Antarctic (A), Gough Island (G), and Spitsbergen (S)

Species	Number sampled	Liver Mean	Liver Range	Kidney Mean	Kidney Range	Locality	Ref.
Adelie penguin *Pygoscelis adeliae*	10	4.7	3.3—6.1	3.6	2.9—4.5	A	90
Chinstrap penguin *Pygoscelis antarctica*	13	4.5	3.6—6.0	3.6	2.8—5.7	A	141
Rockhopper penguin *Eudyptes crestatus*	12	4.1	2.8—5.8	3.8	2.7—6.1	G	130
Macaroni penguin *Eudyptes chrysolophus*	9	4.0	2.4—5.3	3.2	2.6—3.7	A	141
Wandering albatross *Diomedea exulans*	2	6.6	4.9—8.4	5.2	4.7—5.6	G	130
Yellow-nosed albatross *Diomedea chlororhynchos*	9	5.0	3.5—7.0	3.3	2.8—4.9	G	130
Sooty albatross *Phoebetria fusca*	8	6.3	4.1—8.5	4.6	4.2—5.3	G	130
Southern fulmar *Fulmarus glacialoides*	6	4.1	3.4—4.8	4.5	3.3—6.3	A	141
Northern fulmar *Fulmarus glacialis*	10	6.2	5.6—6.7	4.1	3.3—4.7	S	141
Atlantic petrel *Pterodroma incerta*	13	4.9	3.6—13.0	6.1	4.5—9.3	G	130
Kerguelen petrel *Pterodroma brevirostris*	14	6.4	3.4—17.0	4.9	3.8—8.4	G	130
Soft-plumaged petrel *Pterodroma mollis*	18	5.2	3.5—12.0	5.9	4.1—8.0	G	130
Broad-billed prion *Pachyptila vittata*	31	6.4	3.6—8.4	4.4	3.2—6.0	G	130
Great shearwater *Puffinus gravis*	12	5.9	5.1—8.4	6.1	4.7—11.0	G	130
Little shearwater *Puffinus assimilis*	13	7.9	5.5—13.0	5.8	2.8—24.0	G	130
Gray-backed storm petrel *Garrodia nereis*	8	7.6	4.9—15.0	6.5	4.5—8.3	G	130
White-faced storm petrel *Pelagodroma marina*	7	8.5	5.7—11.0	7.0	6.2—8.4	G	130
White-bellied storm petrel *Fregetta grallaria*	8	6.3	3.4—8.1	6.4	5.2—7.5	G	130
Common diving petrel *Pelecanoides urinatrix*	17	6.9	5.3—10.0	5.5	2.8—8.0	G	130
Tristan skua *Catharacta skua hamiltoni*	13	4.2	3.1—5.4	4.6	3.3—6.6	G	130
Brown skua *Catharacta skua lonnbergi*	8	4.6	3.7—5.6	4.5	3.6—5.7	A	141
South polar skua *Catharacta maccormicki*	8	5.8	4.5—7.0	4.6	3.3—6.1	A	141
Glaucous gull *Larus hyperboreus*	11	7.3	5.5—10.0	5.2	4.1—6.8	S	141
Little auk *Alle alle*	9	8.4	6.8—9.4	6.4	5.3—9.7	S	141
Brunnich's guillemot *Uria lomvia*	9	8.2	6.4—9.4	6.9	4.1—11.2	S	141

TABLE 6
Copper Concentrations (μg/g Wet [w] or Dry [d] Weight) in Liver (L), Kidney (K), and Muscle (M) Tissue of Pinnipeds

Species	No. sampled	Tissue	Mean	Range	Locality	Ref.
California sea lion	10	L	86.0	61.0—285.0^d	California (females)	123
Zalophus californianus	10	K	22.4	21.2—52.3^d		
Harbor seal	58	L	—	2.6—17.0^w	German North Sea coast	45
Phoca vitulina	16	K	—	2.3—4.0^w		
Ringed seal	29	L	11.6	4.5—22.3^w	West Greenland	98
Phoca hispida	29	K	10.6	5.0—21.8^w		
	29	M	1.3	1.0—1.6^w		
Harp seal	57	L	24.3^w	—	Gulf of St. Lawrence	161
Phoca groenlandica	56	K	6.7^w	—	(females)	
	50	M	1.8^w	—		
Gray seal	38	L	28.9	6.2—75.0^w	Farne Islands, N.E.	27
Halichoerus grypus	37	K	3.0	1.6—5.0^w	England (females)	
Leopard seal	15	L	44.6	16.5—68.2^w	Antarctic	180
Hydrurga leptonyx	15	M	0.7	0.4—1.2^w		
Weddell seal	2	L	20.4	15.0—25.8^w	Antarctic	189
Leptonychotes weddellii	2	K	8.1	5.1—11.0^w		
	2	M	0.9	0.9—1.0^w		
Crabeater seal	5	L	74.0^d	—	Antarctic	167
Lobodon carcinophagus	5	K	33.3^d	—		
	5	M	5.0^d	—		
Ross seal	20	L	83.0	16.0—255.0^d	Antarctic	117
Ommatophoca rossi						
Elephant seal	1	M	4.1^d	—	Antarctic	43
Mirounga leonina						

reported a negative correlation of copper concentration with length in the narwhal, a trend supported by Honda et al.[89] in the striped dolphin with respect to length and weight in kidney tissue and age in both liver and kidney tissue. The latter study noted a marked decrease in copper concentration in 0- to 1-year-old dolphins, the levels then remaining relatively stable in immature and mature animals. Overall, copper concentration decrease was correlated more strongly both with body length and body weight increase which Honda et al.[89] attributed to greater importance of metabolic turnover which itself is correlated with body size, as opposed to age or exposure time, in determining copper levels. With only one study of copper levels in a baleen whale,[26] it is difficult to make comparisons with the toothed whales.

3. Sirenians and Other Groups

Copper concentrations in the dugong have been reported by Denton et al.[42] and by Miyazaki et al.[128] Levels in liver tissue ranged from 2 to 160 μg/g wet weight[42] which, at the upper extreme, represent the highest values in any marine mammal studied. Kidney values are in the same range (0.5 to 3.2 μg/g wet weight) as other marine mammals, as are values for muscle (0.1 to 1.0 μg/g wet weight) which are also in close agreement between the two studies.[42,128] Denton et al.[42] noted that copper levels in both liver and kidney show a negative correlation with age, although the highest liver copper concentration occurred in a male of at least 33 years of age. Unlike the findings of Honda et al.,[89] the copper levels decreased consistently with age, these decreased concentrations being linked to depigmentation in certain individuals. Dietary levels of copper are considered by the authors to be relatively low in dugongs, although hepatic copper was negatively correlated with increasing concentrations of zinc, cadmium, and iron. Norstrom et al.[144] report relatively high levels of copper in liver tissue of polar bear, concentrations of this element being correlated with

TABLE 7
Copper Concentrations (μg/g Wet Weight) in Liver (L), Kidney (K), and Muscle (M) Tissue of Cetaceans

Species	No. sampled	Tissue	Mean	Range	Locality	Ref.
Beluga	1	L	20.4	—	Baltic Sea	84
Delphinapterus leucas	1	K	3.1	—		
	1	M	1.1	—		
Narwhal	37	L	5.3	2.0—20.3	Pond Inlet, Canada	182
Monodon monoceros	54	K	2.3	1.8—3.5		
	58	M	0.7	0.5—1.2		
White-beaked dolphin	1	L	6.4	—	Kolding Fjord, Denmark	4
Lagenorchynchus albirostris	1	M	1.4	—		
Striped dolphin	57	L	8.1	3.6—15.2	East coast Japan	89
Stenella coeruleoalba	30	K	3.1	1.5—6.1		
	59	M	2.0	1.3—3.4		
Short-finned pilot whale	1	L	6.4	—	North Queensland	41
Globicephala macrorhynchus	1	K	3.7	—		
	1	M	0.7	—		
Harbor porpoise	17	L	7.3	2.7—12.8	East coast Scotland (males)	50
Phocoena phocoena	17	K	3.8	2.6—4.8		
Goose-beaked whale	2	L	5.2	3.2—7.1	Bermuda	104
Ziphius cavirostris	2	K	5.6	3.8—7.3		
	4	M	0.5	—		
Bottlenose whale	1	L	2.8	—	North Sea	84
Hyperoodon ampullatus	1	M	0.6	—		
Bowhead whale	1	L	3.1	—	Alaska	26
Balaena mysticetus	2	K	0.6	0.3—0.9		
	2	M	0.2	—		

zinc levels. Copper concentrations were not found to be correlated with age in the polar bear.[144]

E. CONCLUSIONS

Within each respective subgroup of marine vertebrates covered by this review, there is marked agreement with respect to copper levels regardless of species or geographical location. Highest copper levels in marine vertebrates have been reported in liver tissue of California sea lions (285 μg/g dry weight),[123] kidney tissue of little shearwater *Puffinus assimilis* (24 μg/g wet weight),[130] liver tissue of black marlin (22 μg/g wet weight),[118] and liver tissue of beluga *Delphinapterus leucas* (20.4 μg/g wet weight).[84] It would seem that concentrations of this essential metal are relatively closely regulated, and it is unlikely to show any substantial age-related concentration increase, although its role when at low dietary concentrations, in the absorption of other metals, could be relatively important.[42]

IV. IRON

A. INTRODUCTION

In vertebrates, iron is found most notably as an essential component of several metalloproteins, including hemoglobin and myoglobin and also a number of metalloenzymes.

B. IRON IN MARINE FISH

There are few data on iron levels in marine fish. As an essential element, iron is fairly uniform in its concentration between the few studies which have investigated this metal. Muscle concentrations range from <1 to about 14 μg/g wet weight with mean concentrations

TABLE 8
Iron Concentrations (μg/g Wet[w] or Dry [d] Weight) in Liver (L), Kidney (K), and Muscle (M) Tissue of Pinnipeds

Species	No. sampled	Tissue	Mean	Range	Locality	Ref.
California sea lion	10	L	2000	730—5590[d]	California (females)	123
Zalophus californianus	10	K	448	349—618[d]		
Harbor seal	8	L	—	28—2340[w]	Dutch Wadden Sea	46
Phoca vitulina	2	K	—	31—66[w]		
Weddell seal	2	L	665	389—940[w]	Antarctic	189
Leptonychotes weddellii	2	K	389	159—618[w]		
	2	M	252	237—267[w]		
Ross seal	20	L	519	231—961[d]	Antarctic	117
Ommatophoca rossi						
Elephant seal	1	M	147[w]	—	Antarctic	43
Mirounga leonina						

around 7 μg/g wet weight and less.[34,83,152,179] Liver and kidney concentrations would appear to be somewhat higher; Vas[179] found iron levels up to 12 and 15 μg/g wet weight in kidney and liver tissues, respectively, for tope *Galeorhinus galeus* from Liverpool Bay, northwest England. A summary of the references for iron levels in marine fish is presented in Table 1.

C. IRON IN SEABIRDS

Again, few data exist on the iron levels in seabirds. This metal is unlikely to show marked accumulation effects and is thought to be less of a potential toxic threat when compared to other essential metals. Honda et al.[90] report very high liver iron concentrations in Adelie penguins *Pygoscelis adeliae* (range 233 to 1670 μg/g wet weight) with lower levels in kidney tissue and lower still in muscle tissue. The mean breast feather iron level ranged from 6 to 66 μg/g wet weight. Male Adelie penguins show a redistribution of iron from muscle to liver during a period of starvation after mating resulting in significantly higher hepatic iron concentrations compared to females. Studies on terns[35,92] have reported iron levels lower than those for Adelie penguins with liver concentrations less than 300 μg/g wet weight in crested terns[92] and less than 200 μg/g in common tern *Sterna hirundo*[35] chicks.

D. IRON IN MARINE MAMMALS

1. Pinnipeds

Data for iron levels in seals are presented in Table 8. Comparisons between species are hindered by data presentation either on a dry weight basis[117,123] or on a wet weight basis,[46,189] although liver iron concentrations are higher when compared to kidney and muscle tissues (Table 8). The extremely high iron concentrations in some seal species (Table 8) may reflect increased levels of myoglobin associated with deep diving. No age-accumulation trends have been reported and there is no evidence to suggest that this metal is a pollutant in seals.

2. Cetaceans

Data for iron levels in whales and dolphins are presented in Table 9. As in seals and seabirds, the liver would appear to concentrate iron to the highest level, presumably due to its high blood content, although the few, relatively high, concentrations reported in seals have not been reported for whales and dolphins. Honda et al.[89] described an increase in iron concentration with age in striped dolphin, up to about 8 years, the iron level remaining roughly constant thereafter.

TABLE 9
Iron Concentrations (μg/g Wet Weight) in Liver (L), Kidney (K), and Muscle (M) Tissue of Cetaceans

Species	No. sampled	Tissue	Mean	Range	Locality	Ref.
Striped dolphin	57	L	215	—	East coast Japan	89
Stenella coeruleoalba	30	K	143	40—267		
	59	M	159	47—222		
Short-finned pilot whale	1	L	557	—	North Queensland	41
Globicephala macrorhynchus	1	K	202	—		
	1	M	218	—		
Goose-beaked whale	2	L	500	472—528	Bermuda	104
Ziphius cavirostris	2	K	181	180—182		
	4	M	106	—		

3. Sirenians and Other Groups

Denton et al.[42] reported extremely high liver iron concentrations (up to 21674 μg/g wet weight) in dugongs from North Queensland. Levels in kidney were somewhat lower, (up to 588 μg/g wet weight) while maximum muscle levels reached 82 μg/g wet weight and are generally of the same order as values reported by Miyazaki et al.[128] Iron concentration was shown to be positively correlated with age in liver and muscle tissue.[42] These high iron concentrations in the liver of the dugong were related to high dietary iron levels and much of the iron was found to be stored as hemosiderin. Iron concentration in liver tissue of polar bears was found to average 130 μg/g wet weight and as such is lower than values reported for seals and whales (Tables 8 and 9). No age effects were noted.[144]

E. CONCLUSIONS

Overall, there are relatively few data on iron levels in marine vertebrate groups, although the exceptionally high values reported by Denton et al.[42] of over 21,000 μg/g wet weight in liver tissue of dugongs are notable. As an essential element, it is unlikely that iron will have any real pollution-derived effects, except in severe and localized cases, and levels will tend to be closely regulated. The levels of iron in deep-diving marine mammals may prove to be of interest. The relationship between iron metabolism, iron flux dynamics, and myoglobin storage, for example, may be worthy of further investigation.

V. LEAD

A. INTRODUCTION

Lead has no known biological function or requirement and can be classified as nonessential. Environmental lead derived from anthropogenic sources is likely to have increased markedly in recent decades with the addition of alkyl lead to petrol as an "antiknock" agent.

B. LEAD IN MARINE FISH

Marine fish tend to have low muscle lead levels, generally within the range from limits of detection to 1 μg/g wet weight with values rarely exceeding 2 μg/g wet weight.[10,70,71,74,75,83,112,118,132-135,151-154,168,176,179,186] Where evaluated, liver and kidney lead levels tend to be higher than muscle lead levels.[70,74,118,151,186]

As a nonessential element, one might expect lead to show some accumulation effects with age or size of fish. In black marlin, Mackay et al.[118] reported no such increase in lead concentration with increasing weight, length, or girth in both muscle and liver tissue. In relatively more polluted coastal and estuarine systems, both elevated lead levels and age/

size accumulation effects have been reported. Badsha and Sainsbury[7] noted an increase in lead concentration with increasing weight and length of whiting *Merlangius merlangus* from the Severn estuary, southwest England while Hardisty et al.[82] reported somewhat higher lead concentrations than generally found in other studies in whole fish samples from the same locality with mean lead concentrations of 24 μg/g dry weight in whiting. Generally, only in coastal areas do lead levels show any increase above values which can best be described as low. This increase may reflect anthropogenic inputs of this metal. A summary of the references to lead in marine fish is presented in Table 1.

C. LEAD IN SEABIRDS

Seabirds, in a similar way to marine fish, exhibit generally low levels of lead, rarely exceeding 1 μg/g wet weight in any tissue,[13,21,90,92,120,140,141,145,155,157] although bone has been shown to concentrate lead compared to other tissues[90,93,94,120,145,155,177] and Bull et al.[24] found relatively high liver lead levels (1 to 10 μg/g wet weight) in gulls (*Larus sp.*) found dead on the Mersey estuary, northwest England. The dead gulls showed evidence consistent with poisoning in experimental gulls dosed with lead in the laboratory.[148] Similarly high lead levels have been reported in laughing gulls from Galveston Bay, TX by Hulse et al.,[93] liver tissue values averaging 5.3 μg/g wet weight with a maximum value of 13.8 μg/g wet weight. These levels agree closely with the earlier findings of Munoz et al.[131] but Reid and Hacker[155] noted a significant decrease in lead levels between 1977 and 1980 in this species. The decline was attributed to the reduction in lead emissions to the local environment.[155] The pattern of lead distribution within the bird is less well defined than for most other metals, the distinct gradient of concentrations between organs being less clear cut with lead, except for high levels found in bone.

In studies which include data on seabird bone lead concentrations, the majority give values which fail to approach bone lead levels in feral pigeons in a highly lead-polluted environment of London. Those fall in the range 108 to 669 μg/g dry weight.[95] Turner et al.[177] reported bone lead levels of over 1900 μg/g expressed as 'bone ash' in the silver gull from estuarine regions of New Zealand which are notable, if not directly comparable to other studies, and may be the result of ingestion of lead shot. However, Hutton[94] reported higher bone lead levels in the herring gull *Larus argentatus* (mean 37.7 μg/g dry weight) compared to the great skua (mean 4.5 μg/g dry weight). This difference was attributed to the coastal distribution of the former which would be likely to result in this species being exposed to relatively higher lead concentrations, and the relatively oceanic distribution of the latter, far from any anthropogenic lead sources, although atmospheric transport of this metal is known to be important. Generally, it would appear that avian lead levels tend to be low in oceanic environments while there is some evidence of increased lead concentration in some seabirds with coastal distributions.

D. LEAD IN MARINE MAMMALS

1. Pinnipeds

Data on lead concentrations in various seal tissues are presented in Table 10. It can be seen that levels are generally low and concentrations rarely exceed 1 μg/g wet weight in any tissue,[1,19,27,45,46,51,84,86,98,117,159,161] although bone and teeth tend to accumulate higher concentrations than do the soft organs.[19] Elevated lead levels tend to be found in seals inhabiting relatively industrialized, coastal regions. Lead concentrations of up to 17 μg/g wet weight in liver tissue of gray seals from the east coast of Scotland were reported by Holden,[86] although McKie et al.[119] noted lead levels consistently below the limit of detection (<0.5 μg/g wet weight) for the same species from the same area, a similar trend being noted by Caines[27] in gray seals from the Farne Islands, northeast England. Harbor seals which are found relatively close inshore, tend to show higher lead levels than other more

TABLE 10
Lead Concentrations (μg/g Wet [w] or Dry [d] Weight) in Liver (L), Kidney (K), and Muscle (M) Tissue of Pinnipeds

Species	No. sampled	Tissue	Mean	Range	Locality	Ref.
California sea lion	6	L	1.30[d]	—	California	19
Zalophus californianus	6	K[a]	2.00[d]	—		
	6	M	1.10[d]	—		
Northern fur seal	9	L	0.50	0.20—0.80[w]	Washington coast U.S.	1
Callorhinus ursinus	9	K	1.00	0.80—1.80[w]		
Harbour seal	31	L	—	0.10—0.55[w]	German North Sea coast	45
Phoca vitulina	16	K	—	0.14—0.55[w]		
Ringed seal	29	L	—	<0.01—0.03[w]	West Greenland	98
Phoca hispida	29	K	—	<0.01—0.48[w]		
	29	M	0.04	0.02—0.10[w]		
Harp seal	57	L	0.13[w]	—	Gulf of St. Lawrence	161
Phoca groenlandica	56	K	0.04[w]	—	(females)	
	56	M	0.03[w]	—		
Gray seal	9	L	7.00	<3.0—17.00[w]	East Scotland	86
Halichoerus grypus						
Ross seal	20	L	0.01	0.00—0.08[d]	Antarctic	117
Ommatophoca rossi						
Elephant seal	1	M	7.11[w]	—	Antarctic	43
Mirounga leonina						

[a] Kidney medulla.

pelagic species which may reflect anthropogenic effects with respect to lead. Holden[86] reported lead values from 3 μg/g wet weight up to 12 μg/g wet weight in liver tissue of harbor seals from the east coast of Scotland and England. However, even the lower end of this range is never reached in studies of the same species from similar European coastal sites[45,46,159] which throws Holden's data into doubt. In contrast, Ross seals from the Antarctic exhibit very low lead levels (0 to 0.08 μg/g dry weight), this reflecting the remoteness from any industrialization of the Antarctic.[117] The relatively high lead level in muscle tissue of an elephant seal *Mirounga leonina* (7.11 μg/g wet weight), also from the Antarctic,[43] is difficult to explain while the maximum concentration of 99.2 μg/g wet weight in the muscle tissues of ringed seals *Phoca hispida* from east Arctic Canada, as reported by Fallis,[51] would appear to be erroneous. Ronald et al.[161] reported no bioaccumulation of lead by harp seals while a similar finding is noted by Roberts et al.[159] for the harbor seal.

2. Cetaceans

Representative data on lead concentrations in whale and dolphin tissues are presented in Table 11. Lead concentrations are consistently low (<1 μg/g wet weight) in all studies[26,41,51,84,89,104,182] with levels generally higher in liver tissue > kidney > muscle. This distribution pattern is less well defined for lead than for other metals. Higher lead concentrations have been reported in the harbor porpoise and white-beaked dolphin *Lagenorchynchus albirostris* from the Danish coast when compared to other species.[4] A positive correlation between lead concentration and age, length, and weight in liver and muscle tissue was found by Honda et al.[89] in striped dolphins, the relationship with age being strongest. Lead concentration increased with age up to 1 year then remained fairly constant up to 18 years whereafter a steady increase in lead concentration was noted. The rapid increase in concentration up to 1 year was attributed to lead transfer from the mother via the milk, the 'levelling off' of lead concentration with age between 1 to 18 years was the result of a

TABLE 11
Lead Concentrations (μg/g Wet Weight) in Liver (L), Kidney (K), and Muscle (M) Tissue of Cetaceans

Species	No. sampled	Tissue	Mean	Range	Locality	Ref.
Beluga	1	L	0.36	—	Baltic Sea	84
Delphinapterus leucas	1	K	0.13	—		
	1	M	0.08	—		
Narwhal	37	L	0.03	0.01—0.06	Pond Inlet, Canada	182
Monodon monoceros	54	K	0.02	<0.01—0.08		
	58	M	0.01	<0.01—0.04		
White-beaked dolphin	1	L	4.50	—	Kolding Fjord, Denmark	4
Lagenorchynchus albirostris	1	M	2.20	—		
Striped dolphin	57	L	0.22	0.03—0.64	East coast Japan	89
Stenella coeruleoalba	30	K	0.17	0.01—0.71		
	59	M	0.18	0.04—0.26		
Short-finned pilot whale	1	L	<0.12	—	North Queensland	41
Globicephala macrorhynchus	1	K	<0.12	—		
	1	M	<0.09	—		
Harbor porpoise	4	L	3.50	1.90—5.30	Danish coast	4
Phocoena phocoena	4	M	3.30	1.60—4.70		
Goose-beaked whale	2	L	<0.50	—	Bermuda	104
Ziphius cavirostris	2	K	<0.50	—		
	4	M	<0.50	—		
Bottlenose whale	1	L	0.18	—	North Sea	84
Hyperoodon ampullatus	1	M	0.03	—		
Bowhead whale	1	L	0.12	—	Alaska	26
Balaena mysticetus	2	K	0.78	0.71—0.85		
	2	M	0.22	0.10—0.34		

'dilution effect' of increasing body size and the gradual increase of lead concentration with increasing age above 18 years coincided with the termination of increase in body weight.[89]

3. Sirenians and Other Groups

Lead levels in dugongs were found to be consistently low with values <0.09 μg/g wet weight in liver and kidney tissue[42] while slightly higher in muscle tissue (maximum value 0.25 μg/g wet weight).[42,128] These low levels probably simply reflect the unpolluted nature of the sampling sites. Otters from northern Scotland were found to have lead levels of up to 3.65 and 3.80 μg/g wet weight in liver and kidney tissue respectively with up to 10 μg/g wet weight in hair.[126] These values are comparable to the highest values reported for dolphins (Table 11) and may reflect anthropogenic influences.

E. CONCLUSIONS

If lead levels in marine vertebrates were to exhibit any sign of being the direct result pollution, the coastal environment would be the most likely area in which such trends would be manifest. Anthropogenic sources of this metal would tend to enter the marine environment along coastal and estuarine systems, although atmospheric transport of lead offers an alternative route to the sea.

Lead has no biological function and is likely, therefore, to exhibit age-related accumulation trends similar to those of cadmium and mercury. The levels measured in marine vertebrates, however, tend to be low, often at the limits of detection, although there is evidence to suggest that, especially in coastal environments, lead levels show some degree of elevation and that this lead is of industrial origin. Both Hardisty et al.[82] and Badsha and Sainsbury[7] reported relatively high lead concentrations (>20 μg/g dry weight) in whole fish samples of whiting from the Severn estuary, southwest England while Murray and Portmann[135] noted a maximum lead concentration of 4.2 μg/g wet weight in muscle tissues of thornback rays *Raja clavata* from the Irish Sea. Bull et al.[24] noted relatively high lead levels (up to 10 μg/g wet weight) in liver tissue of gulls found dead from the Mersey estuary, northwest England. Harbor porpoises from the Danish coast have been found to contain lead at a maximum concentration of 5.3 μg/g wet weight in liver tissues.[4] Hard tissues, such as bone and teeth, tend to exhibit the highest lead concentrations; Hutton[94] noted a maximum lead concentration of 78 μg/g dry weight in bone tissue of herring gulls from the Isle of May, Scotland.

VI. MERCURY

A. INTRODUCTION

Mercury has no known biological function and as such can be classified as nonessential. Within the marine biosphere, mercury has been found to exist in both inorganic and organic forms, the relative proportion of these two forms varying in different marine groups.

B. MERCURY IN MARINE FISH

In contrast to other heavy metals, mercury shows well-defined age/size accumulation trends in larger and longer-lived species of pelagic fish. Such fish were the subject of confiscation following the establishment by the U.S. Food and Drug Administration of a maximum permissible level of 5 μg/g wet weight of mercury in fish. This move followed mercury measurements in swordfish and tuna which revealed concentrations above this level and which were initially suggested to be the result of widespread, anthropogenic mercury emission to the marine environment. It is more likely, however, that mercury levels measured are natural and are a consequence of accumulation and storage without metabolic regulation of tissue levels.

Many studies on a wide range of marine fish have reported positive correlations between mercury concentration and a measure of age, weight, or length of fish.[11,29,34,40,48,73,77-79,101,115,116,118,127,158,163,164,175,183,184] This relationship is especially noteworthy when one compares the small number of corresponding relationships reported for the other, nonessential metals cadmium and lead. This may, however, reflect the increased interest in mercury as a potential threat to human health. Despite this tendency for mercury to increase in concentration with increasing size/age of some fish, muscle mercury levels tend to be less than 1 μg/g wet weight with kidney and liver levels slightly higher (liver > kidney > muscle).[10,31,34,55,57,70,73-79,83,101,105,108,115,116,127,132-135,151,153,154,158,166,175,186]

Considerably higher mercury values have been reported in groups, such as marlin, swordfish, tuna, and various shark species.[11,28,29,40,58,118,158,163,164,184] Evidence of slightly elevated mercury levels in fish from inshore, polluted areas is given by Harms[83] and Murray and Norton,[134] although a general trend of decline in mercury levels is noted by the latter study of fish caught around the British Isles and is supported by the findings of Murray and Portmann.[135]

Mercury in fish muscle would appear to be predominantly in the organic form (>80% of total mercury),[10,28,31,101,116,158,173,175,183] although this relationship is different in some large, pelagic species which have muscle organic mercury levels of only 40% or less of the muscle total mercury level.[158,164] This percentage decreases further in liver and kidney tissue.[158,164] The relationship between organic and inorganic mercury in fish tissues is worthy of further investigation in view of the possible biotransformations of this element, from the organic form to the inorganic form which are thought to take place in top predators. A summary of the references to mercury levels in marine fish is presented in Table 1.

C. MERCURY IN SEABIRDS

Total mercury levels in some seabirds from both southern and northern hemisphere studies are presented in Table 12. From this selection of data, it can be seen that total mercury levels vary enormously both within and between species. This general trend is true for all studies and is in contrast to the relatively limited variations in concentration of the essential metals copper and zinc. In addition to increased variation in mercury levels compared to variation in essential metals, the distribution of mercury concentrations within samples from seabird populations tend to be skewed with several high mercury values and, therefore, differ markedly from Gaussian. Muirhead and Furness[130] noted that mercury levels showed the greatest deviation from a Gaussian distribution in a range of seabirds from Gough Island relative to the essential metals copper and zinc and the nonessential metal cadmium. This finding is consistent with mercury accumulation rather than metabolic regulation.

Liver mercury levels tend to be greater than kidney levels which in turn exceed muscle and egg levels.[2,13,32,90,94,129,130,138,140,145,149] The levels of mercury in livers of wandering albatrosses *Diomedea exulans* and sooty albatrosses *Phoebetria fusca* are by far the highest yet reported in any seabird[130] (Table 12). This raises important questions as to what constitutes anthropogenic mercury and what are natural levels; since the above birds were obtained from Gough Island in the south Atlantic Ocean, far from any industrial source of mercury, it can be expected that they represent natural levels, although global transport of this metal cannot be ignored.

Maximum liver mercury concentrations from other studies rarely exceed 20 μg/g even on a dry weight basis, and are usually less than 10 μg/g.[2,20,32,39,61,87,90,94,138,140-143,145,149,177] Feathers have been used to monitor mercury levels in birds to a greater extent than for any other metal e.g.,[44,62,72,185] since mercury bonds strongly to disulfide linkages[33] and mercury levels in feathers are not affected by various vigorous treatments.[5] The results obtained from such studies are difficult to compare, largely due to variation caused by the specific type of feathers analyzed and the stage of molt of the bird at the time feathers were taken.[63] For these reasons, feather mercury values have been reported which both exceed corresponding liver and kidney concentrations[2,61,90,94,145,149,177] and which are less than corresponding liver and kidney concentrations.[61,149] Mercury levels in eggs tend to be low, rarely exceeding 0.5 μg/g wet weight,[2,9,13,47,53,97,140,146] although elevated levels have been reported in royal tern eggs from the Texas coast[103] and in gannet *Sula bassana* eggs from Norway.[53,54] The extrememly high mercury levels reported by Stoneburner et al.[171] in sooty terns from Florida (mean 7.93 μg/g wet weight) are somewhat anomalous and may be erroneous, in that they are about twice as high as levels in common tern eggs thought to be associated with toxic effects and reduced reproductive success.[52]

The form of mercury in seabird tissues has received little attention. Osborn et al.[149] reported mercury being predominantly organic in liver and kidney tissue of three species of seabirds from the north east Atlantic whilst Norheim et al.[143] found the proportion of organic mercury in liver tissue of the south polar skua *Catharacta maccormicki* from the Antarctic to vary from almost 100 to about 20% of the total mercury level. The negative correlation between percentage organic mercury and total mercury in this latter study was found to be statistically significant. Recent work on the relative proportion of organic and inorganic mercury in liver tissue of seabirds from Gough Island has revealed that, in some species, the organic fraction can be as little as 3% of the total mercury level. It would appear that, in such species, eliminatory mechanisms and detoxifying processes may be important in determining the mercury level of internal organs.[174] Age-accumulation trends of mercury are poorly documented, largely due to lack of birds of known age for which metal levels have been determined. Furness and Hutton[61] found a positive, but weak, correlation between liver mercury levels and age in the great skua.

TABLE 12
Total Mercury Concentrations (μg/g Wet Weight) in Liver Tissue of Seabirds from the Antarctic (A), Gough Island (G), and Spitsbergen (S)

Species	Number sampled	Mean	Range	Locality	Ref.
Adelie penguin *Pygoscelis adeliae*	10	<0.1	<0.1—0.2	A	90
Chinstrap penguin *Pygoscelis antarctica*	13	0.5	0.2—0.8	A	141
Rockhopper penguin *Eudyptes crestatus*	12	2.3	1.0—3.7	G	130
Macaroni penguin *Eudyptes chrysolophus*	9	0.9	0.4—1.5	A	141
Wandering albatross *Diomedea exulans*	2	268.0	266.0—271.0	G	130
Yellow-nosed albatross *Diomedea chlororhynchos*	9	7.7	4.8—20.0	G	130
Sooty albatross *Phoebetria fusca*	8	141.0	80.0—227.0	G	130
Southern fulmar *Fulmarus glacialoides*	6	2.9	0.8—6.2	A	141
Northern fulmar *Fulmarus glacialis*	10	2.1	0.6—4.2	S	141
Cape petrel *Daption capense*	7	1.3	—	A	143
Snow petrel *Pagodroma nivea*	7	0.7	—	A	143
Atlantic petrel *Pterodroma incerta*	13	28.0	14.0—53.0	G	130
Kerguelen Petrel *Pterodroma brevirostris*	14	4.6	1.9—6.8	G	130
Soft-plumaged petrel *Pterodroma mollis*	18	21.0	4.0—103.0	G	130
Broad-billed prion *Pachyptila vittata*	31	0.4	0.1—1.1	G	130
Great shearwater *Puffinus gravis*	12	2.0	0.8—6.5	G	130
Little shearwater *Puffinus assimilis*	13	1.2	0.6—1.6	G	130
Common diving petrel *Pelecanoides urinatrix*	17	0.5	0.2—1.5	G	130
Tristan skua *Catharacta skua hamiltoni*	13	7.4	0.9—17.0	G	130
Brown skua *Catharacta skua lonnbergi*	8	7.5	1.0—12.0	A	141
South polar skua *Catharacta maccormicki*	8	2.7	1.7—6.6	A	141
Glaucous gull *Larus hyperboreus*	11	1.6	0.8—2.3	S	141
Little auk *Alle alle*	9	0.5	0.4—0.7	S	141
Brunnich's guillemot *Uria lomvia*	9	0.6	0.3—0.9	S	141

D. MERCURY IN MARINE MAMMALS

1. Pinnipeds

Data on the levels of total mercury in seal tissues are presented in Table 13. There is extensive variation in mercury concentrations both inter- and intraspecifically, a trend which is otherwise only seen to any extent in cadmium. Generally, liver tissue exhibits the highest

TABLE 13
Total Mercury Concentrations (μg/g Wet [w] or Dry [d] Weight) in Liver (L), Kidney (K), and Muscle (M) Tissue of Pinnipeds

Species	No. sampled	Tissue	Mean	Range	Locality	Age (years)	Ref.
California sea lion	10	L	747.0	284.0—1026.0^{d}	California (females)	10—14	123
Zalophus californianus	10	K	28.4	12.1—43.2^{d}		10—14	
Northern fur seal	29	L	—	3.0—19.0^{w}	Pribilof Island, Alaska	2—3	1
Callorhinus ursinus	29	M	—	0.1—0.4^{w}	(males)	2—3	
Walrus	46	L	1.8	0.1—7.3^{w}	Thule Dist., Greenland	1—26	17
Odobenus rosmarus	58	M	<0.1	<0.1—0.1^{w}		1—26	
Harbor seal	31	L	—	1.6—160.0^{w}	German North Sea coast	<1—8	45
Phoca vitulina	16	K	—	1.6—12.5^{w}		<1—8	
Ringed seal	83	L	27.5^{w}	—	West Arctic Canada	12.8[a]	165
Phoca hispida	83	M	0.7^{w}	—		12.8[a]	
Harp seal	57	L	12.7^{w}	—	Gulf of St. Lawrence	6+	161
Phoca groenlandica	56	K	1.0^{w}	—	(females)	6+	
	56	M	0.3^{w}	—		6+	
Gray seal	70	L	27.8	0.2—125.9^{w}	East coast Scotland	1—9	119
Halichoerus grypus	68	K	2.7	0.8—6.7^{w}		1—9	
Bearded seal	56	L	26.2^{w}	—	East Hudson Bay	4.9[a]	165
Erignathus barbatus	55	M	<0.1^{w}	—		4.9[a]	
Leopard seal	15	L	4.5	0.7—12.2^{w}	Antarctic	—	180
Hydrurga leptonyx	15	M	0.1	<0.1—0.5^{w}		—	
Weddell seal	2	L	5.8	3.1—8.5^{w}	Antarctic	Adult	189
Leptonychotes weddellii	2	K	0.7	0.4—1.0^{w}		Adult	
	2	M	0.1	0.1—0.2^{w}		Adult	
Ross seal	20	L	3.4	0.7—19.1^{d}	Antarctic	—	117
Ommatophoca rossi							
Hooded seal	10	L	16.7	2.8—44.4^{w}	West Greenland	—	98
Cystophora cristata	10	M	0.3	0.2—0.5^{w}		—	

[a] Mean age in years.

total mercury concentration, followed by kidney and muscle tissues, in that order.[1,17,27,45,51,56,66,86,98,100,102,119,123,159,161,162,165,178,180,189]

It is not possible within the context of this review, to compare all studies of mercury levels in seals; this would prove difficult when the limitation of data presentation either on wet weight basis on dry weight basis is taken into account. Mercury concentrations in seal tissues are likely to be a reflection of both dietary mercury levels within differing prey types and age accumulation processes. Hence, species such as the walrus *Odobenus rosmarus* which feeds to a large extent on benthic invertebrates, tends to have relatively low mercury levels.[17] Conversely, fish-eating species, such as harbor seals, ringed seals, and gray seals, tend to exhibit relatively high mercury concentrations.[1,27,45,56,66,84,86,98,102,119,156,159,162,165,178] These interspecies differences are complicated by the tendency for mercury concentration to increase with age in seal tissue.[1,18,45,86,106,123,156,159,161,162,165]

Martin et al.[123] found high mercury levels in California sea lions and noted a strong correlation between this metal and selenium (see Section VI.E) and bromine in mothers with normal term pups. This balance broke down with respect to bromine in (younger) mothers with premature pups, suggesting that the balance of these elements is important. The levels of bromine were investigated by Reijnders[156] in harbor seals, although no correlation with age was found for this element and no conclusions were drawn with respect to the role of bromine and mercury toxicity.

The form of mercury in seal tissue has been investigated in several studies. Generally, liver mercury tends to be predominantly in the inorganic form[17,23,56,66,98,102,156,165] while

TABLE 14
Total Mercury Concentrations (μg/g Wet Weight) in Liver (L), Kidney (K), and Muscle (M) Tissue of Cetaceans (Odontoceti)

Species	No. sampled	Tissue	Mean	Range	Locality	Length[a]	Ref.
Beluga	7	L	6.3	2.5—12.1	West Arctic,	3.6	114
Delphinapterus leucas	7	M	0.7	0.6—1.0	Canada	3.6	
Narwhal	37	L	6.1	0.6—13.1	Pond Inlet,	3.7	182
Monodon monoceros	54	K	1.7	0.4—5.7	Canada	3.8	
	58	M	0.9	0.2—1.6		4.3	
Risso's dolphin	1	L	1.2	—	West coast	2.1	190
Grampus griseus	1	K	0.6	—	Scotland	2.1	
White-beaked dolphin	1	L	19.0	—	Kolding Fjord,	2.4	4
Lagenorchynchus albirostris	1	M	2.0	—	Denmark	2.4	
White-dotted dolphin	1	M	1.7	—	Ogasawara,	1.7	6
Stenella attenuata					Japan		
Striped dolphin	45	L	205.0	1.7—485.0	East coast	—	89
Stenella coeruleoalba	20	K	8.7	0.9—17.6	Japan	—	
	51	M	7.0	0.5—15.7		—	
Long-snouted dolphin	2	L	9.5	6.0—13.0	St. Lucia,	1.9	68
Stenella longirostris	2	K	2.5	2.3—2.7	Lesser	1.9	
	2	M	1.1	0.9—1.3	Antilles	1.9	
Gill's bottle-nosed dolphin	1	M	51.8	—	Shizouka,	3.1	6
Tursiops gilli					Japan		
Short-finned pilot whale	5	L	88.7	19.2—157.0	St. Lucia	4.3	68
Globicephala macrorhynchus	4	K	10.0	6.0—14.0	Lesser	4.3	
	5	M	4.0	2.8—5.4	Antilles	4.3	
Long-finned pilot whale	12	M	4.2	3.0—5.2	Wakayama,	3.4	6
Globicephala melaena					Japan		
Finless black porpoise	1	M	0.2	—	Kangawa,	0.7	6
Neophocaena phocaenoides			(fetus)		Japan		
Harbor porpoise	41	L	11.2	—	Deer Island,	1.3	69
Phocoena phocoena	23	K	1.8	—	east Canada	1.3	
	60	M	0.9	—	(males)	1.3	
Bottlenose whale	1	L	0.4	—	North Sea	5.7	84
Hyperoodon ampullatus	1	M	0.3	—		5.7	
Sperm whale	7	M	1.3	1.1—1.6	North Pacific	11.4	136
Physeter catodon							

[a] Mean length in meters.

organic mercury predominates in muscle tissue.[23,98,102,165] This contrasts with mercury form in marine fish tissue where mercury tends to be organic (see Section VI.B). This reduction in the proportion of organic mercury in seal liver has been cited as evidence for a biotransformation of the toxic, organic form into the less toxic, inorganic storage form. The subsequent storage of inorganic mercury would explain the accumulation of mercury in seal tissues with age.

2. Cetaceans

Total mercury data for some whales and dolphins are presented in Table 14 (Odontoceti) and Table 15 (Mysticeti). Both inter- and intraspecies variation in mercury levels is large for this group, in keeping with similar trends for mercury in other marine vertebrate groups. As with mercury concentrations in seals, this variation is likely to reflect both interspecies dietary differences with corresponding differing mercury levels, and age-accumulation trends. The general difference between mercury levels in toothed whales (Table 14) and baleen whales (Table 15) is, however, striking. Reported mercury values for the latter group never

TABLE 15
Total Mercury Concentrations (μg/g Wet Weight) in Liver (L), Kidney (K), and Muscle (M) Tissue of Cetaceans (Mysticeti)

Species	No. sampled	Tissue	Mean	Range	Locality	Length[a]	Ref.
Minke whale	6	L	0.18	0.07—0.41	West	—	98
Balaenoptera acutorostrata	6	M	0.15	0.09—0.25	Greenland	—	
Sei whale	9	M	0.03	0.02—0.07	South Pacific	14.3	136
Balaenoptera borealis							
Fin whale	8	M	0.02	0.01—0.03	South Pacific	20.2	136
Balaenoptera physalus							
Bowhead whale	1	L	<0.01	—	Alaska	9.4	26
Balaena mysticetus	2	K	<0.01	—		9.4	
	2	M	<0.01	—		9.4	

[a] Mean length in meters.

exceed 0.4 μg/g wet weight in liver tissue while the striped dolphin, for example, has been found to contain mercury up to 485 μg/g wet weight in liver tissue.[89] Toothed whales, in general, exhibit liver mercury concentrations in excess of 0.4 μg/g wet weight (Table 14, for example). This marked dichotomy is likely to be the result of clear-cut dietary differences, baleen whales feeding on prey which are generally lower down the marine food chain and which tend to have relatively low mercury concentrations.

The distribution pattern of mercury within the body is similar to that for seabirds and seals. Liver tissue tends to accumulate the highest mercury concentrations with kidney tissue > muscle tissue.[4,12,50,51,67-69,84,89,98,169,182,190] This pattern is less clear in the case of baleen whales with internal organs having fairly uniform, low mercury concentrations.[26,98,136] The tendency for mercury to show a strong positive correlation with age/length of whale[6,50,67,69,89,136] complicates the overall mercury picture and makes direct comparisons between studies difficult. Honda et al.[89] noted a strong positive correlation between mercury concentration and age in liver tissue of striped dolphins, but a leveling off of mercury concentration in kidney and muscle tissue after about 18 years. The correlation of mercury concentration with age in liver, kidney, and muscle tissues was stronger than those with both length and weight. Gaskin et al.[69] found that in harbor porpoises, liver mercury levels correlated well with age, length, and weight whereas, in other tissues, the correlations with weight and length were not so strong as that with age.

As with seals, the form of mercury in whale tissue has been investigated in several studies. Generally, in liver tissue, the majority of mercury is in the inorganic form.[50,67-69,96,98,190] Muscle tissue tends to have a relatively greater proportion of organic mercury compared to inorganic mercury than does liver tissue.[6,67-69,96,98,136] Although the data for mercury form within the various prey organisms of whales are not comprehensive, it is generally thought that the vast majority is in the organic, methyl form. This observed reduction in organic mercury relative to total mercury, especially in liver tissue, is believed to be evidence of a biotransformation of the metal into the less toxic, inorganic form, although the gradual accumulation of small quantities of the latter cannot be ruled out as an alternative mechanism.

3. Sireniens and Other Groups

Denton and Breck[40] determined total mercury levels in two dugongs from North Queensland. Concentrations were found to be low with maximum values of 0.05 μg/g wet weight in both liver and kidney tissue. Mercury levels in liver tissue of polar bears from various areas of the Canadian Arctic were found to average from about 17 to 53 μg/g wet weight with maximum values over 90 μg/g wet weight. The interarea differences in liver mercury concentrations were significant in some cases, and a positive correlation between mercury levels and bear age was also noted.[144]

Otters from the Orkney Islands were found to contain up to 20.3 μg/g wet weight total mercury in liver tissue with a mean value of 4.7 μg/g wet weight. Kidney levels were lower with a maximum of only 3 μg/g wet weight. Hair values were found to range from 9.5 to 29.5 μg/g wet weight.[126]

E. SELENIUM LEVELS AND INTERACTIONS WITH MERCURY IN MARINE VERTEBRATES

The role that selenium plays in mediating the toxicity and determining the accumulation patterns of mercury (and also cadmium) has received much attention since the observations of Parizek and Ostadalova[150] on the prevention by selenium of the deleterious effects of Hg^{2+} on the kidneys and intestine. A complete appraisal of this important topic is not practicable within the context of this review, (see Magos and Webb[121]), although the general trends within the marine vertebrates are presented.

A summary of the references to selenium and other heavy metals, in marine fish is presented in Table 1. Selenium has been shown to accumulate with age/length in several studies of marine fish.[113,116,118,164] Furthermore, a positive correlation between selenium and mercury has been demonstrated in several species[58,101,116,118,164] while Leonzio et al.[113] reported a strong, positive correlation between age and the sum of the selenium and mercury concentrations (nmol/g) in muscle tissue of striped mullet *Mullus barbatus*.

Selenium levels in various seabird tissues are presented in Table 16. Positive, but weak, correlations between selenium levels and age have been found in kidney and, to a lesser extent, liver tissues of great skuas.[61] Significant correlations between selenium and mercury levels have been found by Furness and Hutton[61] in great skuas, by Hutton[94] in herring gulls, and by Norheim[141] in a range of seabird species from both northern and southern hemisphere locations.

Data on selenium levels in seals are presented in Table 17. There is great variation both within and between species. Selenium has been shown to correlate positively with age in several studies.[107,156,161,165] As with marine fish and seabirds, positive correlations between selenium and mercury levels have been found in seal tissues, most notably, liver tissue.[102,106,107,123,156,161,165,178] A similar pattern emerges for whales and dolphins, data for this group being presented in Table 18. Itano et al.[96] noted a general increase in selenium concentration with age in liver and muscle tissue of striped dolphins. Positive correlations between selenium concentration and mercury concentration have been noted in liver and muscle tissues.[6,96,106,107,169] Studies on polar bears from the Canadian Arctic also show that selenium concentration increase with age and that selenium and mercury levels are highly correlated with each other.[144]

In general, the consistent finding that mercury and selenium levels tend to be significantly and positively correlated with each other has been suggested as being indicative of some form of protection by selenium against mercury toxicity. This has been supported by evidence of a molar ratio for mercury:selenium of approximately 1:1 in several studies.[65,106,107,123,165,169,178,182] Conversely, however, some studies, predominantly of marine fish, have reported mercury:selenium molar ratios which show an excess of selenium[28,58,61,101,164] or an excess of mercury.[6,102]

How selenium acts in affording protection against mercury toxicity is debateable, although, in feeding experiments using quail *Coturnix sp.*[64] and minnows *Phoxinus phoxinus*,[37] it has been clearly shown to do so. Sumino et al.[172] suggested that selenite releases methyl mercury from its linkages with proteins and, thus, influences its tissue distribution. In

TABLE 16
Selenium Concentrations (μg/g Wet [w] or Dry [d] weight) in Liver (L), Kidney (K), and Egg (E) Tissue of Seabirds

Species	No. sampled	Tissue	Mean	Range	Locality	Ref.
Chinstrap penguin	13	L	5.4	3.2—7.2^{w}	Antarctic	141
Pygoscelis antarctica						
Macaroni penguin	9	L	22.0	14.0—28.0^{w}	Antarctic	141
Eudyptes chrysolophus						
Southern fulmar	6	L	13.0	11.0—16.0^{w}	Antarctic	141
Fulmarus glacialoides						
Northern fulmar	10	L	3.0	1.4—6.4^{w}	Spitsbergen	141
Fulmarus glacialis						
Wedge-tailed shearwater	25	E	1.2^{w}	—	Hawaii	146
Puffinus pacificus						
Brown pelican	10	L	6.8^{w}	—	Gulf of California	145
Pelecanus occidentalis	10	K	3.5^{w}	—	west coast, U.S.	145
	18	E	0.3	0.2—0.4^{w}		13
Red-footed booby	25	E	0.8^{w}	—	Hawaii	146
Sula sula						
Great skua	10	L	19.7	6.7—34.6^{d}	Foula	61
Catharacta skua skua	9	K	32.8	13.3—89.1^{d}		
Brown skua	8	L	24.0	3.6—34.0^{w}	Antarctic	141
Catharacta skua lonnbergi						
South polar skua	8	L	18.0	2.9—39.0^{w}	Antarctic	141
Catharacta maccormicki						
Herring gull	9	L	7.9	6.9—9.3^{d}	Isle of May	94
Larus argentatus	7	K	14.1	8.6—19.4^{d}		
Kelp gull	34	K	2.0^{w}	—	New Zealand	177
Larus dominicanus						
Glaucous gull	11	L	2.2	1.3—3.8^{w}	Spitsbergen	141
Larus hyperboreus						
Silver gull	29	K	2.2^{w}	—	New Zealand	177
Larus novaehollandiae						
Sooty tern	23	E	1.3^{w}	—	Hawaii	146
Sterna fuscata						
Royal tern	30	E	1.0	0.4—2.1^{w}	Texas	103
Sterna maxima						
Little auk	9	L	2.6	1.5—4.5^{w}	Spitsbergen	141
Alle alle						
Brunnich's guillemot	9	L	1.9	1.1—2.6^{w}	Spitsbergen	141
Uria lomvia						

experimental studies of seal liver tissue, Van de Ven et al.[178] found no evidence of any effect of selenium on the demethylation process of mercury while Martoja and Viale[125] located mercuric selenide storage granules in the liver tissue of the goose-beaked whale *Ziphius cavirostris*. Stoneburner[169] has suggested that the breakdown of the 1:1 mercury:selenium molar relationship may be associated with strandings of whales; a similar breakdown of the relative proportions of elements in California sea lions has been linked with premature births.[123] In this case, the mercury:selenium ratio was near unity for mothers of both full-term and premature pups, but the levels of bromine were severely depressed from the 1:1:1 (mercury:selenium:bromine) ratio in mothers with premature pups.

F. CONCLUSIONS

Of all the heavy metals covered in this review, mercury exhibits the most pronounced age-related concentration changes and variation both within and between species and also

TABLE 17
Selenium Concentrations (μg/g Wet [w] or Dry [d] Weight) in Liver (L), Kidney (K), and Muscle (M) Tissue of Pinnipeds

Species	No. sampled	Tissue	Mean	Range	Locality	Age (years)	Ref.
California sea lion	10	L	260.0	92.0—352.0[d]	California (females)	10—14	123
Zalophus californianus	10	K	22.0	9.2—33.8[d]		10—14	
Harbor seal	8	L	109.0	3.9—350.0[w]	Dutch Wadden Sea	Adult	156
Phoca vitulina	4	K	7.1	2.3—10.0[w]		Adult	
Ringed seal	42	L	15.2	—	West Arctic Canada	12.8[a]	165
Phoca hispida							
Harp seal	16	L	5.8[w]	—	Gulf of St. Lawrence	6+	161
Phoca groenlandica	12	K	2.9[w]	—	(females)	6+	
	13	M	0.6[w]	—		6+	
Gray seal	10	L	38.1	8.6—88.0[w]	Great Britain	—	178
Halichoerus grypus							
Bearded seal	10	L	20.8[w]	—	East Hudson Bay	4.9[a]	165
Erignathus barbatus							
Leopard seal	15	L	3.3	2.6—4.9[w]	Antarctic	—	180
Hydrurga leptonyx	15	M	0.6	0.4—0.7[w]		—	
Weddell seal	1	L	1.0[w]	—	Antarctic	—	180
Leptonychotes weddellii	1	M	0.3[w]	—		—	

[a] Mean age in years.

TABLE 18
Selenium Concentrations (μg/g Wet Weight) in Liver (L), Kidney (K), and Muscle (M) Tissue of Cetaceans

Species	No. sampled	Tissue	Mean	Range	Locality	Length[a]	Ref.
Narwhal	37	L	4.1	0.6—8.0	Pond Inlet, Canada	3.7	182
Monodon monoceros	54	K	3.2	1.7—4.9		3.8	
	58	M	0.4	0.3—0.9		4.3	
White-dotted dolphin	1	M	0.7	—	Ogasawara, Japan	1.7	6
Stenella attenuata							
Striped dolphin	15	L	48.6	—	East coast Japan	—	96
Stenella coeruleoalba	14	K	5.6	—		—	
	26	M	2.8	—		—	
Gill's bottle-nosed dolphin	1	M	13.9	—	Shizouka, Japan	3.1	6
Tursiops gilli							
Short-finned pilot whale	4	L	44.2	22.8—61.6	Cumberland Island,	3.7	169
Globicephala macrorhynchus	4	K	7.2	3.0—10.0	U.S.	3.7	
Long-finned pilot whale	12	M	0.9	0.6—1.3	Wakayama, Japan	3.4	6
Globicephala melaena							
Finless black porpoise	1	M	0.2	—	Kanagawa, Japan	0.7	6
Neophocaena phocaenoides			(fetus)				
Bowhead whale	2	L	0.1	<0.1—0.1	Alaska	9.4	26
Balaena mysticetus	2	K	—	<0.1—0.1		9.4	
	2	M	0.1	<0.1—0.2		9.4	

[a] Mean length in meters.

between different marine groups. Its high toxicity and relatively long biological half-life combine to make this metal the most serious pollution threat from the group of metals covered in this review. Generally, the liver tissue of marine vertebrates accumulates mercury to the highest concentration. The highest reported mercury concentrations in liver tissues of the respective marine groups covered are up to 1026 μg/g dry weight in California sea lions,[123] up to 485 μg/g wet weight in striped dolphins,[89] up to 271 μg/g wet weight in wandering albatrosses,[130] and up to 63 μg/g wet weight in black marlin.[118]

The relationship between inorganic and organic mercury within respective marine groups is of note. The fact that much of the mercury in some marine mammal tissues is largely inorganic whereas that in their prey is thought to be predominantly organic, has been suggested as evidence for a detoxification mechanism for mercury in these long-lived, top predators. It could well be, therefore, that relatively high mercury levels in some species are natural and that some marine animals have become adapted, over an evolutionary time span, to dietary mercury. Whether this is the case for all marine vertebrates is open to speculation, but those species unlikely to encounter large amounts of mercury, via their respective diets, for example, would be especially susceptible to unusual, anthropogenic sources of mercury.

Although much work has been carried out on the levels of mercury and its relationships with other metals, there are still many aspects of its fluxes within the marine biosphere which remain unclear. Direct relationships between prey/predator mercury levels, mercury retention processes, mercury eliminatory pathways and their relative importance in different species, and meaningful data on harmful and deleterious effects of mercury in wild populations are a few of the many areas where further work should be done.

VII. ZINC

A. INTRODUCTION

Zinc can be classified as an essential element since it is a requirement of several metalloproteins, particularly metalloenzymes.

B. ZINC IN MARINE FISH

Zinc concentrations in marine fish tend to be fairly uniform, regardless of species and location. As with copper, this is to be expected of an essential element and levels are likely to be closely regulated. Generally, mean muscle concentrations of zinc are less than 10 μg/g wet weight,[10,15,34,36,49,71,74,75,83,112,118,132-135,137,151-154,168,176,179,187] although values in excess of this level have been reported, but some of these include "whole fish" samples.[3,7,49,82,135,137,186,188]

There are few data on the distribution of zinc within marine fish. It would appear, however, that levels are higher in liver and kidney relative to muscle.[74,118,133,135,151,168,179,186-188] Zinc shows little or no tendency to increase or decrease in concentration with increasing age/length of fish. Positive correlations between zinc and cadmium concentration in marine fish have been noted for black marlin and striped mullet by Mackay et al.[118] and Hornung and Ramelow,[91] respectively.

C. ZINC IN SEABIRDS

Zinc levels in liver and kidney tissue of some seabirds from northern and southern hemisphere studies are presented in Table 19. These data show general agreement between species with little variation which one would expect for an essential element. Comparison between studies is hindered by data presentation on either a wet weight basis or a dry weight basis. Where expressed on a dry weight basis, liver zinc levels tend to be below 200 μg/g,[2,32,35,138,145,157] although higher zinc concentrations have been noted by Hutton[94] in the

TABLE 19
Zinc Concentrations (μg/g Wet Weight) in Liver and Kidney tissue of Seabirds from the Antarctic (A), Gough Island (G), and Spitsbergen (S)

Species	Number sampled	Liver Mean	Liver Range	Kidney Mean	Kidney Range	Locality	Ref.
Adelie penguin *Pygoscelis adeliae*	10	48	32—73	49	30—71	A	90
Chinstrap penguin *Pygoscelis antarctica*	13	36	29—46	31	25—38	A	141
Rockhopper penguin *Eudyptes crestatus*	12	40	27—61	63	41—86	G	130
Macaroni penguin *Eudyptes chrysolophus*	9	41	31—62	46	34—84	A	141
Wandering albatross *Diomedea exulans*	2	53	49—57	48	45—52	G	130
Yellow-nosed albatross *Diomedea chlororhynchos*	9	48	29—59	35	31—42	G	130
Sooty albatross *Phoebetria fusca*	8	67	47—86	56	42—65	G	130
Southern fulmar *Fulmarus glacialoides*	6	42	36—54	47	39—69	A	141
Northern fulmar *Fulmarus glacialis*	10	73	50—95	50	32—96	S	141
Atlantic petrel *Pterodroma incerta*	13	45	33—64	62	50—71	G	130
Kerguelen petrel *Ptreodroma brevirostris*	14	44	29—81	45	35—54	G	130
Soft-plumaged petrel *Ptreodroma mollis*	18	43	30—56	50	36—78	G	130
Broad-billed prion *Pachyptila vittata*	31	44	30—75	36	29—47	G	130
Great shearwater *Puffinus gravis*	12	38	33—45	46	27—88	G	130
Little shearwater *Puffinus assimilis*	13	40	28—54	50	34—66	G	130
Gray-backed storm petrel *Garrodia nereis*	8	42	29—77	28	15—49	G	130
White-faced storm petrel *Pelagodroma marina*	7	34	20—44	39	30—46	G	130
White-bellied storm petrel *Fregetta grallaria*	8	38	28—46	38	35—48	G	130
Common diving petrel *Pelecanoides urinatrix*	17	38	28—51	46	33—78	G	130
Tristan skua *Catharacta skua hamiltoni*	13	22	18—32	37	28—53	G	130
Brown skua *Catharacta skua lonnbergi*	8	32	22—51	42	31—51	A	141
South polar skua *Catharacta maccormicki*	8	35	21—46	40	22—48	A	141
Glaucous gull *Larus hyperboreus*	11	32	26—47	46	37—57	S	141
Little auk *Alle alle*	9	37	31—43	40	32—46	S	141
Brunnich's guillemot *Uria lomvia*	9	35	31—38	39	27—50	S	141

TABLE 20
Zinc Concentrations (μg/g Wet [w] or Dry [d] Weight) in Liver (L), Kidney (K), and Muscle (M) Tissue of Pinnipeds

Species	No. sampled	Tissue	Mean	Range	Locality	Ref.
California sea lion	10	L	220.0	166.0—346.0[d]	California (females)	123
Zalophus californianus	10	K	173.0	146.0—353.0[d]		
Steller sea lion	17	L	47.0	35.5—86.0[w]	Hokkaido, Japan	80
Eumetopias jubata	17	K	27.0	19.8—40.4[w]		
	15	M	31.0	24.3—39.1[w]		
Harbor seal	57	L	—	27.0—56.0[w]	German North Sea coast	45
Phoca vitulina	16	K	—	16.3—32.5[w]		
Ringed seal	29	L	46.0	30.7—67.3[w]	West Greenland	98
Phoca hispida	29	K	46.2	27.9—78.0[w]		
	29	M	22.2	14.2—39.5[w]		
Gray seal	38	L	55.3	30.0—97.5[w]	Farne Islands, England,	27
Halichoerus grypus	37	K	28.4	10.5—75.0[w]	(females)	
Leopard seal	15	L	53.9	32.2—82.8[w]	Antarctic	180
Hydrurga leptonyx	15	M	23.4	14.8—49.3[w]		
Weddell seal	2	L	44.4	41.7—47.0[w]	Antarctic	189
Leptonychotes weddellii	2	K	29.1	27.4—30.7[w]		
	2	M	36.7	33.7—39.6[w]		
Ross seal	20	L	212.4	124.0—406.0[d]	Antarctic	117
Ommatophoca rossi						
Elephant seal	1	M	35.6[w]	—	Antarctic	43
Mirounga leonina						

great skua (range 61 to 497 μg/g dry weight) and by Osborn et al.[149] in the northern fulmar *Fulmarus glacialis* (range 225 to 688 μg/g dry weight). Liver zinc levels have been found to be generally less than 100 μg/g on a wet weight basis.[21,90,92,130,141]

Kidney zinc levels are comparable to those of liver tissue, although high kidney concentrations have been noted in association with high kidney cadmium concentrations, this positive correlation being statistically significant in several studies.[94,129,130,138,141,157] This relationship between zinc and cadmium is thought to involve zinc binding by metallothionein which, by binding cadmium, offers protection against cadmium toxicity.[147] Muscle zinc levels in seabirds tend to be less than corresponding liver and kidney levels.[2,32,90,92,94,138,149,156] Feather zinc levels would appear to be variable, relative to other tissues, although where determined, zinc concentrations in feathers are not inconsiderable.[90,92,94,145,149] Mean zinc levels in egg expressed on a dry weight basis in four seabird species are generally similar,[2,157] the highest mean level of 87 μg/g dry weight being found in eggs of Cory's shearwaters *Calonectris diomedea*.[157] King et al.[103] found egg zinc values of 8.8 to 14.0 μg/g wet weight in royal terns from the Texas coast while Blus et al.[13] found egg zinc levels of 4.3 to 8.3 μg/g wet weight in brown pelicans from South Carolina and Florida.

D. ZINC IN MARINE MAMMALS

1. Pinnipeds

Data on the levels of zinc in some seal tissues are presented in Table 20. As with the essential element copper, there is generally less variation both within and between species for zinc when compared to the nonessential elements. Species from geographically different localities exhibit similar zinc levels with concentrations tending to be less than 90 μg/g wet weight in liver tissue.[27,45,46,51,80,84,86,98,180,189] Highest zinc concentrations tend to be found in liver tissue, although this is not always the case. Some studies have revealed kidney zinc concentrations as high as, or in excess of, corresponding liver concentrations.[80,98,123] Whether

TABLE 21
Zinc Concentrations (μg/g Wet Weight) in Liver (L), Kidney (K), and Muscle (M) Tissue of Cetaceans

Species	No. sampled	Tissue	Mean	Range	Locality	Ref.
Beluga	1	L	32.0	—	Baltic Sea	84
Delphinapterus leucas	1	K	29.5	—		
	1	M	20.0	—		
Narwhal	37	L	38.8	24.0—63.6	Pond Inlet, Canada	182
Monodon monoceros	54	K	41.1	3.8—85.8		
	58	M	17.8	12.4—28.4		
White-beaked dolphin	1	L	24.0	—	Kolding Fjord,	4
Lagenorchynchus albirostris	1	M	13.0	—	Denmark	
Striped dolphin	57	L	44.5	26.5—109.0	East coast Japan	89
Stenella coeruleoalba	30	K	30.1	22.8—41.2		
	59	M	11.4	6.9—20.4		
Short-finned pilot whale	1	L	36.4	—	North Queensland	41
Globicephala macrorhynchus	1	K	38.1	—		
	1	M	15.9	—		
Harbor porpoise	17	L	43.2	18.4—67.6	East coast Scotland	50
Phocoena phocoena	17	K	23.8	19.5—33.1	(males)	
Goose-beaked whale	2	L	49.5	40.0—59.0	Bermuda	104
Ziphius cavirostris	2	K	53.0	52.0—54.0		
	4	M	11.5	—		
Bottlenose whale	1	L	23.0	—	North Sea	84
Hyperoodon ampullatus	1	M	13.5	—		
Bowhead whale	1	L	43.6	—	Alaska	26
Balaena mysticetus	2	K	23.8	19.0—28.5		
	2	M	43.7	36.0—51.4		

these high kidney zinc concentrations are in association with elevated cadmium concentrations incorporated in metallothioneins is not clear, there being little documented evidence of this relationship in this marine group. However, it would appear that high zinc levels correspond to high cadmium levels in kidney tissue in California sea lions.[123] The high zinc levels in liver tissue of Ross seals reported by McClurg[117] may be diet linked, no correlation with cadmium being reported. Zinc concentration has been shown to neither increase nor decrease significantly with age in two studies.[45,80]

2. Cetaceans

Data on zinc levels in whale tissues are presented in Table 21. The similarity in zinc concentrations reported from all studies is striking, zinc showing little variation between species and geographical location.[4,26,41,50,51,84,104,182] Honda et al.[89] noted a negative correlation between liver zinc concentration and age in striped dolphins, although after an initial decrease in concentration, immature and mature dolphins show little variation in zinc concentration with increasing age. No clear relationship between zinc concentration and age could be elucidated by Falconer et al.[50] with respect to harbor porpoises. A positive correlation between kidney zinc and cadmium levels was found by Honda et al.[89] in striped dolphins.

3. Sirenians and Other Groups

Liver zinc levels in dugongs from North Queensland have been shown to range from 58 to 1101 μg/g wet weight with kidney levels (14 to 54 μg/g wet weight) and muscle levels (8 to 28 μg/g wet weight) consistently lower.[42] Muscle zinc levels in dugongs from Sulawesi Island (15 to 30 μg/g wet weight)[128] are in agreement with the findings of Denton et al.[42] The high liver levels of zinc are not thought to reflect unusually high dietary levels, but may be a result of low dietary copper, zinc being absorbed via vacant copper receptor sites in the intestine.[42] A positive correlation between zinc concentration and age was found in liver and kidney tissue; this may explain the positive correlation between zinc and cadmium in these tissues, as the latter exhibits a similar age-related concentration increase.[42] Alternatively, age-related increases in zinc levels may be due to zinc binding by metallothioneins produced to store cadmium which accumulates with age.

The mean liver concentration of zinc in polar bears from the Canadian Arctic was found to be around 63 μg/g wet weight.[144]

E. CONCLUSIONS

As with copper, zinc, being an essential element, varies little in concentration between animals studied and there is general agreement regardless of species and geographical location. Maximum zinc concentrations reported for each group covered by this review are 375.0 μg/g wet weight in liver tissue of black marlin;[118] 688.0 μg/g dry weight in liver tissue of northern fulmar;[149] 406 μg/g dry weight in liver tissue of Ross seal;[117] 109 μg/g wet weight in liver tissue of striped dolphin.[89] Zinc shows little or no age-related accumulation effect, although it has been found to be positively correlated with cadmium levels in several studies.

VIII. OTHER HEAVY METALS

A. INTRODUCTION

Of the many heavy metals for which relatively few data are available for marine vertebrates, the biologically essential metals arsenic, chromium, manganese, and nickel[38] are considered in this section. Each of the above metals is dealt with separately.

B. ARSENIC

Compared to other marine vertebrates, marine fish tend to exhibit the highest arsenic concentrations. Muscle arsenic levels of over 100 μg/g dry weight have been reported by several workers from a range of marine fish.[14,15,160] In contrast, other studies have reported arsenic values which rarely exceed 20 μg/g either on a wet or a dry weight basis for both bony and cartilaginous species.[10,71,75,110,118,187] Bohn[14] noted a positive correlation between arsenic concentration and body weight in several species of marine fish, a relationship also reported by Bohn and Fallis[15] in the shorthorn sculpin *Myoxocephalus scorpius*. Both Bebbington et al.[10] and Bohn and Fallis[15] reported arsenic concentrations in fish which exceeded "standard levels" and "maximum recommended permissible" levels as defined by the National Health and Medical Research Council and the Canadian Food and Drug Directorate.

Arsenic levels in seabirds and marine mammals are generally low when compared to those in marine fish with concentrations rarely exceeding 1 μg/g wet weight in any tissue.[1,13,26,103,119,128,144,145,177]

C. CHROMIUM

Chromium levels in marine fish have been found to be generally less than 1 μg/g wet weight with concentrations often approaching the limits of detection.[71,74,75,132,133,135,152-154] Murray and Portmann[135] reported somewhat higher mean chromium values of 3.8 and 6.4 μg/g wet weight in plaice *Pleuronectes platessa* and herring *Clupea harengus*, respectively, both samples being taken from the Irish Sea. A similarly high chromium concentration of 5.82 μg/g dry weight has been reported by Hornung and Ramelow[91] in muscle tissue of *Saurida undosquamis* from the eastern Mediterranean Sea. Chromium levels in seabirds have been found rarely to exceed 5 μg/g either on a wet or dry weight basis in any tissue.[2,13,92,145] Somewhat higher chromium concentrations have been reported by Custer et al.[35] who noted levels of up to 18.3 μg/g dry weight in liver tissues of common tern chicks, this possibly being the result of local pollution. Anderlini et al.[2] noted similar chromium levels in ashy petrels *Oceanodroma homochroa* from California (mean 12.2 μg/g dry weight, liver tissue) and reported correspondingly high concentrations in bone tissue of seabirds from both the Antarctic and North America.

Chromium levels in marine mammals are generally low, reported values being less than 1 μg/g wet weight in any tissue.[26,27,41-43,46] McClurg[117] reported chromium levels of up to 3.7 μg/g dry weight in liver tissues of Ross seals from the Antarctic.

D. MANGANESE

Manganese concentrations in marine fish tend to be less than 1 μg/g wet weight, often approaching detection limits.[34,71,74,83,152,179] Eustace[49] reported a maximum manganese level of 15 μg/g wet weight in muscle tissue of a seahorse *Hippocampus sp.* from the Derwent estuary, Tasmania. Manganese levels in seabirds tend to be less than 5 μg/g wet weight in any tissue,[90,92,93] although Custer et al.[35] noted manganese concentrations of up to 29 μg/g dry weight in liver tissues of common tern chicks from a somewhat polluted environment.

Manganese concentrations in marine mammals have been found to be generally less than 7 μg/g wet weight in any tissue.[27,41-43,46,89,104,128,144,189] Higher manganese concentrations have been reported by Martin et al.[123] in California sea lions (maximum liver tissue concentration, 24.4 μg/g dry weight) and by McClurg[117] in Ross seals (maximum liver tissue concentration, 39 μg/g dry weight), although direct comparison between such data is hindered by presentation of results either on a wet or a dry weight basis.

E. NICKEL

Nickel concentrations in marine fish tend to be less than 1 μg/g wet weight,[70,71,74,75,83,152] although Vas[179] reported nickel levels of 2 μg/g wet weight in both muscle and kidney tissues of tope from Liverpool Bay, Irish Sea, and Wright[188] noted a mean nickel concentration of 10.8 μg/g wet weight in liver tissues of plaice from the Northumberland coast, northeast England.

Seabirds exhibit similar nickel levels to those of marine fish with concentrations rarely exceeding 1 μg/g even on a dry weight basis.[13,35,90] Elevated nickel levels have been reported in feathers of Adelie penguins from the Antarctic (maximum, 3.04 μg/g fresh weight) by Honda et al.[90] while Anderlini et al.[2] noted mean nickel levels of 17 μg/g dry weight in liver tissues and 16 μg/g dry weight in bone tissues of ashy petrels from California.

Nickel levels in marine mammals tend to be less than 0.5 μg/g wet weight with concentrations close to limits of detection,[26,41-43,89,104] although nickel concentrations of up to 1.25 μg/g wet weight have been reported in muscle tissues of dugong;[128] nickel levels of up to 4.8 μg/g dry weight in liver tissues of Ross seals from the Antarctic have been noted by McClurg.[117]

F. CONCLUSIONS

Based on the relatively few data for these heavy metals in marine vertebrates, it is difficult to make any assessment as to the effects that they may have on this group. Of the four metals considered, arsenic concentrations would appear to show the most variation with some extremely high levels being reported in some species of marine fish (see Section VIII.B). Such wide variation in metal levels is similar to that reported in the nonessential elements and may be worthy of further investigation.

IX. DISCUSSION

Of the metals covered in this review, tissue concentrations of the essential elements copper, iron, and zinc would appear to be closely regulated metabolically and these elements are, therefore, unlikely to be serious pollution threats in general terms. Localized anthropogenic sources of these metals, however, may result in deleterious effects to the marine environment.

Reported values of cadmium show a wide variation within a given study, in association with skewed distribution patterns,[130] and have been shown to increase in concentration with increasing age/size in several studies of all groups covered in this review. There is some evidence, however, that cadmium when compared to the distribution of, and variation in, mercury levels in seabirds, shows signs of being regulated to a small extent.[130]

Mercury concentrations are the most variable of any metal covered in this review and have been shown to exhibit a markedly skewed distribution pattern.[130] Age-related accumulation trends have been widely reported for mercury with metabolic regulation being minimal by comparison to the essential metals copper, iron, and zinc. The trend of a positive correlation between mercury concentration and age/size in marine vertebrates has invariably been reported in terms of total mercury. Given the probable existence of a demethylation process in several species of marine mammals and seabirds which effectively converts a proportion of the dietary organic mercury into the inorganic form, one might expect this correlation to be stronger if inorganic mercury alone was considered. Conversely, a correlation between age/size and organic mercury may prove to be less significant.

In contrast to the amount of work undertaken on cadmium and mercury, there are relatively few data on levels of the nonessential metal lead. Generally, lead concentrations tend to be higher in species inhabiting coastal and estuarine environments[4,24,93,94,131,155] when compared to lead levels in species far from any area of industrialization.[117] Furthermore, coastal seabird fatalities have been reported as being attributable to elevated lead levels.[24] Further detailed work on lead in marine vertebrates would seem worthwhile given its increasing industrial usage and evidence of global increases in lead levels resulting from atmospheric transport (see chapter by Wolff in this volume). Together with lead, cadmium and mercury would appear to represent the most serious pollution risks and warrant further study while further extensive data collection for the essential metals copper, iron, and zinc would seem to be of less value.

Despite a relatively large number of studies into the levels of cadmium and mercury in marine vertebrates, there have been few meaningful assessments of the physiological effects of these metals; for example, relatively few data exist on the toxic effects of cadmium and mercury with respect to reproductive processes, on what concentrations are likely to cause reduced breeding success, and how the respective effects of different metals combine. Geographical variation in metal concentrations in marine vertebrates has tended to be overlooked, both on local and global scales. However, such studies are hindered by the difficulties associated with sampling highly mobile marine species which have relatively extensive ranges. The limitations of comparing metal levels in different species with distinct geographical distributions should be taken into account, as should the potential bias of using dead seabirds and beached and stranded marine mammals for metal analysis.

REFERENCES

1. **Anas, R. E.,** Heavy metals in the northern fur seal *Callorhinus ursinus* and harbour seal *Phoca vitulina richardi, Fish. Bull.*, 72, 133, 1974.
2. **Anderlini, V. C., Connors, P. G., Risebrough, R. W., and Martin, J. H.,** Concentrations of heavy metals in some Antarctic and North American seabirds, in *Proc. Symp. Conservation Problems Antarctica*, Virginia Polytecnic Institute and State University, Blacksburg, VA, 1972, 49.
3. **Andersen, A. T., Dommasnes, A., and Hesthagen, I. H.,** Some heavy metals in sprat *(Sprattus sprattus)* and herring *(Clupea harengus)* from the Inner Oslofjord, *Aquaculture,* 2, 17, 1973.

4. **Andersen, S. H. and Rebsdorff, A.,** Polychlorinated hydrocarbons and heavy metals in harbour porpoise *(Phocoena phocoena)* and whitebeaked dolphin *(Lagenorhynchus albirostris)* from Danish waters, *Aquat. Mammals,* 4, 14, 1976.
5. **Applequist, H., Asbirk, S., and Drabaek, I.,** Mercury monitoring: mercury stability in bird feathers, *Mar. Pollut. Bull.,* 15, 22, 1984.
6. **Arima, S. and Nagakura, K.,** Mercury and selenium content of Odontoceti, *Bull. Jpn. Soc. Sci. Fish.,* 45, 623, 1979.
7. **Badsha, K. S. and Sainsbury, M.,** Uptake of zinc, lead and cadmium by young whiting in the Severn estuary, *Mar. Pollut. Bull.,* 8, 164, 1977.
8. **Bakir, F., Damluji, S. F., Amin-Zaki, L., Martadha, M., Khalidi, A., Al-Rawi, N. Y., Tikriti, Z., Dhahir, H. I., Clarkson, T. W., Smith, J. C., and Doherty, R. A.,** Methyl mercury poisoning in Iraq, *Science,* 181, 230, 1973.
9. **Barrett, R. T., Skaare, J. V., Norheim, G., Vader, W., and Froslie, A.,** Persistent organochlorines and mercury in eggs of Norwegian seabirds 1983, *Environ. Pollut. (A),* 39, 79, 1985.
10. **Bebbington, G. N., Mackay, N. J., Chvojka, R., Williams, R. J., Dunn, A., and Auty, E. H.,** Heavy metals, selenium and arsenic in nine species of Australian commercial fish, *Aust. J Mar. Freshwater Res.,* 28, 277, 1977.
11. **Beckett, J. S. and Freeman, H. C.,** Mercury in swordfish and other pelagic species from the western Atlantic, *Spec. Sci. Rep. Nat. Oceanic Atmos. Adm. U.S. (Fisheries),* 675, 154, 1974.
12. **Bligh, E. G. and Armstrong, F. A. J.,** Marine mercury pollution in Canada, *Int. Counc. Explor. Sea,* ICES C. M. 1971/E34, 1971.
13. **Blus, L. J., Neely, B. S., Lamont, T. G., and Mulhern, B.,** Residues of organochlorines and heavy metals in tissues and eggs of brown pelicans, 1969-73, *Pestic. Monit. J.,* 11, 40, 1977.
14. **Bohn, A.,** Arsenic in marine organisms from West Greenland, *Mar. Pollut. Bull.,* 6, 87, 1975.
15. **Bohn, A. and Fallis, B. W.,** Metal concentrations (As, Cd, Cu, Pb and Zn) in shorthorn sculpins *Myoxocephalus scorpius* (Linnaeus) and Arctic char *Salvelinus alpinus* (Linnaeus) from the vicinity of Strathcona Sound, Northwest Territories, *Water Res.,* 12, 659, 1978.
16. **Borg, K., Wanntorp, H., Erne, K., and Hanko, E.,** Mercury poisoning in Swedish wildlife, *J. Appl. Ecol.,* 3, 171, 1966.
17. **Born, E. W., Kraul, I., and Kristensen, T.,** Mercury, DDT and PCB in the Atlantic walrus *(Odobenus rosmarus rosmarus)* from the Thule District, North Greenland, *Arctic,* 34, 255, 1981.
18. **Botta, J. R., Arsenault, E., and Ryan, H. A.,** Total mercury content of meat and liver from inshore Newfoundland-caught harp seal *(Phoca groenlandica), Bull. Environ. Contam. Toxicol.,* 30, 28, 1983.
19. **Braham, H. W.,** Lead in the California sea lion *(Zalophus californianus), Environ. Pollut.,* 5, 253, 1973.
20. **Braune, B. M.,** Comparison of total mercury levels in relation to diet and moult for nine species of marine birds, *Arch. Environ. Contam. Toxicol.,* 16, 217, 1987.
21. **Brothers, N. P. and Brown, M. J.,** The potential use of fairy prions *(Pachyptila turtur)* as monitors of heavy metal levels in Tasmanian waters, *Mar. Pollut. Bull.,* 18, 132, 1987.
22. **Bryan, G. W.,** Pollution due to heavy metals and their compounds, in, *Marine Ecology,* Vol. 5, Kinne, O., Ed., John Wiley & Sons, New York, 1984, chap. 3.
23. **Buhler, D. R., Claeys, R. R., and Mate, B. R.,** Heavy metals and chlorinated hydrocarbon residues in California sea lions *(Zalophus californianus californianus), J. Fish. Res. Board Can.,* 32, 2391, 1975.
24. **Bull, K. R., Every, W. J., Freestone, P., Hall, J. R., Osborn, D., Cooke, A. S., and Stone, T.,** Alkyl lead pollution and bird mortalities on the Mersey estuary, UK, 1979-1981, *Environ. Pollut. (A),* 31, 239, 1983.
25. **Bull, K. R., Murton, R. K., Osborn, D., Ward, P., and Cheng, L.,** High levels of cadmium in Atlantic seabirds and sea-skaters, *Nature,* 269, 507, 1977.
26. **Byrne, C., Balasubramanian, R., Overton, E. B., and Albert, T. F.,** Concentrations of trace metals in the bowhead whale, *Mar. Pollu. Bull.,* 16, 497, 1985.
27. **Caines, L. A.,** Heavy metal residues in grey seals *(Halichoerus grypus)* from the Farne Islands, *Int. Counc. Explor. Sea, ICES,* C. M. 1978/E: 40, 1978.
28. **Cappon, C. J. and Smith, J. C.,** Mercury and selenium content and chemical form in fish muscle, *Arch. Environ. Contam. Toxicol.,* 10, 305, 1981.
29. **Caputi, N., Edmonds, J. S., and Heald, D. I.,** Mercury content of sharks from south-western Australian waters, *Mar. Pollut. Bull.,* 10, 337, 1979.
30. **Cheng, L., Schulz-Baldes, M., and Harrison, C. S.,** Cadmium in ocean-skaters *Halobates sericeus* (Insecta) and their seabird predators, *Mar. Biol.,* 79, 321, 1984.
31. **Chvojka, R. and Williams, R. J.,** Mercury levels in six species of Australian commercial fish, *Aust. J. Mar. Freshwater Res.,* 31, 469, 1980.
32. **Connors, P. G., Anderlini, V. C., Risebrough, R. W., Gilbertson, M., and Hays, H.,** Investigations of heavy metals in common tern populations, *Can. Field Nat.,* 89, 157, 1975.

33. **Crewther, W. G., Fraser, R. D. B., Lennox, F. G., and Lindley, H.,** The chemistry of keratins, *Adv. Protein Chem.*, 20, 191, 1965.
34. **Cross, F. A., Hardy, L. H., Jones, N. Y., and Barber, R. T.,** Relation between total body weight and concentrations of manganese, iron, copper, zinc and mercury in white muscle of bluefish *(Pomatomus saltatrix)* and a bathyl-demersal fish *Antimora rostrata, J. Fish. Res. Board Can.*, 30, 1287, 1973.
35. **Custer, T. W., Franson, J. C., Moore, J. F., and Myers, J. E.,** Reproductive success and heavy metal contamination in Rhode Island common terns, *Environ. Pollut. (A)*, 41, 33, 1986.
36. **Cutshall, N. H., Naidu, J. R. and Pearcy, W. G.,** Zinc and cadmium in the Pacific hake *Merluccius productus* off the western U.S. coast, *Mar. Biol.*, 44, 195, 1977.
37. **Cuvin, M. L. A. and Furness, R. W.,** Uptake and elimination of inorganic mercury and selenium by minnows *Phoxinus phoxinus, Aquat. Toxicol.*, in press.
38. **Da Silva, J. J. R. F.,** Interaction of the chemical elements with biological systems, in *New Trends in Bioinorganic Chemistry,* Williams, R. J. P. and Da Silva, J. J. R. F., Eds., Academic Press, London, 1978, 449.
39. **Delbeke, K., Joiris, C., and Decadt, G.,** Mercury contamination of the Belgian avifauna 1970—1981, *Environ. Pollut. (B)*, 7, 205, 1984.
40. **Denton, G. R. W. and Breck, W. G.,** Mercury in tropical marine organisms from north Queensland, *Mar. Pollut. Bull.*, 12, 116, 1981.
41. **Denton, G. R. W. and Heinsohn, G. E.,** unpublished data, cited in Reference 42.
42. **Denton, G. W. R., Marsh, H., Heinsohn, G. E., and Burdon-Jones, C.,** The unusual metal status of the dugong *Dugong dugon, Mar. Biol.*, 57, 201, 1980.
43. **Denton, G. W. R., Marsh, H., and Griffiths, D.,** unpublished data, cited in Reference 42.
44. **Doi, R., Ohno, H., and Harada, M.,** Mercury in feathers of wild birds from the mercury-polluted area along the shore of the Shiranui Sea, Japan, *Sci. Total Environ.*, 40, 155, 1984.
45. **Drescher, H. E., Harms, U. and Huschenbeth, E.,** Organochlorines and heavy metals in the harbour seal *Phoca vitulina* from the German North Sea coast, *Mar. Biol.*, 41, 99, 1977.
46. **Duinker, J. C., Hillebrand, M. Th. J., and Nolting, R. F.,** Organochlorines and metals in harbour seals (Dutch Wadden Sea), *Mar. Pollut. Bull.*, 10, 360, 1979.
47. **Dyck, J. and Kraul, I.,** Environmental pollutants and shell thinning in eggs of the guillemot *Uria aalge* from the Baltic Sea and the Faeroes, and a possible relation between shell thickness and sea water salinity, *Dan. Ornithol. Foren. Tidsskr.*, 78, 1, 1984.
48. **Essink, K.,** Monitoring of mercury pollution in Dutch coastal waters by means of the teleostean fish *Zoarces viviparus, Neth. J. Sea Res.*, 19, 177, 1985.
49. **Eustace, I. J.,** Zinc, cadmium, copper and manganese in species of finfish and shellfish caught in the Derwent estuary, Tasmania, *Aust. J. Mar. Freshwater Res.*, 25, 209, 1974.
50. **Falconer, C. R., Davies, I. M., and Topping, G.,** Trace metals in the common porpoise *Phocoena phocoena, Mar. Environ. Res.*, 8, 119, 1983.
51. **Fallis, B.,** unpublished data, cited in Reference 181.
52. **Fimreite, N.,** Mercury contamination of aquatic birds in northwestern Ontario, *J. Wildl. Manage.*, 38, 120, 1974.
53. **Fimreite, N., Brun, E., Froslie, A., Frederichsen, P., and Gundersen, N.,** Mercury in eggs of Norwegian seabirds, *Astarte*, 7, 71, 1974.
54. **Fimreite, N., Kveseth, N., and Brevik, E. M.,** Mercury, DDE and PCBs in eggs from a Norwegian gannet colony, *Bull. Environ. Contam. Toxicol.*, 24, 142, 1980.
55. **Forrester, C. R., Ketchen, K. S., and Wong, C. C.,** Mercury content of spiny dogfish *(Squalus acanthias)* in the Strait of Georgia, British Columbia, *J. Fish. Res. Board Can.*, 29, 1487, 1972.
56. **Freeman, H. C. and Horne, D. A.,** Mercury in Canadian seals, *Bull. Environ. Contam. Toxicol.*, 10, 172, 1973.
57. **Freeman, H. C., Horne, D. A., McTague, B., and McMenemy, M.,** Mercury in some Canadian Atlantic coast fish and shellfish, *J. Fish. Res. Board Can.*, 31, 369, 1974.
58. **Freeman, H. C., Shum, G., and Uthe, J. F.,** The selenium content in swordfish *(Xiphias gladius)* in relation to total mercury content, *J. Environ. Sci. Health,* A13, 235, 1978.
59. **Friberg, L., Piscator, M., Nordberg, G., and Kjellstrom, T.,** *Cadmium in the Environment*, 2nd. ed., CRC Press, Boca Raton, FL, 1974, chap. 8.
60. **Fujiki, M., Tajima, S., and Omori, A.,** The transition of the mercury contamination in Minamata District, *Jpn. J. Hyg.*, 27, 115, 1972.
61. **Furness, R. W. and Hutton, M.,** Pollutant levels in the great skua *Catharacta skua, Environ. Pollut.*, 19, 261, 1979.
62. **Furness, R. W. and Hutton, M.,** Pollutants and impaired breeding success of great skuas *Catharacta skua* in Britain, *Ibis,* 122, 88, 1980.
63. **Furness, R. W., Muirhead, S. J., and Woodburn, M.,** Using bird feathers to measure mercury in the environment: relationships between mercury content and moult, *Mar. Pollut. Bull.*, 17, 27, 1986.

64. **Ganther, H. E., Goudie, C., Sunde, M. L., Kopecky, M. J., Wagner, P., Oh, S-H., and Hoekstra, W. G.,** Selenium: relation to decreased toxicity of methylmercury added to diets containing tuna, *Science,* 175, 1122, 1972.
65. **Ganther, H. E. and Sunde, M. L.,** Effect of tuna fish and selenium on the toxicity of methylmercury: a progress report, *J. Food Sci.,* 39, 1, 1974.
66. **Gaskin, D. E., Frank, R., Koldrinet, M., Ishida, K., Walton, C. J., and Smith, M.,** Mercury, DDT and PCB in harbour seals *(Phoca vitulina)* from the Bay of Fundy and Gulf of Maine, *J. Fish. Res. Board Can.,* 30, 471, 1973.
67. **Gaskin, D. E., Ishida, K., and Frank, R.,** Mercury in harbour porpoises *(Phocoena phocoena)* from the Bay of Fundy region, *J. Fish. Res. Board Can.,* 29, 1644, 1972.
68. **Gaskin, D. E., Smith, G. J. D., Arnold, P. W., Louisy, M. V., Frank, R., Holdrinet, M., and McWade, J. W.,** Mercury, DDT, dieldrin and PCB in two species of Odontoceti (Cetacea) from St. Lucia, Lesser Antilles, *J. Fish. Res. Board Can.,* 31, 1235, 1974.
69. **Gaskin, D. E., Stonefield, K. I., Suda, P., and Frank, R.,** Changes in mercury levels in harbour porpoises from the Bay of Fundy, Canada, and adjacent waters during 1969—1977, *Arch. Environ. Contam. Toxicol.,* 8, 733, 1979.
70. **Gilmartin, M. and Revelante, N.,** The concentration of mercury, copper, nickel, silver, cadmium and lead in the northern Adriatic anchovy *Engraulis encrasicholus* and sardine *Sardina pilchardus, Fish. Bull.,* 73, 193, 1975.
71. **Glover, J. W.,** Concentrations of arsenic, selenium and ten heavy metals in school shark *Galeorhinus australis* (Macleay) and gummy shark *Mustelus antarcticus* Gunther from south-eastern Australian waters, *Aust. J. Mar. Freshwater Res.,* 30, 505, 1979.
72. **Gochfeld, M.,** Mercury levels in some seabirds of the Humboldt Current, Peru, *Environ. Pollut. (A),* 22, 197, 1980.
73. **Greig, R. A. and Krzynonek, J.,** Mercury concentrations in three species of tunas collected from various oceanic waters, *Bull. Environ. Contam. Toxicol.,* 22, 120, 1979.
74. **Greig, R. A. and Wenzloff, D. R.,** Trace metals in finfish from the New York Bight and Long Island Sound, *Mar. Pollut. Bull.,* 8, 198, 1977.
75. **Greig, R. A., Wenzloff, D. R., and Peare, J. B.,** Distribution and abundance of heavy metals in finfish, invertebrates and sediments collected at a deepwater disposal site, *Mar. Pollut. Bull.,* 7, 185, 1976.
76. **Greig, R. A., Wenzloff, D., and Shelpuk, C.,** Mercury concentrations in fish, north Atlantic offshore waters—1971, *Pestic. Monit. J.,* 9, 15, 1975.
77. **Greig, R. A., Wenzloff, D., Shelpuk, C., and Adams, A.,** Mercury concentrations in three species of fish from north Atlantic offshore waters, *Arch. Environ, Contam. Toxicol.,* 5, 315, 1977.
78. **Hall, A. S., Teeny, F. M. and Gauglitz, E. J.,** Mercury in fish and shellfish of the northeast Pacific. II. Sablefish *Anoplopoma fimbria, Fish. Bull.,* 74, 791, 1976.
79. **Hall, A. S., Teeny, F. M., Lewis, L. G., Hardman, W. H., and Gauglitz, E. J.,** Mercury in fish and shellfish of the northeast Pacific. I. Pacific halibut *Hippoglossus stenolepis, Fish. Bull.,* 74, 783, 1976.
80. **Hamanaka, T., Itoo, T., and Mishima, S.,** Age-related change and distribution of cadmium and zinc concentrations in the Steller sea lion *(Eumetopias jubata)* from the coast of Hokkaido, Japan, *Mar. Pollut. Bull.,* 13, 57, 1982.
81. **Hamanaka, T., Kato, H., and Tgujita, T.,** Cadmium and zinc in ribbon seal *Histriophoca fasciata* in the Okhotsk Sea, *Res. Inst. Natl. Pac. Fish. Spec. Vol.,* 547, 1977.
82. **Hardisty, M. W., Kartar, S., and Sainsbury, M.,** Dietary habits and heavy metal concentrations in fish from the Severn estuary and Bristol Channel, *Mar. Pollut. Bull.,* 5, 61, 1974.
83. **Harms, U.,** The levels of heavy metals (Mn, Fe, Co, Ni, Cu, Zn, Cd, Pb, Hg) in fish from onshore and offshore waters of the German Bight, *Z. Lebensm. Unters. Forsch.,* 157, 125, 1975.
84. **Harms, U., Drescher, H. E., and Huschenbeth, E.,** Further data on heavy metals and organochlorines in marine mammals from German coastal waters, *Meeresforsch,* 26, 153, 1978.
85. **Havre, G. N., Underdal, B., and Christiansen, C.,** Cadmium concentrations in some fish species from a coastal area in southern Norway, *Oikos,* 24, 155, 1973.
86. **Holden, A. V.,** The accumulation of oceanic contaminants in marine mammals. *Rapp. P. V. Reun. Cons. Int. Explor. Mer,* 169, 353, 1975.
87. **Holt, G., Froslie, A., and Norheim, G.,** Mercury, DDE and PCB in the avian fauna in Norway 1965—1976, *Acta Vet. Scand. Suppl.,* 70, 1, 1979.
88. **Honda, K., Nasu, T., and Tatsukawa, R.,** Seasonal changes in mercury accumulation in the black-eared kite *Milvus migrans lineatus, Environ. Pollut. (A),* 42, 325, 1986.
89. **Honda, K., Tatsukawa, R., Itano, K., Miyazaki, N. and Fujiyama, T.,** Heavy metal concentrations in muscle, liver and kidney tissue of striped dolphin *Stenella coeruleoalba* and their variations with body length, weight, age and sex, *Agric. Biol. Chem.,* 47, 1219, 1983.

90. **Honda, K., Yamamoto, Y., Hidaka, H., and Tatsukawa, R.,** Heavy metal accumulations in Adelie penguins *Pygoscelis adeliae* and their variations with the reproductive process, *Mem. Natl. Inst. Polar Res. Spec. Issue,* 40, 443, 1986.
91. **Hornung, H. and Ramelow, G. J.,** Distribution of Cd, Cr, Cu, and Zn in eastern Mediterranean fishes, *Mar. Pollut. Bull.,* 18, 45, 1987.
92. **Howarth, D. M., Hulbert, A. J., and Horning, D.,** A comparative study of heavy metal accumulation in tissues of the crested tern *Sterna bergii* breeding near industrialised and non-industrialised areas, *Aust. Wildl. Res.,* 8, 665, 1981.
93. **Hulse, M., Mahoney, J. S., Schroder, G. D., Hacker, C. S., and Pier, S. M.,** Environmentally acquired lead, cadmium and manganese in the cattle egret *Bubulcus ibis* and the laughing gull *Larus atricilla, Arch. Environ. Contam. Toxicol.,* 9, 65, 1980.
94. **Hutton, M.,** Accumulation of heavy metals and selenium in three seabird species from the United Kingdom, *Environ. Pollut. (A),* 26, 129, 1981.
95. **Hutton, M. and Goodman, G. T.,** Metal contamination of feral pigeons *Columba livia* from the London area: part 1 — tissue accumulation of lead, cadmium and zinc, *Environ. Pollut. (A),* 22, 207, 1980.
96. **Itano, K., Kawai, S., Miyazaki, N., Tatsukawa, R., and Fujiyama, T.,** Mercury and selenium levels in striped dolphins caught off the Pacific coast of Japan, *Agric. Biol. Chem.,* 48, 1109, 1984.
97. **Jensen, S., Johnels, A. G., Olsson, M., and Westermark, T.,** The avifauna of Sweden as indicators of environmental contamination with mercury and chlorinated hydrocarbons, *Proc. 15th Int. Ornithol. Congr.,* Leiden, 1972, 455.
98. **Johansen, P., Kapel, F. O. and Kraul, I.** Heavy metals and organochlorines in marine mammals from Greenland. *Int. Counc. Explor. Sea ICES C. M.* 1980/E: 32, 1980.
99. **Johnels, A. G. and Westermark, T.,** Mercury contamination of the environment in Sweden, in *Chemical Fallout. Current Research on Persistent Pesticides,* Miller, M. W. and Berg, G. G., Eds., Charles C Thomas, Springfield, IL, 1969, 221.
100. **Jones, D., Ronald, K., Lavigne, D. M., Frank, R., Holdrinet, M., and Uthe, J. F.,** Organochlorine and mercury residues in the harp seal *(Pagophilus groenlandicus), Sci. Total Environ.,* 5, 181, 1976.
101. **Kai, N., Ueda, T., Takeda, M., and Kataoka, A.,** On mercury and selenium in tuna fish tissues. VIII. The levels of mercury and selenium in albacore from the Indian Ocean, *J. Shimonoseki Univ. Fisheries,* 31, 69, 1983.
102. **Kari, T. and Kauranen, P.,** Mercury and selenium contents of seals from fresh and brackish waters in Finland, *Bull. Environ. Contam. Toxicol.,* 19, 273, 1978.
103. **King, K. A., Lefever, C. A., and Mulhern, B. M.,** Organochlorine and metal residues in royal terns nesting on the central Texas coast, *J. Field Ornithol.,* 54, 295, 1983.
104. **Knap, A. H. and Jickells, T. D.,** Trace metals and organochlorines in the goosebeaked whale, *Mar. Pollut. Bull.,* 14, 271, 1983.
105. **Knauer, G. A. and Martin, J. H.,** Mercury in a marine pelagic food chain, *Limnol. Oceanogr.,* 17, 868, 1972.
106. **Koeman, J. H., Peeters, W. H. M., Koudstaal-Hol, C. H. M., Tjioe, P. S., and De Goeij, J. J. M.,** Mercury-selenium correlations in marine mammals, *Nature,* 245, 385, 1973.
107. **Koeman, J. H., Van de Ven, W. S. M., de Goeij, J. J. M., Tjioe, P. S., and Van Haaften, J. L.,** Mercury and selenium in marine mammals and birds, *Sci. Total Environ.,* 3, 279, 1975.
108. **Kureishy, T. W., George, M. D., and Gupta, R. S.,** Total mercury content in some marine fish from the Indian Ocean, *Mar. Pollut. Bull.,* 10, 357, 1979.
109. **Lantzy, R. J. and Mackenzie, F. T.,** Atmospheric trace metals: global cycles and assessment of man's impact, *Geochim. Cosmochim. Acta,* 43, 511, 1979.
110. **LeBlanc, P. J. and Jackson, A. L.,** Arsenic in marine fish and invertebrates, *Mar. Pollut. Bull.,* 4, 88, 1973.
111. **Lee, S. S., Mate, B. R., Von der Trenck, K. T., Rimerman, R. A., and Buhler, D. R.,** Metallothionein and the subcellular localization of mercury and cadmium in the California sea lion, *Comp. Biochem. Physiol.,* 57C, 45, 1977.
112. **Leonzio, C., Bacci, E., Focardi, S., and Renzoni, A.,** Heavy metals in organisms from the northern Tyrrhenian Sea, *Sci. Total Environ.,* 20, 131, 1981.
113. **Leonzio, C., Focardi, S., and Bacci, E.,** Complementary accumulation of selenium and mercury in fish muscle, *Sci. Total Environ.,* 24, 249, 1982.
114. **Lutz and Armstrong, F. A. J.,** unpublished data, cited in Reference 181.
115. **Lyle, J. M.,** Mercury concentrations in four Carcharhinid and three hammerhead sharks from coastal waters of the Northern Territory, *Aust. J. Mar. Freshwater Res.,* 35, 441, 1984.
116. **Lyle, J. M.,** Mercury and selenium concentrations in sharks from northern Australian waters, *Aust. J. Mar. Freshwater Res.,* 37, 309, 1986.
117. **McClurg, T. P.,** Trace metals and chlorinated hydrocarbons in Ross seals from Antarctica, *Mar. Pollut. Bull.,* 15, 384, 1984.

118. **Mackay, N. J., Kazacos, M. N., Williams, R. J., and Leedow, M. I.,** Selenium and heavy metals in black marlin, *Mar. Pollut. Bull,* 6, 57, 1975.
119. **McKie, J. C., Davies, I. M., and Topping, G.,** Heavy metals in grey seals *(Halichoerus grypus)* from the east coast of Scotland, *Int. Counc. Explor. Sea, ICES,* 1980/E: 41, 1980.
120. **Maedgen, J. L., Hacker, C. S., Schroder, G. D., and Weir, F. W.,** Bioaccumulation of lead and cadmium in the royal tern and sandwich tern, *Arch. Environ. Contam. Toxicol.,* 11, 99, 1982.
121. **Magos, L. and Webb, M.,** The interaction of selenium with cadmium and mercury, *CRC Crit. Rev. Toxicol.,* 8, 1, 1980.
122. **Maher, W. A.,** Selenium in marine organisms from St. Vincent's Gulf, South Australia, *Mar. Pollut. Bull.,* 14, 35, 1983.
123. **Martin, J. H., Elliott, P. D., Anderlini, V. C., Girvin, D., Jacobs, S. A., Risebrough, R. W., Delong, R. L., and Gilmartin, G. W.,** Mercury-selenium-bromine imbalance in premature parturient California sea lions, *Mar. Biol.,* 35, 91, 1976.
124. **Martin, J. H. and Flegal, A. R.,** High copper concentrations in squid livers in association with elevated levels of silver, cadmium and zinc, *Mar. Biol.,* 30, 51, 1975.
125. **Martoja, R. and Viale, D.,** Accumulation de granules de seleniure mercurique dans le foie d'Odontocetes (Mammiferes, Cetaces): un mecanisme possible de detoxication du methyl-mercure par le selenium, *C.R. Acad. Sci. Ser. D,* 285, 109, 1977.
126. **Mason, C. F. and Reynolds, P.,** Organochlorine residues and metals in otters from the Orkney Islands, *Mar. Pollut. Bull.,* 19, 80, 1988.
127. **Menasveta, P. and Siriyong, R.,** Mercury content of several predacious fish in the Andaman Sea, *Mar. Pollut. Bull.,* 8, 200, 1977.
128. **Miyazaki, N., Itano, K., Fukushima, M., Kawai, S., and Honda, K.,** Metals and organochlorine compounds in the muscle of dugong from Sulawesi Island, *Sci. Rep. Whales Res. Inst.,* 31, 125, 1979.
129. **Muirhead, S. J.,** The Accumulation, Storage and Elimination of Metals and Organochlorines in the Great Skua *Catharacta skua* and Metal Accumulation in Atlantic Procellariiformes, Ph.D. thesis, University of Glasgow, 1986.
130. **Muirhead, S. J. and Furness, R. W.,** Heavy metal concentrations in the tissues of seabirds from Gough Island, South Atlantic Ocean, *Mar. Pollut. Bull.,* in press.
131. **Munoz, R. V., Hacker, C. S. and Gesell, T. F.,** Environmentally acquired lead in the laughing gull *Larus atricilla, J. Wildl. Dis.,* 12, 139, 1976.
132. **Murray, A. J.,** Metals, organochlorine pesticides and PCB residue levels in fish and shellfish landed in England and Wales during 1974, *Aquat. Environ. Monit. Rep., MAFF Direct. Fish. Res.,* Lowestoft, 2, 1979.
133. **Murray, A. J.,** Metals, organochlorine pesticides and PCB residue levels in fish and shellfish landed in England and Wales during 1975, *Aquat. Environ. Monit. Rep., MAFF Direct. Fish. Res.,* Lowestoft, 5, 1981.
134. **Murray, A. J. and Norton, M. G.,** The field assessment of effects of dumping wastes at sea. 10: analysis of chemical residues in fish and shellfish from selected coastal regions around England and Wales, *Fish. Res. Tech. Rep., MAFF Direct. Fish. Res.,* Lowestoft 69, 1982.
135. **Murray, A. J. and Portmann, J. E.,** Metals and organochlorine pesticide and PCB residues in fish and shellfish in England and Wales in 1976 and trends since 1970, *Aquat. Environ. Monit. Rep., MAFF Direct. Fish. Res.,* Lowestoft, 10, 1984.
136. **Nagakura, K., Arima, S., Kurihara, M., Koga, T., and Fujita, T.,** Mercury content of whales, *Bull. Tokai Reg. Fish. Res. Lab.,* 78, 41, 1974.
137. National Marine Fisheries Service, 1975 microconstituent resource survey, unpublished data, cited in, *Medical and Biological Effects of Environmental Pollutants: Zinc,* National Research Council, Subcommittee on Zinc, University Park Press, Baltimore, 1979, 88.
138. **Nicholson, J. K.,** The comparative distribution of zinc, cadmium and mercury in selected tissues of the herring gull *(Larus argentatus), Comp. Biochem. Physiol.,* 68C, 91, 1981.
139. **Nicholson, J. K. and Osborn, D.,** Kidney lesions in pelagic seabirds with high tissue levels of cadmium and mercury, *J. Zool. London,* 200, 99, 1983.
140. **Noble, D. G. and Elliott, J. E.,** Environmental contaminants in Canadian seabirds, 1968-1984: trends and effects, *Technical report series No. 13,* Canadian Wildlife Service, Ottawa, 1986.
141. **Norheim, G.,** Levels and interactions of heavy metals in seabirds from Svalbard and the Antarctic, *Environ. Pollut.,* 47, 83, 1987.
142. **Norheim, G. and Kjos-Hanssen, B.,** Persistent chlorinated hydrocarbons and mercury in birds caught off the west coast of Spitsbergen, *Environ. Pollut. (A),* 33, 143, 1984.
143. **Norheim, G., Somme, L., and Holt, G.,** Mercury and persistent chlorinated hydrocarbons in Antarctic birds from Bouvetoya and Dronning Maud Land, *Environ. Pollut. (A),* 28, 233, 1982.
144. **Norstrom, R. J., Schweinsberg, R. E., and Collins, B. T.,** Heavy metals and essential elements in livers of the polar bear *(Ursus maritimus)* in the Canadian Arctic, *Sci. Total Environ.,* 48, 195, 1986.

145. **Ohlendorf, H. M., Anderson, D. W., Boellstorff, D. E., and Mulhern, B. M.,** Tissue distribution of trace elements and DDE in brown pelicans, *Bull. Environ. Contam. Toxicol.*, 35, 183, 1985.
146. **Ohlendorf, H. M. and Harrison, C. S.,** Mercury, selenium, cadmium and organochlorines in eggs of three Hawaiian seabird species, *Environ. Pollut. (B)*, 11, 169, 1986.
147. **Osborn, D.,** A naturally occurring cadmium and zinc binding protein from the liver and kidney of *Fulmarus glacialis,* a pelagic North Atlantic seabird, *Biochem. Pharmacol.*, 27, 822, 1978.
148. **Osborn, D., Every, W. J., and Bull., K. R.,** The toxicity of trialkyl lead compounds to birds, *Environ. Pollut. (A),* 31, 261, 1983.
149. **Osborn, D., Harris, M. P. and Nicholson, J. K.,** Comparative tissue distribution of mercury, cadmium and zinc in three species of pelagic seabirds, *Comp. Biochem. Physiol,* 64C, 61, 1979.
150. **Parizek, J. and Ostadalova, I.,** The protective effect of small amounts of selenite in sublimate intoxication, *Experientia,* 23, 142, 1967.
151. **Perttila, M., Tervo, V., and Parmanne, R.,** Heavy metals in Baltic herring and cod, *Mar. Pollut. Bull.*, 13, 391, 1982.
152. **Plaskett, D. and Potter, I. C.,** Heavy metal concentrations in the muscle tissue of 12 species of teleost from Cockburn Sound, Western Australia, *Aust. J. Mar. Freshwater Res.*, 30, 607, 1979.
153. **Portmann, J. E.,** The levels of certain metals in fish from coastal waters around England and Wales, *Aquaculture,* 1, 91, 1972.
154. **Portmann, J. E.,** Chemical monitoring of residue levels in fish and shellfish landed in England and Wales during 1970—73, *Aquat. Environ. Monit. Rep., MAFF Direct. Fish. Res.*, Lowestoft, 1, 1979.
155. **Reid, M. and Hacker, C. S.,** Spatial and temporal variation in lead and cadmium in the laughing gull *Larus atricilla, Mar. Pollut. Bull.*, 13, 387, 1982.
156. **Reijnders, P. J. H.,** Organochlorine and heavy metal residues in harbour seals from the Wadden Sea and their possible effects on reproduction, *Neth. J. Sea Res.*, 14, 30, 1980.
157. **Renzoni, A., Focardi, S., Fossi, C., Leonzio, C., and Mayol, J.,** Comparison between concentrations of mercury and other contaminants in eggs and tissues of Cory's shearwater *Calonectris diomedea* collected on Atlantic and Mediterranean islands, *Environ. Pollut. (A)*, 40, 17, 1986.
158. **Rivers, J. B., Pearson, J. E., and Shultz, C. D.,** Total and organic mercury in marine fish, *Bull. Environ. Contam. Toxicol.*, 8, 257, 1972.
159. **Roberts, T. M., Heppleston, P. B., and Roberts, R. D.,** Distribution of heavy metals in tissues of the common seal, *Mar. Pollut. Bull.*, 7, 194, 1976.
160. **Robertson, D. E., Rancitelli, L. A., Langford, J. C., and Perkins, R. W.,** Battelle-Northwest contribution to the IDOE Baseline Study, in *Baseline Studies of Pollutants in the Marine Environment,* NSF-IDOE Workshop, Brookhaven, New York, 1972, 209.
161. **Ronald, K., Frank, R. J., and Dougan, J.,** Pollutants in harp seals *(Phoca groenlandica).* II. Heavy metals and selenium, *Sci. Total Environ.*, 38, 153, 1984.
162. **Sergeant, D. E. and Armstrong, F. A. J.,** Mercury in seals from Eastern Canada, *J. Fish. Res. Board Can.*, 30, 843, 1973.
163. **Shomura, R. S. and Craig, W. L.,** Mercury in several species of billfishes taken off Hawaii and southern California, *Spec. Sci. Rep. Natl. Oceanic Atmos. Adm. U.S. (Fisheries),* 675, 160, 1974.
164. **Shultz, C. D. and Ito, B. M.,** Mercury and selenium in blue marlin *Makaira nigricans* from the Hawaiian islands, *Fish. Bull.*, 76, 872, 1979.
165. **Smith, T. G. and Armstrong, F. A. J.,** Mercury and selenium in ringed and bearded seal tissues from Arctic Canada, *Arctic,* 31, 75, 1978.
166. **Sorentino, C.,** Mercury in marine and freshwater fish of Papua New Guinea, *Aust. J. Mar. Freshwater Res.*, 30, 617, 1979.
167. **Steinhagen-Schneider, G.,** Cadmium and copper levels in seals, penguins and skuas from the Weddell Sea in 1982/83, *Polar Biol.*, 5, 139, 1986.
168. **Stevens, J. D. and Brown, B. E.,** Occurrence of heavy metals in the blue shark *Prionace glauca* and selected pelagic in the N.E. Atlantic Ocean, *Mar. Biol.*, 26, 287, 1974.
169. **Stoneburner, D. L.,** Heavy metals in tissues of stranded short-finned pilot whales, *Sci. Total Environ.*, 9, 293, 1978.
170. **Stoneburner, D. L. and Harrison, C. S.,** Heavy metal residues in sooty tern tissues from the Gulf of Mexico and north central Pacific Ocean, *Sci. Tot. Environ.*, 17, 51, 1981.
171. **Stoneburner, D. L., Patty, P. C., and Robertson, W. B.,** Evidence of heavy metal accumulations in sooty terns, *Sci. Total Environ.*, 14, 147, 1980.
172. **Sumino, K., Yamamoto, R., and Kitamura, S.,** The role of selenium against methylmercury toxicity, *Nature,* 268, 73, 1977.
173. **Suzuki, T., Miyama, T., and Toyama, C.,** The chemical form and bodily distribution of mercury in marine fish, *Bull. Environ. Contam. Toxicol.*, 10, 347, 1973.
174. **Thompson, D. R.,** Total and organic mercury concentrations in liver tissue of seabirds, in *Seabird Food and Feeding Ecology; Proc. Third Int. Conf. Seabird Group,* Tasker, M. L., Ed., 1988, 44.

175. **Thomson, J. D.,** Mercury concentrations of the axial muscle tissues of some marine fishes of the continental shelf adjacent to Tasmania, *Aust. J. Mar. Freshwater Res.*, 36, 509, 1985.
176. **Topping, G.,** Heavy metals in fish from Scottish waters, *Aquaculture,* 1, 373, 1973.
177. **Turner, J. C., Solly, S. R. B., Mol-Krijnen, J. C. M., and Shanks, V.,** Organochlorine, fluorine and heavy metal levels in some birds from New Zealand estuaries, *N.Z. J. Sci.*, 21, 99, 1978.
178. **Van de Ven, W. S. M., Koeman, J. H., and Svenson, A.,** Mercury and selenium in wild and experimental seals, *Chemosphere,* 8, 539, 1979.
179. **Vas, P.,** Observations on trace metal concentrations in a Carcharhinid shark *Galeorhinus galeus* from Liverpool Bay, *Mar. Pollut. Bull.*, 18, 193, 1987.
180. **Wagemann, R.,** unpublished data, cited in Reference 181.
181. **Wagemann, R. and Muir, D. C. G.,** Concentrations of heavy metals and organochlorines in marine mammals of northern waters: overview and evaluation, *Can. Tech. Rep. Fish. Aquat. Sci.*, No. 1279, 1984.
182. **Wagemann, R., Snow, N. B., Lutz, A., and Scott, D. P.,** Heavy metals in tissues and organs of the narwhal *(Monodon monoceros), Can. J. Fish. Aquat. Sci.*, 40 (Suppl. 2), 206, 1983.
183. **Walker, T. I.,** Effects of species, sex, length and locality on the mercury content of school shark *Galeorhinus australis* (Macleay) and gummy shark *Mustelus antarcticus* Guenther from south-eastern Australian waters, *Aust. J. Mar. Freshwater Res.*, 27, 603, 1976.
184. **Watling, R. J., McClurg, T. P., and Stanton, R. C.,** Relation between mercury concentration and size in the mako shark, *Bull. Environ. Contam. Toxicol.*, 26, 352, 1981.
185. **Westermark, T., Odsjo, T., and Johnels, A. G.,** Mercury content of bird feathers before and after Swedish ban on alkyl mercury in agriculture, *Ambio,* 4, 87, 1975.
186. **Wharfe, J. R. and Van Den Broek, W. L. F.,** Heavy metals in macroinvertebrates and fish from the lower Medway estuary, Kent, *Mar. Pollut. Bull.*, 8, 31, 1977.
187. **Windom, H., Stickney, R., Smith, R., White, D., and Taylor, F.,** Arsenic, cadmium, copper, mercury and zinc in some species of north Atlantic finfish, *J. Fish. Res. Board Can.*, 30, 275, 1973.
188. **Wright, D. A.,** Heavy metals in animals from the north east coast, *Mar. Pollut. Bull.*, 7, 36, 1976.
189. **Yamamoto, Y., Honda, K., Hidaka, H., and Tatsukawa, R.,** Tissue distribution of heavy metals in Weddell seals *(Leptonychotes weddellii), Mar. Pollut. Bull.*, 18, 164, 1987.
190. **Zonfrillo, B., Sutcliffe, R., Furness, R. W., and Thompson, D. R.,** Notes on a Risso's dolphin from Argyll with analyses of its stomach contents and mercury levels, *Glasgow Nat.*, 21, 297, 1988.

Chapter 10

THE USE OF SEABIRDS AS MONITORS OF HEAVY METALS IN THE MARINE ENVIRONMENT

P. M. Walsh

TABLE OF CONTENTS

I. INTRODUCTION

In attempts to quantify geographical variations in heavy metal levels in marine environments and assess the impact of anthropogenic inputs of metals, analysis of levels in biological indicator organisms offers several advantages over seawater or sediment analysis.[1] The choice of indicator depends on many variables, such as ability to regulate tissue concentrations or tendency to accumulate high concentrations of specific metals; macroinvertebrates are generally favored (see also Chapter 6).[1,2]

Although seabirds may average out (''integrate'') very localized or short-term variations in contaminant levels of marine food chains, and in some cases (notably organochlorines and mercury) accumulate higher tissue concentrations than in food items, they may offer no overall advantage over other marine biological indicators for metal analysis.[3] However, seabirds have a high conservation profile, and mercury has been implicated as a cause of mortality or of reduced breeding success in terrestrial, freshwater, and coastal populations of birds.[4-8] Mercury concentrations in eggs and body tissues of seabirds have thus been analyzed on a large scale since the mid 1960s, particularly in the northern hemisphere,[9-19] and levels of cadmium, lead, copper, zinc, and other metals are increasingly being measured in seabirds.[20-31]

Most studies have not been specifically designed to quantify environmental variations in metal levels but are primarily concerned with possible adverse effects of elevated metal levels on seabirds. However, a few studies have specifically aimed to use seabirds as indicators of geographical variation,[32-37] while seabird feathers have been used for assessment of historical variations in mercury inputs to the sea.[38,39] The establishment of baseline concentrations of metals for seabird tissues is difficult, given geological, dietary, taxonomic, and other likely influences on 'natural' levels, but there is a growing database of metal levels in populations remote from sources of localized pollution.[28,30,31,40-42]

The present review aims primarily to draw attention to factors which need to be taken into account in designing programs to monitor geographical or temporal variations in marine levels of heavy metals using seabird tissues, or in interpreting existing work.

II. METABOLIC REGULATION OF METALS

Of the metals most often analyzed in avian tissues, copper and zinc are biologically essential, with important roles in, for example, enzyme function.[43] Seabird data support the idea that tissue concentrations of these metals are regulated (see also Chapter 9).[10,20,30] Copper concentrations are relatively constant between species and populations, with arithmetic means ranging from 2.3 to 18.5 (mainly 4 to 8) μg/g wet weight in liver and 1.0 to 7.8 (3.5 to 6.5) μg/g in kidney.[28,32,40,42] Similarly, zinc concentrations vary only 19 to 110 (mainly 30 to 50) μg/g in liver, 18 to 78 (25 to 50) μg/g in kidney.[24,32,42]

Further evidence for regulation of copper and zinc is provided by relatively low intrapopulation variation (Table 1).[20] Tissue concentrations of environmental contaminants are often not normally distributed in population samples,[44,45] with arithmetic means disproportionately influenced by a few individuals exhibiting high concentrations. However, copper and zinc tend to be normally distributed in seabirds,[20,46] as evident from comparison of means with the meager data available on median concentrations (Table 1).

For mercury and lead, coefficients of variation (CV) and skewness tend to be high (Table 1),[46] implying a lack of regulation, and concentrations vary widely between species and populations (see also Chapter 9).[42] Mean concentrations of mercury in adult seabirds range from 0.06 to 268 μg/g wet weight in liver (Figure 1),[31,41] at least 0.06 to 5.4 μg/g in kidney,[41,47] and 0.05 to 11.1 μg/g in egg contents.[19,48] More limited data for lead indicate mean concentrations <0.01 to 5.3 μg/g in liver and <0.01 to 2.1 μg/g in kidney.[41,49]

TABLE 1
Coefficients of Variation (100 SD/Mean) and "Skewness Ratios" (Arithmetic Mean/Median) for Metal Concentrations in Seabird and Wader Tissues (Averages of Individual Species Means)

Species	Location	Tissue	Cd	Cu	Hg	Pb	Zn
		Coefficient of Variation					
14 Seabird species[31]	Gough Island	Liver	38	27	57	—	21
		Kidney	30	26	—	—	18
2 Wader species[45]	Netherlands	Liver	202	21	—	105	16
		Kidney	114	27	—	267	10
Brown pelican[14,54]	South Carolina	Egg	53	17	37	78	14
		Skewness ratio					
14 Seabird species	Gough Island	Liver	0.99	1.05	1.19	—	1.02
		Kidney	1.04	1.02	—	—	1.00
2 Wader species	Netherlands	Liver	1.73	1.08	—	1.88	1.02
		Kidney	3.18	1.03	—	1.23	1.00
Brown pelican	South Carolina	Egg	1.31	0.96	1.02	1.42	1.01

Cadmium concentrations also vary widely between populations, with means <0.1 to 32 μg/g wet weight in liver and 1.5 to 138 μg/g in kidney (Figure 1).[31,49,50] However, intraspecific variation and skewness tend to be less than for mercury and lead in seabirds, though not, apparently, in waders (Table 1) or seaducks,[51,52] despite lower concentrations in the latter groups. It may be that, as an evolutionary consequence of long-term exposure to proposed high 'natural' levels of cadmium in oceanic food chains,[21,53] seabirds have some capacity to regulate their tissue levels of cadmium.

Available data for other metals imply metabolic regulation of the essential metals manganese and iron (e.g., coefficients of variation 12 to 27%, skewness ratios of 1.00 to 1.04 in waders),[45,49] but not of arsenic, chromium, and nickel (e.g., CVs 69 to 157%, skewness 1.2 to 1.79 in brown pelican *Pelecanus occidentalis* eggs).[14]

Metabolic regulation decreases the potential for detecting variations in environmental levels of essential metals using seabirds as sampling agents.[28] However, regulatory ability may not always be sufficient to cope with acute local pollution, allowing some potential for monitoring,[50] at least qualitatively. Such a 'breakdown' of regulation might be evident as an increase in skewness and intrapopulation variation of metal concentrations.

At the other extreme, marked geographical trends may be seen for tissue concentrations of some toxic metals despite apparent evidence for regulation. For example, mercury can show low intrapopulation variability and skewness in seabird eggs (Table 1) and nestling feathers.[35,37] In such cases, it may be that the tissue or age-class is a particularly good indicator of localized or short-term exposure to metals (see Sections IV and VI), and thus less likely to reflect temporal or spatial heterogeneities in exposure within a population.

III. CHOICE OF SPECIES: TAXONOMIC AND ECOLOGICAL FACTORS

An ideal seabird monitor of heavy metal pollution should be resident within a restricted radius of its breeding colony throughout the year. Its diet should be well known and should show no systematic variation between sampling locations; species feeding on few prey species are particularly suitable. However, the possibility that seasonal movements of prey species

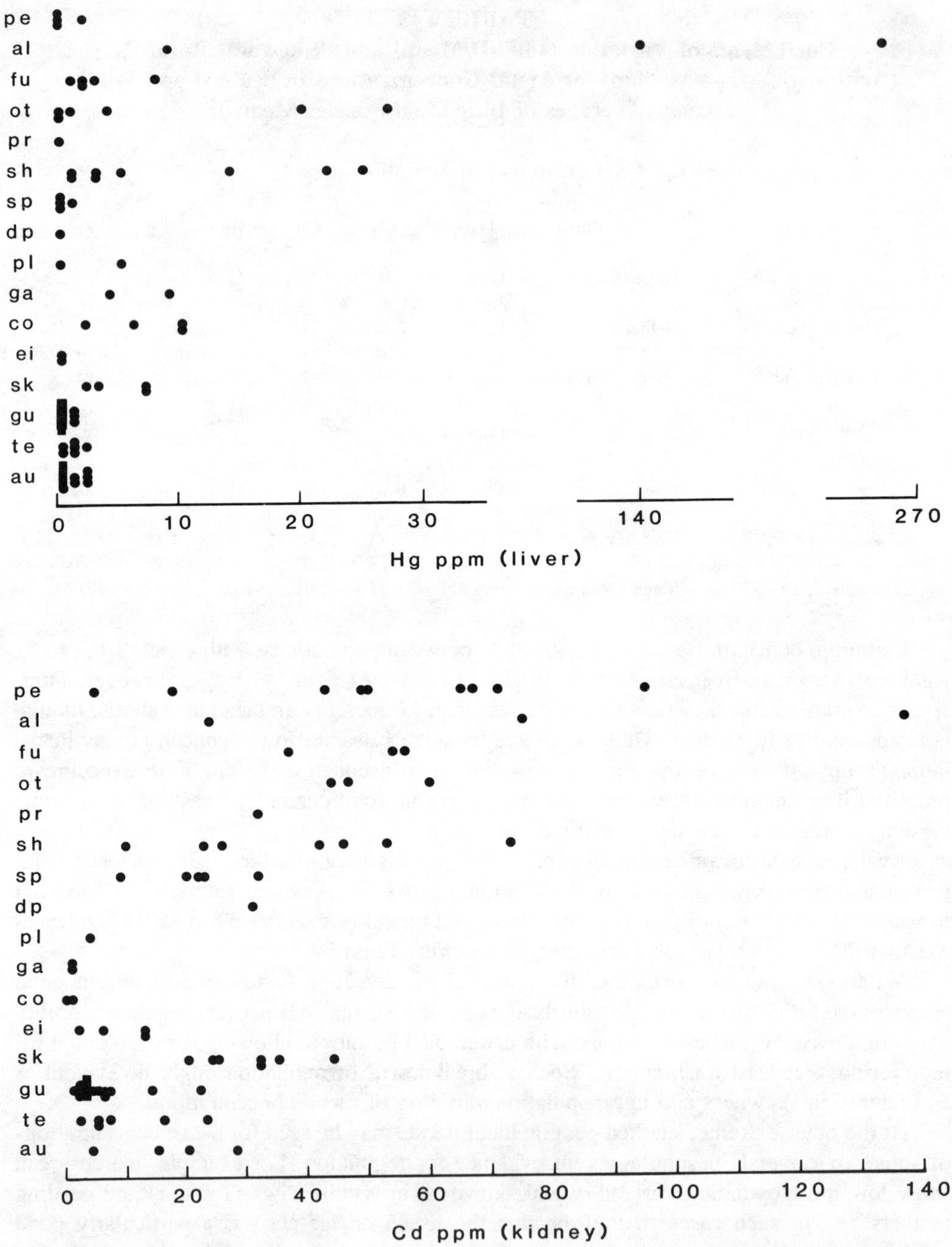

FIGURE 1. Variation of liver mercury and kidney cadmium concentrations between seabird taxa (ppm = parts per million or μg/g wet weight). Each point represents an arithmetic population mean based on 2+ individuals (small square = 2, large square = 4 means). Pe = penguins (Spheniscidae), al = albatrosses (Diomedeidae), fu = fulmars *Fulmarus*, ot = other tubenoses *(Pagodroma, Pterodroma)*, pr = broad-billed prion *Pachyptila vittata*, sh = shearwaters *(Puffinus, Calonectris)*, sp = storm petrels (Hydrobatidae), dp = common diving petrel *Pelecanoides urinatrix*, pl = brown pelican, ga = northern gannet *Sula bassana*, co = cormorants *Phalacrocorax*, ei = common eider *Somateria mollissima*, sk = skuas *Catharacta*, gu = gulls *(Larus, Rissa)*, terns *Sterna* and auks (Alcidae). Data from References 11, 14, 15, 17, 19—21, 23—33, 40, 41, 46, 47, 49, 51, 57, 59—64, 95.

could affect metal levels should be noted.[3] Furthermore, reliability of metal analyses, and ability to discriminate environmental variation, will be improved if the seabird species, or their prey, tend to accumulate relatively high tissue concentrations of the specific metal. The seabird species should also be relatively common and accessible,[3] and sufficiently widespread to provide good coverage of the proposed region being monitored for geographical variation.

Most seabirds meet only some of the above criteria. For many seabirds, however, diet, seasonal movements, and molt are reasonably well known,[38] so some allowance can be made for influences of these variables. Since different species tend to feed at varying distances from land, it is also possible in theory to select species which will reflect metal levels in either inshore or offshore waters.[3] Sampling at the breeding colony narrows down the sampling radius, particularly in the case of eggs or nestling tissues, which more closely reflect metal uptake from local foraging than do tissues of adults.[13,15,18,37] For example, eggs of common murres *Uria aalge*, which feed on relatively few species of small fish, [55] seem to be particularly good indicators of coastal variations in mercury levels.[13]

Variations in heavy metal concentrations between seabird species potentially reflect many factors, including feeding and migratory habits, body size, lifespan, molt strategy, and taxonomic influences on physiology. Inferences about geographical variations in environmental metal levels based on tissue analyses of different species are therefore open to question, though interspecific comparisons may shed light on relative metal loads in different food chains. Apparent tendencies for particular taxonomic or ecological groups to accumulate high levels of particular metals may also recommend suitable indicator species.

Figure 1 summarizes most of the published work on mercury and cadmium concentrations in liver and kidney tissues, respectively, of adult seabirds. For mercury, much of the variation reflects the well-known tendency for concentrations to be highest in species feeding on fish (or on other seabirds).[11,12,15,25,35,47] However, markedly high concentrations have been found in some procellariform species, notably albatrosses and gadfly petrels, from Gough Island,[31] where no association with diet was evident. In the latter case, a possible influence of molt-related or other constraints on mercury elimination was suggested.[31] Marked interspecific variation is also evident for cadmium, with a tendency for concentrations to be highest in species feeding on pelagic invertebrates, and in skuas feeding on pelagic seabirds.[3,21,25,30,31,41,56] For both cadmium and mercury, pelagic seabirds might seem ideal candidates for assessment of possible global oceanic pollution. However, difficulties posed by such factors as age accumulation (Section VI) and, in the case of cadmium, the apparent lack of a suitable tissue for historical analysis,[39] would need to be overcome.

Interspecific variation in zinc concentrations is comparatively slight, and in general follows the trends shown by cadmium (Section VII). For copper, the only notable feature is the occurrence of concentrations of 100 to 270 μg/g wet weight or more in liver tissue of common eiders from northern Europe,[3,30,51,57] 20 to 30 times higher than in "true" seabirds. Copper concentrations also tend to be high in other duck species,[52,58] though nowhere near as high as in eiders. It would appear either that eiders are physiologically unable to regulate their liver concentrations of copper, or that their dietary uptake of copper exceeds their regulatory capacity. Among wholly marine or coastal birds, eiders may thus be the best potential indicator of coastal pollution by copper.

The highest concentrations of lead occur in inshore or estuarine seabirds, notably laughing gulls *Larus atricilla*,[22,33] though this may partly reflect a bias towards sites known or suspected to be lead polluted. Interspecific differences for other metals, including silver, arsenic, chromium, iron, manganese, and nickel, are unclear in view of the paucity of data, and possible variations in analytical reliability.[14,20,23,32,41,57,59]

IV. VARIATION OF METAL CONCENTRATIONS BETWEEN TISSUES

Table 2 summarizes published data on relative concentrations of mercury, cadmium, lead, copper, and zinc in the most frequently analyzed tissues of adult seabirds. Such tissue/tissue ratios, although highly variable between studies, indicate the tissues which tend to accumulate high concentrations of particular metals and which may thus be 'good' indicators of metal uptake from the environment. The latter assumption has not been tested critically in seabirds, but various studies have looked for parallel variation of metal levels in different seabird tissues.[23-25,30,46,59,65,66] However, many multiple-tissue studies have made little attempt to investigate intertissue correlations, and no quantitative information is available on the relationship between concentrations in particular seabird tissues and the total metal burden in those tissues or in the whole body.

Tissues tending to have low concentrations could still reflect environmental variations, but analytical difficulties may be encountered. Also, such low concentrations might imply a saturation of metal-accumulatory ability. Conversely, some tissues may be more prone to the influence, and potential bias, of age accumulation of metals (Section VI), though it is sometimes possible to choose a specific tissue to reflect either long-term or recent exposure.[52,67-69]

Note that in Table 2, the water-content figures used for any necessary wet-weight/dry-weight conversions are very approximate, and may vary between studies,[19,25,51,59] for analytical, physiological, or taxonomic reasons. The figure for 10% water content of feathers may be a good average for full-grown feathers, but will vary with collecting conditions and will often be higher in chicks: the use of surface-cleaned,[70] dried samples is recommended. A 10% water content in bone is rather notional, and will depend on the degree of adhering soft tissue. Intertissue comparisons of metals on a wet-weight basis are rather meaningless, however, except in terms of ease of analysis. Wet-weight comparisons between eggs of different species, e.g., in attempts at toxicological assessment, are also questionable, as the water content tends to be highest in species hatching altricial (naked, poorly developed) chicks.[71] For example, fresh eggs of northern gannets and common eiders averaged 83.7 and 66.4% water, respectively, in Canadian samples,[19] representing a two-fold difference in relative dry-tissue content.

Compared to inorganic mercury, which accumulates mainly in the kidney in experimentally dosed animals, methylmercury distributes more evenly between the soft tissues.[67,72,73] Tissue concentrations in seabirds fit this pattern, with a liver/kidney ratio close to unity, although liver concentrations are usually highest. Liver and kidney concentrations are often significantly correlated,[25,59,66] and this correlation is also evident between populations (Table 2 data). However, kidney tissue alone might provide an underestimate of exposure, as the liver/kidney ratio tends to be higher at high liver and kidney concentrations of mercury. Liver and pectoral muscle concentrations are also significantly, but more variably, correlated between populations, again with some evidence of an exponential increase in liver concentration. The apparent position of liver as the best single soft tissue to indicate mercury exposure is further supported by the high degree of average correlation with other tissue concentrations in nestling common terns *Sterna hirundo*.[66]

Mercury shows a particularly strong tendency to be incorporated in feathers, becoming firmly bound to keratin during feather growth. Once growth is completed, feathers are physiologically isolated from other tissues, and their mercury content is remarkably resistant to further change under the influences of heat, ultraviolet light, and other forms of weathering,[74] with little evidence for external contamination under field conditions.[75] Additional advantages of feathers as mercury monitors are the tendency for concentrations to be high relative to other tissues and the ease with which live birds can be sampled.

TABLE 2
Between-Tissue Ratios of Metal Concentrations in Seabird Adults

			Ratio							
			l/k[a]	l/pm	l/br	l/bo	l/f	k/l	bo/l	f/l
Hg	md, n[b]	d[c]	1.12, 18	4.9, 19	6.4, 11	21, 6	0.35, 25	0.88, 18	0.05, 6	2.8, 25
	range		0.69—4.8	1.0—18.5	1.3—21.8	5.4—36	0.15—95	0.20—1.4	0.03—0.2	0.01—6.6
	md, n	w[c]	1.35, 18	5.9, 19	9.0, 11	7.0, 6	0.12, 25	0.73, 18	0.14, 6	8.5, 25
	range		0.82—5.8	1.1—22.2	1.8—30.4	1.8—12.0	0.05—32	0.17—1.2	0.08—0.6	0.03—20
Cd	md, n	d	0.17, 62	13.0, 26	6.1, 14	19.1, 14	>25.1, 7	5.9, 62	0.06, 14	<0.04, 7
	range		0.03—0.3	1.9—108	0.9—>276	0.97—577	2.7—304	1.9—35.2	0.002—1	0.003—0.4
	md, n	w	0.20, 62	15.6, 26	8.5, 14	6.4, 14	>8.4, 7	4.9, 62	0.17, 14	<0.11, 7
	range		0.03—0.4	2.3—130	1.2—>385	0.32—193	0.89—101	1.6—29.3	0.01—3.1	0.01—1.1
Pb	md, n	d	0.89, 12	1.44, 9	0.92, 11	0.97, 8	1.79, 4	1.12, 12	1.31, 8	0.57, 4
	range		0.01—2.1	0.87—3.3	0.04—3.3	0.37—3.1	<0.11—3	0.47—69	0.33—2.7	0.33—>9.3
	md, n	w	1.06, 12	1.73, 9	1.29, 11	0.32, 8	0.60, 4	0.93, 12	3.9, 8	1.70, 4
	range		0.02—2.5	1.04—4.0	0.53—3.4	0.12—1.0	<0.04—1	0.39—58	0.98—>7	1.0—>28
Cu	md, n	d	1.04, 36	1.39, 17	2.6, 3	2.4, 5	1.66, 3	0.95, 36	0.43, 5	0.60, 3
	range		0.55—2.5	0.39—17	2.41—3.8	1.65—22	1.09—5.9	0.40—1.8	0.04—0.6	0.17—0.9
	md, n	w	1.25, 36	1.67, 17	3.7, 3	0.78, 5	0.55, 3	0.79, 36	1.28, 5	1.81, 3
	range		0.66—3.0	0.47—20	3.36—5.3	0.55—7.5	0.36—2.0	0.33—1.5	0.13—1.8	0.51—2.7
Zn	md, n	d	0.81, 38	2.4, 19	2.1, 12	0.90, 10	1.54, 7	1.26, 38	1.12, 10	0.67, 7
	range		0.49—1.3	1.13—5.3	1.10—6.3	0.45—1.5	0.33—3.7	0.75—2.0	0.64—2.2	0.26—3.0
	md, n	w	0.98, 38	2.9, 19	2.9, 12	0.30, 10	0.49, 7	1.05, 38	3.3, 10	2.0, 7
	range		0.59—1.6	1.36—7.6	1.53—8.8	0.15—0.5	0.11—1.2	0.62—1.7	1.94—6.6	0.80—9.0

Note: Data from References 20, 22—29, 31—33, 41, 46, 47, 49, 51, 59, 65.

[a] Tissues = liver (1), kidney (k), pectoral muscle (pm), brain (br), bone (bo), feather (f).

[b] md = median, sample size, and range of population mean ratios.

[c] w = wet, d = dry tissue; conversions assume water content of kidney and muscle 75%, liver 70%, brain 78.5%, feather 10%, bone ?10%.

Although feathers have been shown experimentally to incorporate mercury in a dose-dependent fashion,[67] the relationship of feather to other tissue concentrations in seabirds is poorly documented. In part, this reflects the paucity of multitissue studies, but it also reflects the great variation of mercury concentrations among feathers of individual birds. Feathers replaced early in a molt cycle incorporate the highest concentrations of mercury, newly mobilized after accumulation in soft tissues between molts.[47,70] This is most strikingly shown by the tendency for mercury concentrations to decrease linearly along such feather sequences as the primaries, corresponding to the order in which feathers have been dropped and renewed.[47,70,76] Thus comparisons between studies are greatly complicated unless feathers from a similar position in the molt sequence are analyzed. However, suggestions that the mercury exposure of individual migratory birds on their wintering and breeding grounds can thus be compared, on the basis of where a particular feather was grown,[76,77] are based on the erroneous assumption that feather concentrations largely reflect recent dietary intake.

For environmental monitoring using feathers of adults, the effect of molt-influences on mercury levels of individual feathers can be minimized by pooling small body feathers from a defined plumage area.[70] If nestlings are used, little variation due to growth sequence can be expected; however, part-grown feathers tend to have disproportionately high concentrations of mercury.[37]

Dietary methylmercury is dose-dependently transferred to the contents of the egg,[67] and there is abundant evidence from seabirds of a relationship to environmental exposure.[13,15,16,29,78] Mercury accumulates particularly in the egg-white proteins,[67] which derive from serum proteins,[71] and egg concentrations thus apparently more closely reflect mercury from recent dietary uptake than from accumulated tissue stores.[18,65,69,79] (There is also evidence that the ovalbumin fraction of egg white has a specific affinity for dietary mercury, while the globulin fraction tends to accumulate low levels of ''nondietary'' mercury.[80]) Eggs may thus provide a particularly good indicator of mercury exposure in the vicinity of a breeding colony in the immediate prelaying season.[18,69]

Cadmium concentrations are invariably highest in seabird kidney, the critical organ for chronic exposure.[25,67] Nevertheless, liver may be a better indicator of recent exposure,[52,68] and may thus be an appropriate organ for short-term, localized monitoring. Liver/kidney ratios >1 indicate acute exposure to high cadmium doses,[67] which could occur in cases of severe coastal pollution. A possible advantage of liver over kidney is that elevated levels of cadmium can, by inducing tubular dysfunction, lead to an actual lowering of kidney concentrations,[67] though this has not been demonstrated in seabirds. Kidney and liver concentrations of cadmium tend to be significantly correlated both between and within populations.[24,30,46,59,62] No clear correlations with other tissues are evident from population means, but significant liver/kidney/pancreas and kidney/duodenum correlations have been noted in Atlantic puffins *Fratercula arctica*.[24]

Information on cadmium concentrations in seabird feathers is conflicting, with population means ranging from 0.02 to 27μg/g and little evidence of a relationship to other tissue levels.[24,32,41,56,65,81] The highest concentrations have been noted in pelagic seabirds (notably blue-gray ternlets *Procelsterna cerulea*),[56] but some pelagic populations have shown very low concentrations.[24,65] Nevertheless, the potential use of feathers in assessing historical variation in exposure to cadmium, and the possible influences of external contamination and molt sequence, merit further study.

Experimental studies showing little transfer of cadmium to eggs are confirmed by seabird data, with concentrations usually <0.7 (mainly 0.003 to <0.05) μg/g wet weight.[67,14,15,20,27,29,33,35,41,49,62] Such an apparent barrier to cadmium transfer might break down if other tissues became overloaded,[25] possibly accounting for the anomalously high concentrations reported in one study of sooty tern *Sterna fuscata* eggs from Hawaii (75.0 μg/g).[26] There is limited evidence that cadmium concentrations in egg shells may be relatively high, with means of 1.39 and 1.75 μg/g recorded for two laughing gull populations on the Texas coast (vs. <0.028 μg/g in egg contents).[33]

Although recorded lead concentrations vary rather little between seabird tissues, the avian kidney tends to accumulate higher levels than other soft tissues and may thus be a good indicator of recent exposure to lead.[67] However, it has been suggested that liver may be a better short-term monitoring tissue,[52] and that increased liver/kidney ratios may indicate recent acute exposure.[82] Current or recent exposure can also be monitored by lead concentrations in blood, most readily assessed by indirect techniques based on inhibition of certain enzymes by lead.[67] In birds, measurement of aminolevulinic acid dehydratase activity is particularly reliable,[67] but has not been applied to seabirds.

Lead accumulates most strongly in bone, which thus provides probably the best measure of lifetime or chronic exposure.[67,83] However, for assessment of localized or recent uptake of lead, particularly in migratory species, the use of soft tissues is preferable.[22] Although lead concentrations can be high in seabird feathers, interpretation is difficult as there is evidence from waders of heavy contamination by secretory products.[75] However, this still allows some potential for indirect monitoring of lead exposure over known periods. Little lead is transferred to eggs, which have invariably low concentrations in seabirds (usually <0.4 μg/g), often below detection limits.[20,22,26,29,33,49,54,81]

Copper concentrations tend to be highest in seabird kidney and, especially, liver. Significant liver/kidney correlations have been noted in individual studies,[30,59] and are also evident between species and populations. Common eiders are exceptional in showing a marked (8-fold to 17-fold) excess of liver over kidney concentration.[30,57] No liver/pectoral muscle correlation is evident between populations, but the liver/muscle ratio increases with increasing liver concentration. Concentrations in seabird eggs are comparatively low (means 0.15 to 1.8 μg/g).[15,20,34,41,81]

Zinc concentrations are relatively high in kidney, bone, liver, and feathers. There is a significant liver/kidney correlation between seabird populations, and at Gough Island kidney/liver ratios of zinc tended to be highest in species having high kidney concentrations of cadmium.[31] Zinc concentrations are lower in pectoral than in femoral and gastrocnemius muscle, probably reflecting differences in the content of zinc metalloenzymes.[41,46] Pancreatic concentrations are also high,[24,59] and correlate with feather levels in brown pelicans.[59] In waders, the zinc concentration of feathers is apparently 'fixed' in the shaft and highly variable in the vane, but how this variation relates to zinc exposure is unclear.[75] Mean concentrations of zinc in seabird eggs range from 1.5 to 22 μg/g.[15,20,29,34,41]

A significant feather/liver correlation of arsenic concentrations has been noted for estuarine gulls in New Zealand,[23] but concentrations may only be meaningful in feathers of known age, in view of external, probably secretory, contamination noted in waders.[75] Feather concentrations of iron, manganese, and nickel are poorly documented, though nickel levels may be comparatively high.[41,84] Concentrations of chromium in bone tissue of Antarctic and Pacific procellariform species exceeded liver and pectoral muscle concentrations in the ratio 5/2/1 (dry-weight terms),[20] and concentrations may also be high in the kidney and salt glands.[14,32] Liver tends to have the highest concentration of iron,[32,41,57] while manganese concentrations tend to be highest in liver, salt-gland, pancreas, and bone.[32,41,49] Silver, cobalt, and nickel dry-weight concentrations in bone/liver/pectoral muscle followed approximate ratios of 60/1/2, 12/1/1, and 3/2/1, respectively, in several procellariforms,[20] while common eiders can accumulate disproportionately high levels of silver in liver.[57] In seabird egg contents, mean concentrations of 0.13 to 0.30 μg/g have been noted for arsenic,[14,34] 0.01 to 0.55 for chromium,[20] 5.8 to 6.0 for iron,[41] 0.04 to 1.2 for manganese,[41,49] and 0.02 μg/g for nickel.[14]

V. PHYSIOLOGICAL AND SEASONAL INFLUENCES

Insufficient attention is often paid to possible physiological or seasonal influences on the metal content of seabird tissues.[85] Seasonal changes in tissue concentrations of essential metals, notably zinc, have been documented in house sparrows *Passer domesticus*,[86] starlings *Sturnus vulgaris*,[87] and waders.[3] Some such changes are associated with molt and can be accompanied by elevated concentrations of nonessential metals (e.g., cadmium and mercury). Seasonal variation in fat or protein content of tissues, or in enzyme activity, may also be important.[24,87] For example, an increase in metalloprotein content may increase the number of binding sites available to essential metals (e.g., zinc) and their toxic competitors (e.g., cadmium).[24,88] The most pessimistic view is that metal concentrations in tissues "depend as much on physiology as they do on exposure."[85]

In Atlantic puffins at a colony in southeast Scotland,[85] liver mercury concentrations declined significantly from a mean of 1.73 μg/g dry weight in March (prelaying) to 0.59 μg/g in July (chick rearing), kidney cadmium from 15.5 to 4.48 μg/g. No change was seen for zinc, probably reflecting only partial coverage of the annual cycle (excluding molt, when peak concentrations would be expected).[85] A proposed peak of cadmium during early to mid molt may have contributed to differences in cadmium concentrations between three pelagic seabird species at St. Kilda, sampled in June but molting at different seasons.[24]

Fairy prions *Pachyptila turtur* in Tasmania showed no variation in cadmium or zinc concentrations between breeding and nonbreeding seasons,[36] although tissue concentrations of copper and lead differed significantly. A significant difference in zinc and iron concentrations in greater crested terns *Sterna bergii* between two New South Wales colonies was considered an artifact of sampling date, reflecting seasonal declines.[32]

Seasonal variation in mercury concentrations, largely reflecting mobilization and incorporation into plumage,[70] is well known, but not always taken into account when 'soft-tissue' levels are presented. Body burdens of mercury (excluding plumage) in Bonaparte's gulls *Larus philadelphia* declined by 93% during autumn molt in southeast Canada.[89] Average body burdens in Japanese black-eared kites *Milvus migrans* declined by somewhere in the range 44 to 53% during molt, and concentrations in liver by up to 64%, in kidney by up to 59%, and in other tissues by 28 to 78%.[90] Soft-tissue comparisons between populations of adult seabirds are thus ideally based on samples collected immediately before molt, when mercury concentrations are highest and least variable.

Mercury incorporated into feathers is stably bound, and concentrations in individual feathers thus remain constant.[74,89] However, given that mercury concentrations tend to be highest in feathers renewed early in a molt sequence,[47,70] the average mercury concentration of the plumage can, in theory, alter. If individual feathers tend to be replaced at the same stage of each molt period, average concentrations will be reduced after such high-mercury feathers are dropped but before they are regrown. As molt proceeds, this effect may be reversed as high-mercury feathers are replaced and low-mercury feathers dropped. While the occurrence and potential magnitude of such effects remain to be confirmed, it is likely that feather mercury concentrations will be most consistent in samples taken outside the molt period.

No significant correlation between mercury concentration in eggs and season date was found for common terns in Ontario, but mercury in breast muscle of five adults increased significantly with date during the breeding season.[69] Since egg proteins are largely derived from blood serum,[71] this discrepancy was taken to imply an equilibrium between mercury in eggs (but not muscle), blood, and diet.

Egg laying had little influence on metal concentrations in female Adélie penguins *Pygoscelis adeliae* in Antarctica.[41] Efficiency of transfer to the egg was greatest for mercury, but the total egg content was negligible relative to adult burdens. Females had higher zinc concentrations in bone than males; zinc concentrations are also highest in female house sparrows, and peak during egg laying, apparently related to alkaline phosphatase activity associated with calcium mobilization.[86] Male Adélie penguins took the first, about 20-d incubation shift, and muscle concentrations of iron appeared to decline during this starvation period, while liver concentrations increased, implying redistribution of iron. This probably contributed to the higher iron concentrations in livers of males (987 μg/g wet weight) than of females (478 μg/g).[41]

Although evidence from seabird studies is lacking, sex differences in tissue (especially bone) concentrations of lead can be expected. Reproductively active female mallard *Anas platyrhynchos* and Barbary doves *Streptopelia 'risoria'* accumulate lead more rapidly than males, resulting in order-of-magnitude differences in bone concentrations of experimentally dosed birds.[67] Increased synthesis of calcium-binding protein during eggshell formation apparently increases intestinal absorption of both calcium and lead.[67,91]

Possible influences of starvation should be taken into account in metal analyses of seabirds found dead. Even if a metal ''does not appear to be freely mobilized in starved birds'' (mercury in liver),[11] changes in tissue weight may alter metal concentrations without altering tissue burdens. Thus in northern gannets found dead around British coasts, a twofold difference in liver concentrations of mercury and zinc between east and west coasts was largely offset by a twofold difference in liver weights, reflecting the emaciated condition of corpses in the west coast sample.[10,15] Stress associated with slow death from injury or starvation could also influence metal concentrations. For example, lead, which follows calcium during bone formation,[67] might be released with calcium during stress, as known for mammals.[83]

VI. AGE-RELATED VARIATION

Variations in the age composition of a sample of breeding or other 'adult' seabirds could potentially contribute to apparent differences in tissue concentrations of metals between species, populations, or years. If a species can be aged at all using morphological characters, this is usually possible only in the first few, mainly nonbreeding, years of life. Sampling of known-age adults, therefore, usually requires a population of birds which have been banded as chicks.

The metal levels of few such populations have been studied, greatly limiting information on possible accumulation of metals over adult life. In 13 3- to 12-year-old breeding great skuas *Catharacta skua* from Shetland, cadmium concentrations showed statistically significant positive correlations with age in kidney, liver, and pectoral muscle.[25,62] This was considered to reflect the long biological half-life of cadmium and its binding to a metallothionein-like protein and subsequent retention in the kidney.[25] Mercury concentrations in primary feathers, kidney, and liver also increased with age in great skuas,[62] though nonsignificantly.[25] In fact, both metals showed great variation between birds of the same age. Moreover, a further analysis of cadmium and mercury in liver, kidney, and muscle of Shetland great skuas found only weak age-accumulatory tendencies, only that for cadmium in muscle ($n = 27$ adults) being significant.[92,93]

No age accumulation of cadmium, mercury, or zinc was detected in 2- to 11-year-old herring gulls *Larus argentatus* from southeast Scotland.[25,46] For cadmium, this was thought possibly to reflect a lower dietary exposure than in great skuas.[46] Recaptures of black guillemots *Cepphus grylle* in the Baltic Sea likewise indicated no systematic variation of (plumage) mercury concentrations with age.[38] It may be that elimination of mercury into feathers during molt,[70] almost entirely as methylmercury,[39] is usually efficient enough to prevent age accumulation of this particularly toxic form. However, there is also evidence for demethylation of methylmercury by seabirds and subsequent storage of inorganic mercury in the liver.[94] This suggests that age accumulation of mercury is likely to occur in species having particularly high liver concentrations of inorganic mercury, e.g., certain albatrosses *Diomedea* with slow molt patterns.[31,94]

Evidence for accumulation of metals in the first few months or years posthatching is much more readily available. While of little direct relevance to adult seabirds, such studies help elucidate the relative merits of different tissues or age classes as metal-monitors.

Several studies have compared cadmium and lead concentrations in tissues of adult and nestling (or newly fledged) laughing gulls and terns (Figure 2). For cadmium, age accumulation is most marked in kidney, with adult:chick ratios varying from 13 in a laughing gull study to 387 in royal tern.[27,49] Ratios range from 1.9 to 148 in liver and only 1 to 7.3 in other soft tissues, with conflicting evidence for accumulation in bone. Liver and kidney concentrations of cadmium in first-year common eiders from Denmark were about half those in older birds.[51] Age accumulation of lead is most evident in bone, with adult:chick ratios of 1.8 to 152 (Figure 2).[33,49] For both cadmium and lead, discrepancies between studies presumably reflect differences in temporal availability of metals, in physiological responses or, perhaps, in analytical accuracy (especially at low tissue concentrations).

Surprisingly little has been published on mercury levels in chicks. An adult:prefledgling ratio of 13.3 was recorded for liver tissue of great cormorants *Phalacrocorax carbo* in the Netherlands,[95] and adult:first year ratios of 6.9, 6.4, and 10 for pectoral muscle of Baltic great cormorants, common murres, and black guillemots, respectively.[77] Limited data for brown pelicans in the eastern U.S. suggest an adult:nestling ratio of 1.6 to 9.2.[14] Plumage mercury concentrations of northern gannets average about three to five times higher in adults than in well-grown chicks.[37]

Very few data are available for other metals,[14,32] though significant adult:prefledgling differences were recorded in greater crested terns from New South Wales for manganese (ratio 1.3) and iron (2.2) in liver.[32]

To summarize, while metal concentrations may differ considerably between chicks and adults, evidence for gradual accumulation of metals in adults is very limited. It seems likely that most metals reach a dynamic equilibrium in adult seabirds, while others, such as cadmium, and perhaps inorganic mercury, may increase in some tissues of some species throughout adult life. Chicks of known age may thus provide the most reliable indication of between-colony variations in metal exposure, assuming that tissue concentrations fall within the range of analytical reliability. A further advantage, that chicks will reflect dietary uptake within a comparatively restricted radius of a breeding colony, should be noted.

VII. CORRELATIONS BETWEEN METALS, AND EVIDENCE FOR DETOXIFICATION MECHANISMS

Correlations between concentrations of different metals in seabird tissues are of interest for several reasons. Such correlations may indicate antagonistic, 'detoxifying' interactions between certain essential and nonessential metals. Clues may also be provided to the mechanisms by which metals are distributed between tissues and accumulated. In addition, if several metals tend to coaccumulate for physiological or toxicological reasons, only one of those metals may be amenable to monitoring of environmental variation.

There is particularly strong evidence in seabirds for a relationship between tissue concentrations of zinc and cadmium. Significant positive correlations have been noted for kidney and liver tissue (usually both) between individual Cory's shearwaters *Calonectris diomedea*,[29] great skuas, and herring gulls.[25,46] Similar relationships occur both within and between species from Spitsbergen, Antarctica, and Gough Island.[30,31] The zinc/cadmium relationship is particularly strong in kidney, and is summarized in Figure 3 over a range of seabird populations.

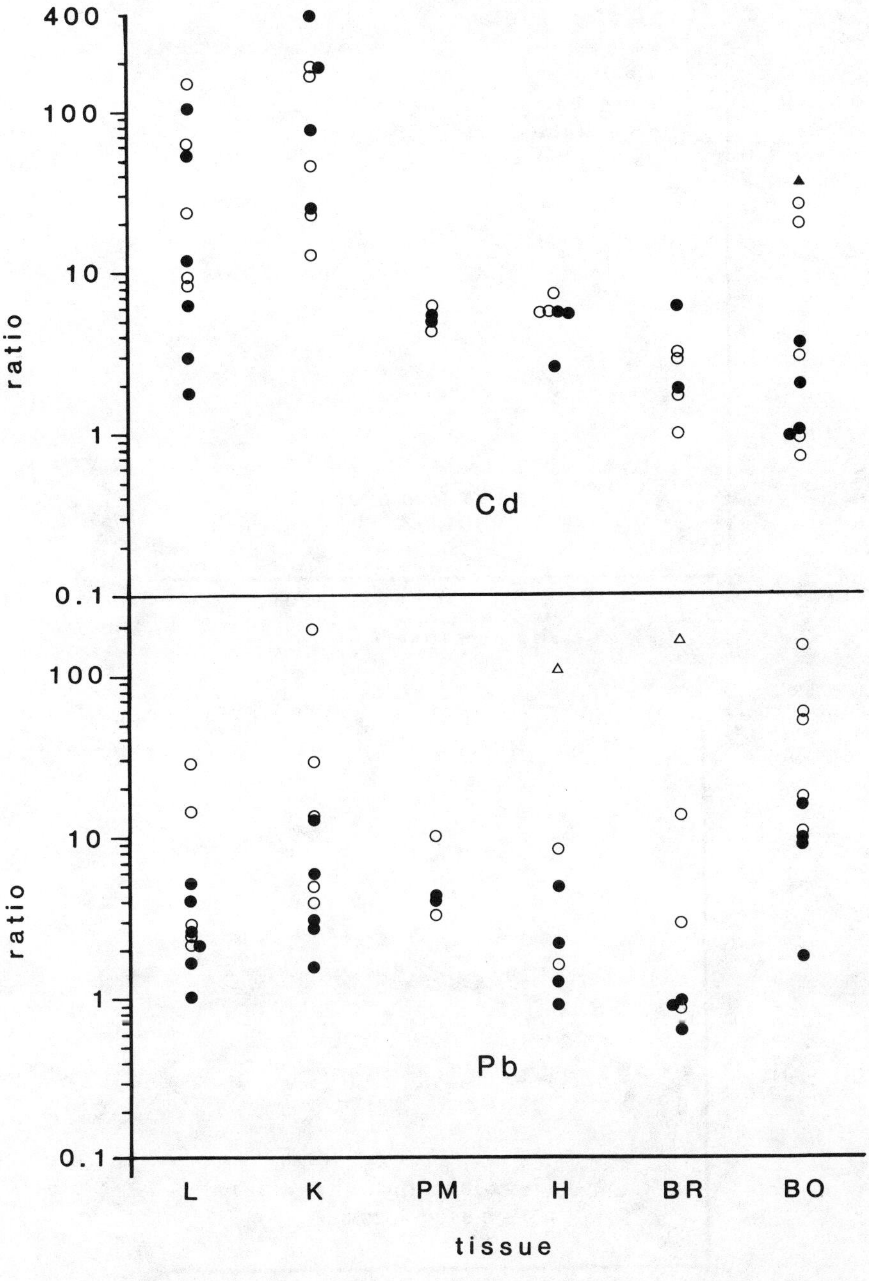

FIGURE 2. Adult:chick ratios of cadmium and lead concentrations (wet weight) in seabird tissues: laughing gulls,[22,33,49] greater crested,[32] royal *Sterna maxima* and sandwich terns *S. sandvicensis*.[27] Filled circles = well-feathered or newly fledged chicks, open circles = downy chicks; triangles = minimum ratio. Tissues: liver (L), kidney (K), pectoral muscle (PM), heart (H), brain (BR), and bone (BO). Note logarithmic scale of y-axes.

Most of the cadmium and zinc in seabird kidney and liver appears to be bound to a low molecular weight, metallothionein-like protein.[25,88] Such proteins can be experimentally induced in mammals and birds by dosing with heavy metals, including cadmium and zinc.[67] Increased synthesis of cadmium-induced metallothionein at the same time increases the number of binding sites available to zinc, thereby accounting for parallel accumulation of

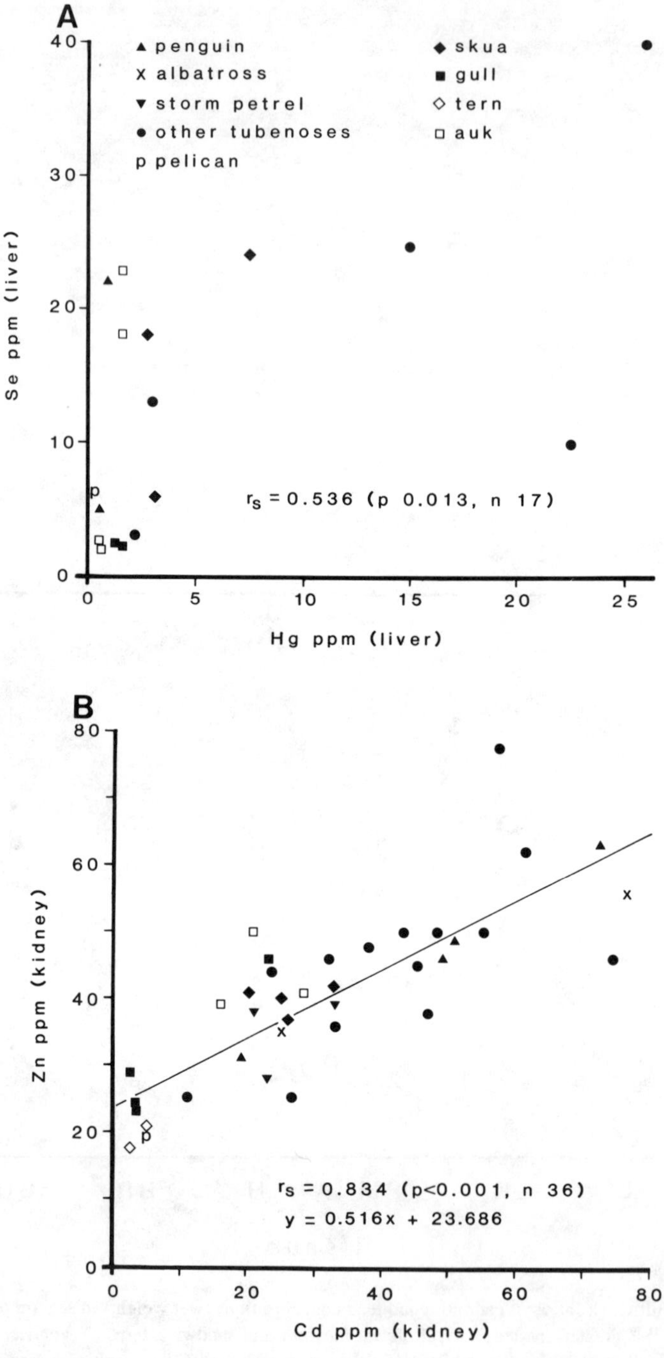

FIGURE 3. Selenium/mercury and zinc/cadmium relationship in liver and kidney tissue, respectively, of seabird adults (ppm = parts per million or μg/g wet weight). r_s = Spearman rank correlation coefficient, n = number of species or population means (based on 4+ individuals) plotted. Other significant ($p < 0.05$) Spearman rank correlations were obtained for: Zn/Cd (liver) all species (n = 52), tubenoses (25); Zn/Cd (kidney) penguins (4), tubenoses (19), skuas/gulls/terns (10), skuas/gulls/terns/auks (12); Cu/Cd (K) skuas to terns (9), skuas to auks (11); Cu/Zn (L) skuas (3), gulls (3), skuas to terns (8), skuas to auks (10); Cu/Zn (K) skuas to terns (6); Zn/Hg (L) all species (41); Se/Hg (K) all species (8); Se/Cd (L) all species (16), skuas (3), skuas to terns (12); Hg/Cd (L) all species (43), skuas/gulls (14), skuas to terns (16), skuas to auks (18). Data from References 20, 23—26, 28—32, 36, 40, 41, 46, 57, 59, 108.

the two metals. There is considerable evidence that, by binding cadmium, metallothionein helps prevent its toxic interactions with other proteins.[67] Nevertheless, kidney damage associated with high concentrations of cadmium has been noted in pelagic seabirds,[96] consistent with the known nephrotoxicity of metallothionein-bound cadmium.[67]

The primary role of metallothionein may in fact be as a zinc store,[97] in which case cadmium binding could represent a toxic interference with zinc metabolism.[88,96] It is also possible that seasonal variation in zinc requirements, e.g., for molt, could result in increased retention of dietary cadmium at particular seasons.[85,88] Zinc itself has a protective influence against various toxic effects of cadmium,[68,98] though the relative importance of zinc-induced metallothionein synthesis and outcompetition of cadmium for other binding sites is unclear.

Cadmium ingestion in experimental mallard can result in increased kidney concentrations of both zinc and copper, possibly related to metallothionein synthesis.[99] Significant correlations of zinc/copper in liver and cadmium/copper in kidney have been reported in brown pelican.[59] However, in Gough Island seabirds and over a range of populations (Figure 3), there is only limited evidence of such a relationship.[31,92]

Although inorganic mercury is known to bind to metallothionein in mammalian kidney,[67] mercury was not associated with the metallothionein-like protein in northern fulmar *Fulmarus glacialis* liver and great skua kidney.[25,88] This was suspected to reflect a preponderance in seabird tissues of methylmercury, which may not be able to bind to metallothionein.[46,85]

Methylmercury does in fact often comprise the bulk of mercury in seabirds, e.g., up to 97% in liver.[24,30,47] Being lipid soluble, unlike inorganic mercury, methylmercury can readily cross the 'blood/brain' and other membrane barriers, and is thus particularly toxic to the central nervous system.[67] Demethylation to the less toxic inorganic form has been shown to occur in experimental mammals,[100] and has also been inferred in marine mammals.[42] In South Polar skuas *Catharacta maccormicki* and a range of Gough Island seabirds, the percentage of mercury present as methylmercury is negatively correlated with the total mercury concentration in liver tissue.[30,94] This may indicate that seabirds are capable of demethylating organic mercury,[94] as has been suggested for freshwater and predatory species,[73,101] though experimental evidence in birds is apparently lacking. However, while the tissue concentration of methylmercury is the critical factor from a toxicological viewpoint, variations in dietary exposure to mercury are likely to be most clearly reflected by tissue levels of total mercury.

More prevalent may be a detoxification mechanism involving mercury and selenium, a nonmetallic element. Selenium has a protective effect against both inorganic mercury and methylmercury in experimental animals, including birds,[67,102] and approximate 1:1 molar (or atomic) ratios and increments have been noted between mercury and selenium in marine mammals.[103] A direct linkage between mercury and selenium, involving sulfur, has been postulated, the resulting complex being stable and nontoxic.[67,104,105] Nevertheless, this interaction may increase retention and accumulation of mercury,[105] especially in liver and kidney. There is also some evidence for involvement of selenium in demethylation of methylmercury.[67,105]

The mercury/selenium relationship is not as clearcut in seabirds as in marine mammals,[65,104] but this is not necessarily evidence that no protective effect is operating. Certainly there is rarely a direct 1:1 relationship, but the mean selenium:mercury molar ratio is always >1 (6.0 to 69 in kidney and 1.1 to 70 in liver over a range of species).[25,26,29,30,59] Perhaps reflecting this 'surplus' of selenium, correlations between the two elements in liver and kidney of different seabird populations are not very clearcut (Figure 3). Nonetheless, significant positive correlations have also been noted within populations.[25,29,30] A positive correlation has also been noted in eggs of little terns *Sterna albifrons*,[106] but no relationship is apparent between species, though again there is usually a (1.2- to 26-fold) molar excess of selenium over mercury.[14,26,29,34,35] However, the toxicological significance of selenium in eggs is unclear, given that dietary selenium increases the transfer of dietary methylmercury to eggs and that the two bind preferentially to egg white and yolk, respectively.[80]

Significant selenium/cadmium correlations have been noted in kidney or liver of several seabirds,[25,30] possibly related to a protective role of selenium against cadmium.[107] Significant correlations also occur in kidney, but not liver, over a range of populations (Figure 3). A strong correlation between molar liver concentrations of combined cadmium plus mercury and selenium plus zinc in species from Spitsbergen and the Antarctic was interpreted as suggesting integration of detoxifying interactions between these elements.[30]

Although evidence from seabirds is lacking, coaccumulation of zinc and lead in bone, unrelated to dietary levels of zinc, can apparently occur in feral rock doves *Columba livia*.[98] It is unclear whether this reflects the known antagonistic interaction between these metals or an influence on both metals of active bone metabolism during eggshell formation.[86,98] An antagonistic interaction also exists between lead and calcium, since intestinal absorption of lead is greatest from calcium-deficient diets.[67]

Various other metal/metal correlations, of doubtful toxicological or physiological significance, have been reported in seabirds.[14,50,59] There is an overall, though highly variable, positive correlation between cadmium and mercury in liver of different populations (Figure 3), but this largely reflects the tendency for recorded concentrations of these metals to be higher in procellariforms and skuas than in gulls, terns, and auks.

VIII. DOSE-RESPONSE RELATIONSHIPS

The use of seabirds as metal-monitors depends on the assumption that environmental variations will be reflected by levels of heavy metals in tissues. This assumption has not, however, received as much scrutiny in seabirds as in the case of other proposed marine biological indicators.[1,2] At the very least, tissue concentrations should reflect dietary uptake of the particular metal being monitored. The evidence in seabirds tends to be rather anecdotal, being based on inadequately quantified associations with dietary levels of metals or with known sources of metal pollution.[11,15,18,25,27,31,33,35,36,38,41,49,50]

Experimental studies of non-seabird species confirm that increased dietary doses of mercury, cadmium, and lead are reflected in increased tissue concentrations (particularly in specific 'target' organs).[67] In some studies, tissue levels have increased in direct proportion to dietary levels over a particular dose range,[109] but dose responses are often nonlinear, with a lower ratio of tissue to dietary concentration at higher doses. For cadmium, there is some evidence of an increased dose response at higher doses, but there are also studies suggesting up to 75% decreases in response.[67] Similarly, for lead, decreases in dose response of up to 60% have been noted.[110] Mercury may tend to show a more linear tissue/dose relationship,[109] although there is also evidence for a decrease in dose response of up to 20% in eggs.[111] For analogy with natural rates of metal uptake, however, a critical appraisal of dose responses over lower, chronic dose ranges is needed.

Even assuming that a coherent picture can be deduced from existing experimental studies, applying such dose-response relationships to free-living seabirds is far from straightforward. In particular, the exposure of individual seabirds to heavy metals over periods of years or even decades, with associated seasonal variations in uptake and elimination rates, may invalidate comparisons with experimental dosing over weeks or months. Seabird nestlings would be more amenable to direct comparison, but experimental dosing would be far more informative. The most practicable approach, however, might be a quantitative assessment of tissue concentrations in relation to dietary uptake for nestlings at several colonies varying in exposure to heavy metals. Given such information, attempts at using seabirds to quantify environmental variations in heavy metals, at least in terms of biological availability, would be based on firmer ground.

IX. CURRENT EVIDENCE FOR GEOGRAPHICAL VARIATION IN EXPOSURE OF SEABIRDS TO HEAVY METALS

Particularly striking patterns of geographical variation are evident from studies of mercury in single or closely related species in European and North Atlantic waters. Common murres provide a particularly good example. Eggs collected from 16 colonies around Britain and Ireland in the early 1970s showed a 20-fold variation in mean mercury concentrations, with lowest concentrations (0.8 to 1.9 μg/g dry weight or about 0.25 to 0.6 μg/g wet weight) from northwest Scotland around to northeast England but markedly high concentrations (about 1.9 to 4.9 μg/g wet weight) at five Irish Sea colonies.[13] Anthropogenic inputs of mercury to the Irish Sea, in industrial and other waste, are high and are probably exacerbated by low rates of water exchange; mercury levels in seawater and in fish are also high.[112] Eggs from other parts of the murre's range are lower in mercury, with means of 0.12 μg/g at a Canadian colony,[19] about 0.20 to 0.25 in the Faroes, [13,113] <0.07 to 0.13 in northern Norway, [12,18] and slightly higher levels of 0.25 to 0.54 μg/g in the Baltic Sea.[77,113]

Elevated mercury levels in the enclosed Baltic were also reflected in a study of murre feather concentrations: north Baltic samples averaged 3.3 μg/g, south Baltic 2.6, Faroes 1.2, and Greenland (thick-billed murres *Uria lomvia*) 1.1 μg/g.[38] Black guillemots showed a similar pattern, which in both species was also evident, though less marked, even in feathers from pre-1900 museum specimens.[38] In addition to historical trends revealed by auk feather samples,[38,39] there is limited evidence of a decline in Baltic mercury levels within the 1970s, possibly reflecting reduced mercury discharges from Swedish and Finnish chlorine-alkali and paper/pulp industries.[38,114,115]

Other seabirds showing clear geographical variations in mercury levels include northern gannets (mean egg concentrations about 0.6 to 1.7 μg/g wet weight at British and Irish colonies, again peaking in the Irish Sea) and herring gulls (a fairly even trend of increasing egg concentrations from 0.09 μg/g in northern Norway to 1.28 in the south).[12,15,16] Herring gull and common tern eggs from seven regions of the German North Sea coast averaged about 0.2 to 0.5 and about 0.3 to 0.7 μg/g respectively, except for pronounced peaks of 1.55 and 4.58 μg/g at the Elbe estuary.[48]

Further south, mercury concentrations in eggs and adult tissues of Cory's shearwaters from three Mediterranean colonies were consistently higher than at an Atlantic colony (Selvagen),[29] e.g., egg/liver concentrations about 1.3 to 1.9/14.9 to 25.9 vs. about 0.5/3.8 μg/g wet weight. High levels of mercury in the Mediterranean, reflecting both anthropogenic and geological influences,[29] are also reflected in eggs of Audouin's gulls *Larus audouinii* (means 0.76 to 1.32 μg/g).[116,117]

Assessment of variations in seabird mercury levels on a broader, global scale, and of the relative roles of human and natural inputs of mercury,[20,31,65] is difficult, given differences in species and food chains between studies and the current lack of data on historical levels in pelagic seabirds. However, it has been suggested that Antarctic species (Antarctic fulmar *Fulmarus glacialoides* and skuas *Catharacta*) may be exposed to higher mercury levels than comparable species from Spitsbergen (northern fulmar and glaucous gull *Larus hyperboreus*).[61] Comparison between pelagic and inshore populations of seabirds further suggests that mercury in oceanic food chains may be largely natural in origin, despite high tissue concentrations in some pelagic species and evidence for coastal mercury pollution.

The tendency for cadmium levels to be relatively high in pelagic species also suggests a natural origin.[21,24,26,31,53,56,96] This is supported by data for individual species, e.g., kidney concentrations in Atlantic puffins from St. Kilda (about 29 μg/g wet weight) about 12 times higher than from the more coastal Isle of May population.[85] Similarly, for Cory's shearwaters, kidney cadmium concentrations were higher in Atlantic (Selvagen) adults (about 53.5 μg/g) than at Mediterranean colonies (about 10.7 to 46.9 μg/g).[29] However, cadmium variations between broad oceanic regions are unclear.

Anthropogenic inputs of cadmium should be more marked in inshore waters, but little information is available for coastal seabirds. A comparison of laughing gull tissue concentrations between Galveston Bay and Matagorda Bay, Texas found no consistent variation, though prefledgling kidney concentrations were significantly higher in the more polluted Galveston Bay.[33] Fairy prions apparently reflected dumping of industrial waste off the coast of Tasmania, with significantly higher tissue concentrations of cadmium at a nearby colony than at colonies more remote from the dumping site.[36] However, cadmium concentrations in greater crested terns in New South Wales and in estuarine gulls in New Zealand showed no relationship to degree of industrialization.[23,32]

There is some evidence of geographical variation in lead levels of coastal seabirds, reflecting local pollution. Tissue concentrations were significantly higher in laughing gulls from Galveston Bay than from Matagorda Bay, Texas.[33] The mean liver concentration of 5.3 μg/g wet weight in Galveston adults is apparently the highest yet reported for living seabirds, but declined (to 1.1 μg/g) between 1977 and 1980, as did other tissue levels.[33] In Britain, many birds, including gulls, were poisoned by trimethyllead in the Mersey Estuary during 1977 to 1981, the gull casualties having liver concentrations of about 1 to 8 μg/g.[82] However, despite the apparently high contribution of anthropogenically derived atmospheric lead to surface oceanic waters,[2,20] little or no information is available for pelagic seabirds.

Copper and zinc concentrations vary little between seabird populations, as may be expected from their metabolic regulation (Section II). Geographical differences can be evident for zinc,[85] but may largely reflect variation in exposure to cadmium (Section VII).[31] However, copper and zinc concentrations in nestling common terns from Rhode Island, U.S., apparently reflected industrial pollution, and showed the same localized patterns of variation as metals in food items.[50]

Variations in tissue concentrations of arsenic, chromium, iron, manganese, and nickel in some cases appear to reflect coastal pollution,[20,23,32,50,59] but interpretation is difficult at present, as natural levels and the suitability of particular tissues as metal monitors are particularly unclear.

REFERENCES

1. **Phillips, D. J. H.,** The use of biological indicator organisms to monitor trace metal pollution in marine and estuarine environments — a review, *Environ. Pollut.*, 13, 281, 1977.
2. **Bryan, G. W.,** Pollution due to heavy metals and their compounds, In *Marine Ecology*, Vol. 5, Part 3, Kinne, O., Ed., J. Wiley & Sons, Chichester, 1984, 1289.
3. Natural Environment Research Council, *Contaminants in Marine Top Predators*, NERC, Swindon, England, 1983.
4. **Kurland, L. T., Faro, S. N., and Siedler, H.,** Minamata disease, *World Neurol.*, 1, 370, 1960.
5. **Henriksson, K., Karppanen, E., and Helminen, M.,** High residue of mercury in Finnish white-tailed eagles, *Ornis Fenn.*, 43, 38, 1966.
6. **Borg, D., Wanntorp, H., Erne, K., and Hanko, E.,** Alkyl mercury poisoning in terrestrial Swedish wildlife, *Viltrevy*, 6, 301, 1969.
7. **Koeman, J. H., Vink, J. A. J., and de Goeij, J. J. M.,** Causes of mortality in birds of prey and owls in the Netherlands in the winter of 1968—1969, *Ardea*, 57, 67, 1969.
8. **Fimreite, N.,** Mercury contamination of aquatic birds in northwestern Ontario, *J. Wildl. Manage.*, 38, 120, 1974.
9. Natural Environment Research Council, *The Sea Bird Wreck in the Irish Sea Autumn 1969*, NERC, London, 1971, 14.
10. **Parslow, J. L. F., Jefferies, D. J., and Hanson, H. M.,** Gannet mortality incidents in 1972, *Mar. Pollut. Bull.*, 4, 41, 1973.
11. **Dale, I. M., Baxter, M. S., Bogan, J. A., and Bourne, W. R. P.,** Mercury in seabirds, *Mar. Pollut. Bull.*, 4, 77, 1973.

12. **Fimreite, N., Brun, E., Frøslie, A., Frederichsen, P., and Gundersen, N.,** Mercury in eggs of Norwegian seabirds, *Astarte,* 7, 71, 1974.
13. **Parslow, J. L. F. and Jefferies, D. L.,** Geographical variation in pollutants in guillemot eggs, *Annu. Rep. Inst. Terr. Ecol. 1974,* 28, 1975.
14. **Blus, L. J., Neely, B. S., Jr., Lamont, T. G., and Mulhern, B.,** Residues of organochlorines and heavy metals in tissues and eggs of brown pelicans, 1969—73, *Pestic. Monit. J.,* 11, 40, 1977.
15. **Parslow, J. L. F. and Jefferies, D. J.,** Gannets and toxic chemicals, *Br. Birds,* 70, 366, 1977.
16. **Holt, G., Frøslie, A., and Norheim, G.,** Mercury, DDE, and PCB in the avian fauna in Norway 1965—1976, *Acta Vet. Scand. Suppl.,* 70, 1, 1979.
17. **Delbeke, K., Joiris, C., and Decadt, G.,** Mercury contamination of the Belgian avifauna 1970—1981, *Environ. Pollut. Ser. B,* 7, 205, 1984.
18. **Barrett, R. T., Skaare, J. U., Norheim, G., Vader, W. and Frøslie, A.,** Persistent organochlorines and mercury in eggs of Norwegian seabirds 1983, *Environ. Pollut. Ser. A,* 39, 79, 1985.
19. **Noble, D. G. and Elliott, J. E.,** Environmental Contaminants in Canadian Seabirds, 1968—1985: trends and effects, Tech. Rep. Ser. No. 13, Canadian Wildlife Service, Ottawa, 1986.
20. **Anderlini, V. C., Connors, P. G., Risebrough, R. W., and Martin, J. H.,** Concentrations of heavy metals in some Antarctic and North American sea birds, in *Proc. Colloq. Conservation Problems in Antarctica,* Parker, B. C., Ed., Allen Press, Lawrence, KS, 1972, 49.
21. **Bull, K. R., Murton, R. K., Osborn, D., Ward, P., and Cheng, L.,** High levels of cadmium in Atlantic seabirds and sea-skaters, *Nature,* 269, 507, 1977.
22. **Munoz, R. V., Jr., Hacker, C. S., and Gesell, T. F.,** Environmentally acquired lead in the laughing gull, *Larus atricilla, J. Wildl. Dis.* 12, 139, 1976.
23. **Turner, J. C., Solly, S. R. B., Mol-Krijnen, J. C. M., and Shanks, V.,** Organochlorine, fluorine, and heavy-metal levels in some birds from New Zealand estuaries, *N.Z. J. Sci.,* 21, 99, 1978.
24. **Osborn, D., Harris, M. P., and Nicholson, J. K.,** Comparative tissue distribution of mercury, cadmium and zinc in three species of pelagic seabirds, *Comp. Biochem. Physiol.,* 64C, 61, 1979.
25. **Hutton, M.,** Accumulation of heavy metals and selenium in three seabird species from the United Kingdom, *Environ. Pollut. Ser. A,* 26, 129, 1981.
26. **Stoneburner, D. L. and Harrison, C. S.,** Heavy metal residues in sooty tern tissues from the Gulf of Mexico and north central Pacific Ocean, *Sci. Total Environ.,* 17, 51, 1981.
27. **Maedgen, J. L., Hacker, C. S., Schroder, G. D., and Weir, F. W.,** Bioaccumulation of lead and cadmium in the royal tern and sandwich tern, *Arch. Environ. Contam. Toxicol.,* 11, 99, 1982.
28. **Steinhagen-Schneider, G.,** Cadmium and copper levels in seals, penguins and skuas from the Weddell Sea in 1982/1983, *Polar Biol.,* 5, 139, 1986.
29. **Renzoni, A., Focardi, S., Fossi, C., Leonzio, C., and Mayol, J.,** Comparison between concentrations of mercury and other contaminants in eggs and tissues of Cory's shearwater *Calonectris diomedea* collected on Atlantic and Mediterranean islands, *Environ. Pollut. Ser. A,* 40, 17, 1986.
30. **Norheim G.,** Levels and interactions of heavy metals in sea birds form Svalbard and the Antarctic, *Environ. Pollut.,* 47, 83, 1987.
31. **Muirhead, S. J. and Furness, R. W.,** Heavy metal concentrations in the tissues of seabirds from Gough Island, South Atlantic Ocean, *Mar. Pollut. Bull.,* 19, 278, 1988.
32. **Howarth, D. M., Hulbert, A. J., and Horning, D.,** A comparative study of heavy metal accumulation in tissues of the crested tern, *Sterna bergii,* breeding near industrialized and non-industrialized areas, *Aust. Wildl. Res.,* 8, 665, 1981.
33. **Reid, M. and Hacker, C. S.,** Spatial and temporal variation in lead and cadmium in the laughing gull, *Larus atricilla, Mar. Pollut. Bull.,* 13, 387, 1982.
34. **King, K. A., Lefever, C. A., and Mulhern, B. M.,** Organochlorine and metal residues in royal terns nesting on the central Texas coast, *J. Field Ornithol.,* 54, 295, 1983.
35. **Ohlendorf, H. M. and Harrison, C. S.,** Mercury, selenium, cadmium and organochlorines in eggs of three Hawaiian seabird species, *Environ. Pollut. Ser. B,* 11, 169, 1986.
36. **Brothers, N. P. and Brown, M. J.,** The potential use of fairy prions *(Pachyptila turtur)* as monitors of heavy metal levels in Tasmanian waters, *Mar. Pollut. Bull.,* 18, 132, 1987.
37. **Walsh, P. M.,** Geographical Variation in Mercury Levels of North Atlantic Gannets *Sula bassana,* presented at 3rd Int. Conf. Seabird Group, Cambridge, February 12 to 14, 1988.
38. **Appelquist, H., Drabaek, I., and Asbirk, S.,** Variation in mercury content of guillemot feathers over 150 years, *Mar. Pollut. Bull.,* 16, 244, 1985.
39. **Furness, R. W., Thompson, D. R., and Walsh, P. M.,** Evidence from biological samples for historical changes in global metal pollution, chapter in this volume.
40. **Schneider, R., Steinhagen-Schneider, G., and Drescher, H. E.,** Organochlorines and heavy metals in seals and birds from the Weddell Sea, in *Antarctic Nutrient Cycles and Food Webs: Proc. 4th SCAR Symp. Antarct. Biol.,* Siegfried, W. R., Condy, P. R., and Laws, R. M., Eds., Springer-Verlag, Berlin, 1985, 652.

41. **Honda, K., Yamamoto, Y., Hidaka, H., and Tatsukawa, R.,** Heavy metal accumulations in Adélie penguin, *Pygoscelis adeliae,* and their variations with the reproductive processes, *Mem. Natl. Inst. Polar Res., Spec. Issue,* 40, 443, 1986.
42. **Thompson, D. R.,** Metal levels in marine vertebrates, chapter in this volume.
43. **Fisher, G. L.,** Function and homeostasis of copper and zinc in mammals, *Sci. Total Eviron.,* 4, 373, 1975.
44. **Talbot, V. and Simpson, C.,** The validity of using arithmetic means to summarize environmental pollution data, *Chem. Aust.,* 50, 156, 1983.
45. **Blomqvist, S., Frank, A., and Petersson, L. R.,** Metals in liver and kidney tissues of autumn-migrating dunlin *Calidris alpina* and curlew sandpiper *Calidris ferruginea* staging at the Baltic Sea, *Mar. Ecol. Prog. Ser.,* 35, 1, 1987.
46. **Nicholson, J. K.,** The comparative distribution of zinc, cadmium and mercury in selected tissues of the herring gull *(Larus argentatus), Comp. Biochem. Physiol.,* 68C, 91, 1981.
47. **Braune, B. M.,** Comparison of total mercury levels in relation to diet and molt for nine species of marine birds, *Arch. Environ. Contam. Toxicol.,* 16, 217, 1987.
48. **Becker, P. H., Büthe, A., and Heidmann, W.,** Decreasing contaminants in shorebird eggs on the German North Sea coast?, *J. Orn.,* 129, 104, 1988.
49. **Hulse, M., Mahoney, J. S., Schroder, G. D., Hacker, C. S., and Pier, S. M.,** Environmentally acquired lead, cadmium, and manganese in the cattle egret, *Bubulcus ibis,* and the laughing gull, *Larus atricilla, Arch. Environ. Contam. Toxicol.,* 9, 65, 1980.
50. **Custer, T. W., Franson, J. C., Moore, J. F., and Myers, J. E.,** Reproductive success and heavy metal contamination in Rhode Island common terns, *Environ. Pollut. Ser. A,* 41, 33, 1986.
51. **Karlog, O., Elvestad, K., and Clausen, B.,** Heavy metals (cadmium, copper, lead and mercury) in common eiders *(Somateria mollissima)* from Denmark, *Nord. Vet. Med.,* 35, 448, 1983.
52. **Di Giulio, R. T. and Scanlon, P. F.,** Heavy metals in tissues of waterfowl from the Chesapeake Bay, USA, *Environ. Pollut. Ser. A,* 35, 29, 1984.
53. **Murton, R. K., Osborn, D. and Ward, P.,** Are heavy metals pollutants in Atlantic seabirds?, *Ibis,* 120, 106, 1978.
54. **Blus, L. J., Belisle, A. A., and Prouty, R. M.,** Relations of the brown pelican to certain environmental pollutants, *Pestic. Monit. J.,* 7, 181, 1974.
55. **Cramp, S., Ed.,** *The Birds of the Western Palearctic,* Vol. IV, Oxford University Press, Oxford, 1985, 174.
56. **Cheng, L., Schulz-Baldes, M. and Harrison, C. S.,** Cadmium in ocean-skaters, *Halobates sericeus* (Insecta), and in their seabird predators, *Marine Biol,* 79, 321, 1984.
57. **Lande, E.,** Heavy metal pollution in Trondheimsfjorden, Norway, and the recorded effects on the fauna and flora, *Environ. Pollut.,* 12, 187, 1977.
58. **Ohlendorf, H. M., Lowe, R. W., Kelly, P. R., and Harvey, T. E.,** Selenium and heavy metals in San Francisco Bay diving ducks, *J. Wildl. Manage.,* 50, 64, 1986.
59. **Ohlendorf, H. M., Anderson, D. W., Boellstorff, D. E., and Mulhern, B. M.,** Tissue distribution of trace elements and DDE in brown pelicans, *Bull. Environ. Contam. Toxicol.,* 35, 183, 1985.
60. **Norheim, G., Sømme, L., and Holt, G.,** Mercury and persistent chlorinated hydrocarbons in Antarctic birds from Bouvetøya and Dronning Maud Land, *Environ. Pollut. Ser. A,* 28, 233, 1982.
61. **Norheim, G. and Kjos-Hanssen, B.,** Persistent chlorinated hydrocarbons and mercury in birds caught off the west coast of Spitsbergen, *Environ. Pollut. Ser. A,* 33, 143, 1984.
62. **Furness, R. and Hutton, M.,** Pollutant levels in the great skua *Catharacta skua, Environ. Pollut.,* 19, 261, 1979.
63. **Burger, J., Gochfeld, M., and Leck, C.,** Waterbirds on Raritan Bay: a preliminary analysis of their distribution and heavy metal levels, *Bull. N.J. Acad. Sci.,* 29, 133, 1984.
64. **Osborn, D., Young, W. J., and Gore, D. J.,** Pollutants in auks from the 1983 North Sea bird wreck, *Bird Study,* 31, 99, 1984.
65. **Stoneburner, D. L., Patty, P. C., and Robertson, W. B., Jr.,** Evidence of heavy metal accumulations in sooty terns, *Sci. Total Environ.,* 14, 147, 1980.
66. **Gochfeld, M.,** Tissue distribution of mercury in normal and abnormal young common terns, *Mar. Pollut. Bull.,* 11, 362, 1980.
67. **Scheuhammer, A. M.,** The chronic toxicity of aluminium, cadmium, mercury, and lead in birds: a review, *Environ. Pollut.,* 46, 263, 1987.
68. **Friberg, L., Piscator, M., Nordberg, G. F., and Kjellstrom, T.,** *Cadmium in the Environment,* 2nd ed., CRC Press, Cleveland, OH, 1974.
69. **Gilbertson, M.,** Seasonal changes in organochlorine compounds and mercury in common terns of Hamilton Harbour, Ontario, *Bull. Environ. Contam. Toxicol.,* 12, 726, 1974.
70. **Furness, R. W., Muirhead, S. J., and Woodburn, M.,** Using bird feathers to measure mercury in the environment: relationships between mercury content and moult, *Mar. Pollut. Bull.,* 17, 27, 1986.

71. **Romanoff, A. L. and Romanoff, A. J.,** *The Avian Egg,* J. Wiley and Sons, New York, 1949.
72. **Friberg, L. and Vostal, J. V., Eds.,** *Mercury in the Environment,* CRC Press, Cleveland, OH.
73. **Fimreite, N.,** Accumulation and effects of mercury on birds, in *The Biogeochemistry of Mercury in the Environment,* Nriagu, J. O., Ed., Elsevier/North-Holland, Amsterdam, 1979, 601.
74. **Appelquist, H., Asbirk, S., and Drabaek, I.,** Mercury monitoring: mercury stability in bird feathers, *Mar. Pollut. Bull.,* 15, 22, 1984.
75. **Goede, A. A. and de Bruin, M.,** The use of bird feather parts as a monitor for metal pollution, *Environ. Pollut. Ser. B,* 8, 281, 1984.
76. **Gochfeld, M.,** Mercury levels in some seabirds of the Humboldt Current, Peru, *Environ. Pollut. Ser. A,* 22, 197, 1980.
77. **Jensen, S., Johnels, A. G., Olsson, M., and Westermark, T.,** The avifauna of Sweden as indicators of environmental contamination with mercury and chlorinated hydrocarbons, in *Proc. XVth Int. Ornithol. Congr., Leiden,* 1972, 455.
78. **Becker, P. H., Ternes, W., and Rüssel, H. A.,** Pollutants in eggs of birds breeding at the German North Sea coast. II. Mercury, *J. Orn.,* 126, 253, 1985.
79. **Fimreite, N., Kveseth, N., and Brevik, E. M.,** Mercury, DDE, and PCBs in eggs from a Norwegian gannet colony, *Bull. Environ. Contam. Toxicol.,* 24, 142, 1980.
80. **Magat, W. and Sell, J. L.,** Distribution of mercury and selenium in egg components and egg-white proteins, *Proc. Soc. Exp. Biol. Med.,* 161, 458, 1979.
81. **Parslow, J. L. F., Jefferies, D. J., and French, M. C.,** Ingested pollutants in puffins and their eggs, *Bird Study,* 19, 18, 1972.
82. **Bull, K. R., Every, W. J., Freestone, P., Hall, J. R., and Osborn, D.,** Alkyl lead pollution and bird mortalities on the Mersey Estuary, UK, 1979-1981, *Environ. Pollut. Ser. A,* 31, 239, 1983.
83. **Bagley, G. E. and Locke, L. N.,** The occurrence of lead in tissues of wild birds, *Bull. Environ. Contam. Toxicol.,* 2, 297, 1967.
84. **Doi, R. and Fukuyama, Y.,** Metal content in feathers of wild and zoo-kept birds from Hokkaido, 1976—78, *Bull. Environ. Contam. Toxicol.,* 31, 1, 1983.
85. **Osborn, D.,** The significance of metal residues in wild animals, in *Proc. Int. Conf. Management and Control of Heavy Metals in the Environment, 1979,* CEP Consultants, Edinburgh, 1979, 187.
86. **Ojanen, M., Haarakangas, H., and Hyvärinen, H.,** Seasonal changes in bone mineral content and alkaline phosphatase activity in the house sparrow *(Passer domesticus* L.), *Comp. Biochem. Physiol.,* 50A, 581, 1975.
87. **Osborn, D.,** Seasonal changes in fat, protein and metal content of the liver of the starling *Sturnus vulgaris, Environ. Pollut.,* 19, 145, 1979.
88. **Osborn, D.,** A naturally occurring cadmium and zinc binding protein from the liver and kidney of *Fulmarus glacialis,* a pelagic North Atlantic seabird, *Biochem. Pharmacol.,* 27, 822, 1978.
89. **Braune, B. M. and Gaskin, D. E.,** A mercury budget for the Bonaparte's gull during autumn moult, *Ornis Scand.,* 18, 244, 1987.
90. **Honda, K., Nasu, T., and Tatsukawa, R.,** Seasonal changes in mercury accumulation in the black-eared kite, *Milvus migrans lineatus, Environ. Pollut. Ser. A,* 42, 325, 1986.
91. **Edelstein, S., Fullmer, C. S., and Wasserman, R. H.,** Gastrointestinal absorption of lead in chicks: involvement of the cholecalciferol endocrine system, *J. Nutr.,* 114, 692, 1984.
92. **Muirhead, S. J.,** The Accumulation, Storage and Elimination of Metals and Organochlorines in the Great Skua *Catharacta skua skua* and Metal Accumulation in Atlantic Procellariformes, Ph. D. thesis, University of Glasgow, Glasgow, 1986.
93. **Furness, R. W.,** *The Skuas,* Poyser, Calton, England, 1987.
94. **Thompson, D. R.,** Total and organic mercury concentrations in liver tissue of seabirds, *Proc. Int. Conf. Seabird Group,* Cambridge, 1988, 44.
95. **Koeman, J. H., van Velzen-Blad, H. C. W., de Vries, R., and Vos, J. G.,** Effects of PCB and DDE in cormorants and evaluation of PCB residues from an experimental study, *J. Reprod. Fert. Suppl.,* 19, 353, 1973.
96. **Nicholson, J. K. and Osborn, D.,** Kidney lesions in pelagic seabirds with high tissue levels of cadmium and mercury, *J. Zool., London,* 200, 99, 1983.
97. **Bremner, I. and Davies, N. T.,** The induction of metallothionein in rat liver by zinc injection and restriction of food intake, *Biochem. J.,* 149, 733, 1975.
98. **Hutton, M. and Goodman, G. T.,** Metal contamination of feral pigeons *Columba livia* from the London area: part 1 — tissue accumulation of lead, cadmium and zinc, *Environ. Pollut. Ser. A,* 22, 207, 1980.
99. **Di Giulio, R. T. and Scanlon, P. F.,** Effects of cadmium ingestion and food restriction on energy metabolism and tissue metal concentrations in mallard ducks *(Anas platyrhynchos), Environ. Res.,* 37, 433, 1985.
100. **Norseth, T. and Clarkson, T. W.,** Studies on the biotransformation of ^{203}Hg-labelled methyl mercury chloride in rats, *Arch. Environ. Health,* 21, 717, 1970.

101. **Norheim, G. and Frøslie, A.,** The degree of methylation and organ distribution of mercury in some birds of prey in Norway, *Acta Pharmacol. Toxicol.,* 43, 196, 1978.
102. **Stoewsand, G. S., Bache, C. A., and Lisk, D. J.,** Dietary selenium protection of methylmercury intoxication of Japanese quail, *Bull. Environ. Contam. Toxicol.,* 11, 152, 1974.
103. **Koeman, J. H., Peeters, W. H. M., Koudstaal-Hol, C. H. M., Tjioe, P. S., and de Goeij, J. J. M.,** Mercury-selenium correlations in marine mammals, *Nature,* 245, 385, 1973.
104. **Koeman, J. H., van de Ven, W. S. M., de Goeij, J. J. M., Tjioe, P. S., and van Haaften, J. L.,** Mercury and selenium in marine mammals and birds, *Sci. Total Environ.,* 3, 279, 1975.
105. **Beijer, K. and Jernelöv, A.,** Ecological aspects of mercury-selenium interactions in the marine environment, *Environ. Health Perspect.,* 25, 43, 1978.
106. **Focardi, S., Leonzio, C., and Renzoni, A.,** Chlorinated hydrocarbons and trace metals in eggs of *Larus argentatus michahellis* Naumann and *Sterna albifrons albifrons* Pallas, in *Atti Congr. Soc. Ital. Ecol., 1° Salsomaggiore Terme, Parma,* 1980, 411.
107. **Parizek, J., Ostadalova, I., Kalouskova, J., Babicky, A. and Benes, J.,** The detoxifying effects of selenium: interrelations between compounds of selenium and certain metals, in *Newer Trace Elements in Nutrition,* Mertz, W. and Cornatzer, W. E., Eds., Marcel Dekker, New York, 1971, 85.
108. **Thingstad, G.,** Environmental contaminants in some adult *Larus fuscus fuscus* from Tarva, an islet on the coast of Sørtrøndelag, *Fauna Norv. Ser. C Cinclus,* 9, 55, 1986.
109. **Scheuhammer, A. M.,** Chronic dietary toxicity of methylmercury in the zebra finch, *Poephila guttata, Bull. Environ. Contam. Toxicol.,* 40, 123, 1988.
110. **Osborn, D., Every, W. J., and Bull, K. R.,** The toxicity of trialkyl lead compounds to birds, *Environ. Pollut. Ser. A,* 31, 261, 1983.
111. **Tejning, S.,** Biological effects of methyl mercury dicyandiamide-treated grain in the domestic fowl *Gallus gallus* L., *Oikos Suppl.,* 8, 1, 1967.
112. **Gardner, D.,** Mercury in fish and waters of the Irish Sea and other United Kingdom fishing grounds, *Nature,* 272, 49, 1978.
113. **Dyck, J. and Kraul, I.,** Environmental pollutants and shell thinning in eggs of the guillemot *Uria aalge* from the Baltic Sea and the Faeroes, and a possible relation between shell thickness and sea water salinity, *Dan. Ornithol. Foren. Tidsskr.,* 78, 1, 1984.
114. **Odsjø, T.,** Toxic chemicals in sedentary and migratory birds in Fennoscandia and the Baltic area, *Ornis Fennica,* 52, 74, 1975.
115. **Johnels, A., Tyler, G., and Westermark, T.,** A history of mercury levels in Swedish fauna, *Ambio,* 8, 160, 1979.
116. **Lambertini, M.,** Mercury levels in *Larus audouinii* and *Larus argentatus michahellis* breeding in Capraia Island (Tyrrhenian Sea), *Riv. Ital. Ornithol. Milano,* 52, 75, 1982.
117. **Bijleveld, M. F. I. J., Goeldlin, P., and Mayol, J.,** Persistent pollutants in Audouin's gull *(Larus audouinii)* in the western Mediterranean: a case study with wide implications?, *Environ. Conserv.,* 6, 139, 1979.

Chapter 11

EVIDENCE FOR HISTORICAL CHANGES IN GLOBAL METAL POLLUTION: SNOW AND ICE SAMPLES

Eric Wolff

TABLE OF CONTENTS

I. THE RECORD IN SNOW AND ICE

Snowcover contains within it a sample of the contents of the atmosphere at the time the snow fell, modified by processes occuring in the snowflake and subsequently in the snowpack. Snow can remain frozen for a few days, for a single winter, for decades or centuries (e.g., Alpine glaciers), or for thousands of years in permanent ice caps and ice sheets. Where it does persist, it contains a potentially unique record of changes in the atmosphere. In the extreme cases, in central Antarctica and Greenland, the ice is several kilometers thick and it may be possible to obtain useful information over more than 500,000 years.

In practice, melting and movement of water and impurities within the snow will confuse the picture in all but the highest, coldest mountain glaciers and the polar ice caps. Ice cores, particularly from Antarctica and Greenland, have been studied extensively for many parameters.

Useful information is enclosed in the ice in three main ways: (1) The isotopic composition of the ice itself is determined mainly by changes in climatic factors. (2) Aerosol material is incorporated into snowflakes or deposits onto the snow surface. (3) As the snow sinks deeper into the ice sheet, it is compressed to form solid ice with enclosed bubbles of air which record the content of stable gases in the air. Each of these types of information has been used. For instance, the oxygen isotope composition of ice from Vostok in Central Antarctica has been determined from an ice core extending 150,000 years into the past.[1] This has allowed temperature changes to be inferred over a complete glacial cycle. The CO_2 content of the same core[2] has shown a close correspondence between changes in temperature and in carbon dioxide levels. Increases in the atmospheric concentration of CO_2 over the industrial period have also been clearly demonstrated in ice cores.[3] From the aerosol component, anions, such as sulfate and nitrate have been particularly well studied, both in long records[4] and short ones.[5,6] It is in the aerosol component that the metal record is to be found. There are parts of Antarctica and Greenland where a meter of snow falls each year, so that it is possible in cores of a hundred meters depth to measure pollutant concentrations throughout the industrial period with seasonal resolution. Research into heavy metals in snow has been carried out for over 20 years. It might, therefore, be expected that there would now be complete and reliable records of heavy metal concentrations in snow over a period of at least the last 200 years. Unfortunately, technical problems have prevented this promise being fulfilled. As a result, the story to be told here is at best patchy and at worst misleading.

II. PROBLEMS OF ANALYSIS AND CONTAMINATION

The main problems arise from the very low concentrations of metals involved. For many metals in the polar ice sheets, concentrations are in the low-pg/g or sub-pg/g range. These levels are lower than those in any other natural waters (except, in some cases, deep sea water), and lower than those obtainable in many laboratory water purification systems.

This has led to two major difficulties. The first is to find an analytical technique capable of achieving such low detection limits. The second has been to collect, treat, and analyze samples without introducing contamination. Only in the last 5 years have some analysts solved these problems to an extent where reasonable data have been produced. The literature is, therefore, full of unreliable data that can trap the unsuspecting outsider. Some criteria can be given for assessing data, and any work not meeting these criteria is suspect.

A. ANALYSIS

The ideal technique would allow metals to be analyzed at the pg/g level and below with the minimum of pretreatment. In practice, most analytical methods require a preconcentration step. The techniques that have provided published data generally considered reliable are

FIGURE 1. Stringent clean techniques should be used in the collection of snow and ice samples in the field.

isotope dilution mass spectrometry (IDMS) (for Pb alone)[7] and atomic absorption spectrometry using a simple preconcentration onto tungsten wires.[8,9] Anodic stripping voltammetry has been used with some success.[10]

Many techniques used in the past appeared suitable only because they were successfully determining concentrations that were elevated by contamination. Newer techniques, such as inductively coupled plasma-mass spectrometry (ICP-MS), laser fluorescence spectrometry, and laser photoionization analysis hold out promise for the future. They may be able to offer lower detection limits, multielement analysis or simple isotopic analyses.

B. CONTAMINATION

Contamination can be introduced at every stage from before the snow falls until the analysis is complete. Solutions have been found to many of the possible sources of contamination but areas of uncertainty remain.

Contamination can occur in the atmosphere from nearby local sources, such as a manned base. In this case, concentrations subsequently measured in the snow will not have any global or regional significance. There is currently some discussion about the extent of contamination from local sources in the Antarctic,[11] and this is dealt with later. We have little knowledge of the distance traveled by local emissions. However, sampling sites should be as far as possible from manned bases and other sources, and upwind of them. The air and the snow surface can also be affected by activities at the sampling site itself. Again, samples should be taken well upwind of all activities, and contaminating items (such as leaded fuels) should be eliminated from field operations.

During sampling, contamination can be added by the sampling personnel, who should, therefore, wear at least full clean-room clothing, and by sampling tools (Figure 1). These may range from plastic bottles, saws, and chisels to ice drills capable of reaching deep into the ice sheet. It is desirable that the cleanest possible materials are used (e.g., stainless steel, polyethylene, fluorocarbons). However, it is recognized as virtually impossible to collect samples in the field totally cleanly (especially where deep drills are used). It is, therefore, far more important to collect samples that are large enough for subsampling or removal of

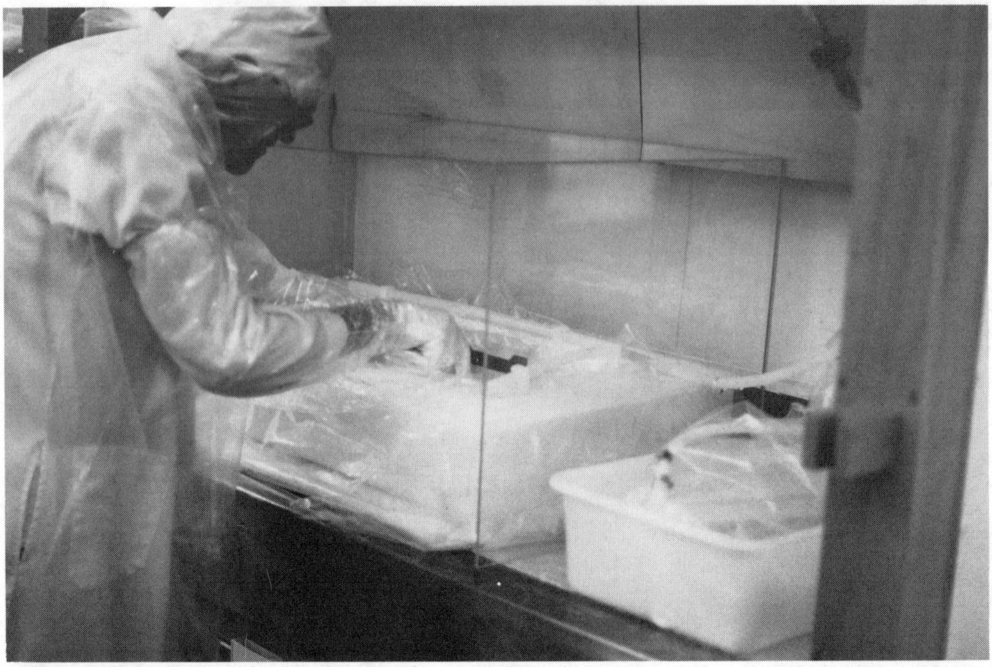

FIGURE 2. Extreme precautions are required during work in the clean laboratory to subsample Antarctic ice cores. (Photo courtesy of C. Boutron.)

contaminated outer layers to be possible at a later stage. Crucial to success is that samples should be kept below $-10°C$ at all times during transport to avoid the movement of contaminants from the outside to the center of the sample.

Once the sample is returned to the home laboratory, scrupulous cleanliness is still vital, but it is easier to monitor sampling tools and reagents for signs of contamination. It is now considered mandatory[12] to take samples across cores or blocks, from the outside in, to see whether contamination on the outside has reached the center. Successive samples can be taken either using a clean subsampler,[9] or by chiseling successive veneers from samples[7] (Figure 2). Only if a concentration plateau is reached before the inner sample can the central value of the sample be considered acceptable. In solid ice, even some cores drilled with drilling fluid have been decontaminated successfully using these techniques.[13]

In the laboratory, blanks should be obtained for all reagents and procedures (using ultrapure water of known content). In this way, a total procedural blank can be derived and subtracted from obtained results.

Only a very few published papers meet the criteria described above, and only their data are considered reliable and discussed here.

III. A BRIEF HISTORY OF HEAVY METAL ANALYSIS IN SNOW

The potential of ice cores to provide data on heavy metals became clear with the publication of a landmark paper by Clair Patterson and co-workers in 1969.[14] The major achievement of this work was to present a temporal curve of Pb in Greenland snow from 800 B.C. to the present. It provided the first clear evidence of a massive pollution of the northern hemisphere by lead aerosol. Details of it have been criticized since,[15-17] and in some respects, it does not fulfil the criteria given above. Nonetheless, its painstaking description of suitable clean procedures have made it the standard by which later work has been judged, and the findings have not been challenged substantially.

During the 1970s, there followed a period in which a number of other investigators attempted to emulate Patterson's[14] achievement for Pb and other metals. It is now clear that, although each made valiant efforts, none really solved the problems of contamination and analysis. The data that cannot now be considered acceptable are discussed and enumerated in two earlier reviews.[18,19] The finding of both these reviews can best be summarized by stating that, with one exception,[20] no data on heavy metals in polar snow produced in the years 1970 to 1982 inclusive can be considered reliable. The findings of this period had been highly confusing, suggesting that many metals were greatly enriched, apparently by natural processes, in polar ice.

In the early 1980s, spurred on by criticism and by new data[20] from Patterson's group, a concensus started to emerge as to the way forward by the two groups most active in this field, in Grenoble and Cambridge. Both groups acknowledged that their previous data were wrong[18,19] and both adopted the techniques described above to reduce contamination.[7,9]

Subsequently, the Grenoble group has made considerable progress on Pb in the Antarctic, providing a convincing record of this metal over a full glacial cycle,[13,21,22] thus elucidating the role of natural sources of lead in snow. They have also published data on present-day concentrations of Pb.[11] The Cambridge group, concentrating to date on the simpler problem of near-surface snow, has provided the best values for present-day snow in both Greenland and Antarctica for a set of four metals.[9,23] Further rapid progress seems likely in the next 2 to 3 years.

IV. DATA

A. Pb IN ANTARCTIC SNOW

Because of the intense environmental interest, Pb is the metal for which the most data are available, and it is pleasing to report a reasonably coherent picture.

A total of 20 pieces at various depths in cores from Vostok and Dome C in Central Antarctica have been analyzed using IDMS methods.[13,21,22,24] Both cores were highly contaminated on the outside, so that an elaborate chiseling procedure was used to obtain profiles across the cores and to produce clean inner subsamples. The resulting record of Pb over the last 155,000 years is shown in Figure 3. The main features are large increases in Pb concentration at the end of the last and penultimate glaciations. It is possible to use Al concentrations as a marker of windblown, continental dust (called here the crustal input). Using ratios of metals, such as Pb, to Al in average crustal rocks and soils, the input of Pb from the crustal source can be estimated (to within a factor 3). Similarly, using Na and part of the sulfate contribution as markers for marine and volcanic input, we can calculate their contributions to the Pb concentration. The peaks in concentration in the glacial maxima correspond to high values of Al, due to an influx of terrestrial dusts, and can be almost completely accounted for by this source. The volcanic contribution, though highly uncertain, may be a significant part of the input (up to 50%) in the low concentration interglacial periods. The marine contribution, even allowing for enrichment in the sea-surface microlayer, is always negligible.

The concentration of Pb in Antarctic snow before the industrial era is, therefore, seen to be about 0.5 pg/g, made up of natural crustal and volcanic Pb. There is no longer any room for exotic natural enrichment mechanisms in this value.

Changes in Pb concentration over the last 200 years are much harder to assess. Boutron and Patterson[7] attempted to produce a record of these concentrations. However, the same authors now believe that only some of the values presented there were correct.[11] In the absence of any firm reasons to accept or reject the values, it must be concluded that the time variations of Pb in Antarctic snow over the last 200 years are essentially unknown.

For present day snow, various values have been presented. For the Antarctic Peninsula,

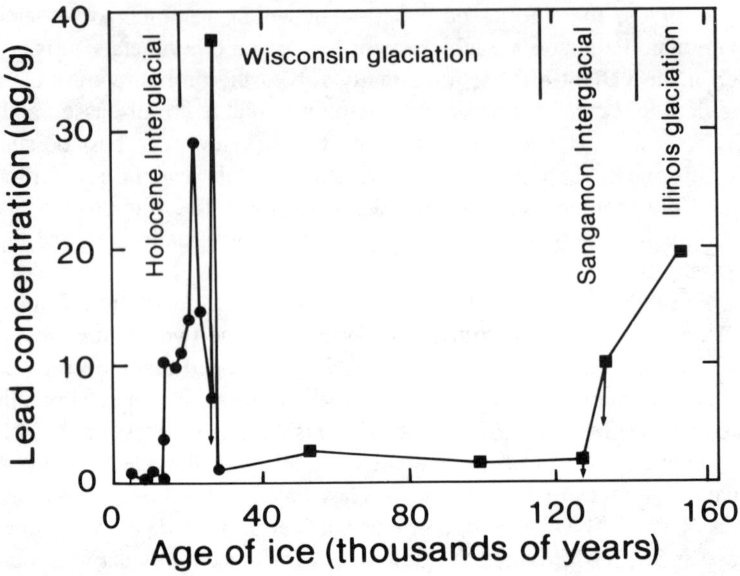

FIGURE 3. Variations in Pb concentrations at Dome C (circles) and Vostok (squares), Antarctica during the last 155,000 years.[22] Arrows mark samples where no clear central plateau was obtained, so that these values are upper limits. (Figure courtesy of C. Boutron/La Recherche, from Boutron C., *La Recherche*, 198, 454, April, 1988.)

two sets of samples covering only a part year gave concentrations of 6.3 pg/g[9] and 5.5 pg/g.[25] In east Antarctica, surface snow values ranging from 2.3 to 7.4 pg/g are reported.[11] The authors believe that some of these sites are affected by contamination from local bases,[11] and that only the value of 2.3 pg/g has a global significance. It is indeed possible that, in the pristine Antarctic environment, even small point sources could have a large influence. However, the observed pattern of concentrations does not seem consistent with this explanation alone. Further study of the emissions from bases, and of their effect at distance, is needed to clarify this point.

In summary, the preindustrial Pb concentration of about 0.5 pg/g can be accounted for entirely by known natural sources. The present-day concentration represents an increase of between 5- and 12-fold over this value. However, the time-course of the increase and the global significance of the finding are not yet established.

B. Pb IN GREENLAND SNOW

The record of Pb in Greenland snow produced in 1969,[14] supplemented by further data on ancient snow[20] is presented in Figure 4. There have been criticisms of the data,[15-17] including the fact that samples came from a number of different sites within a distance of 80 km from Camp Century in north Greenland. In addition, it is now recognized that some samples in the same paper from Byrd Station, Antarctica, did show contamination, so it must be possible that some of the Greenland samples were contaminated also, especially by other operations in the Camp Century area. However, the reported general increase has stood the test of time, even if the detailed time sequence is less certain. Other investigations were able to confirm the value of about 200 pg/g for snowfall at various Greenland sites in the 1970s.[15,26,27] The value of about 1 pg/g for ancient ice is close to that expected from natural crustal material.

A more recent paper,[23] investigating snow that fell in 1983 to 1984, found a lower average value of 28 pg/g for a complete year's snowfall at Dye 3 in south Greenland. This lower modern value may be in part a first sign of the smaller usage of leaded gasoline

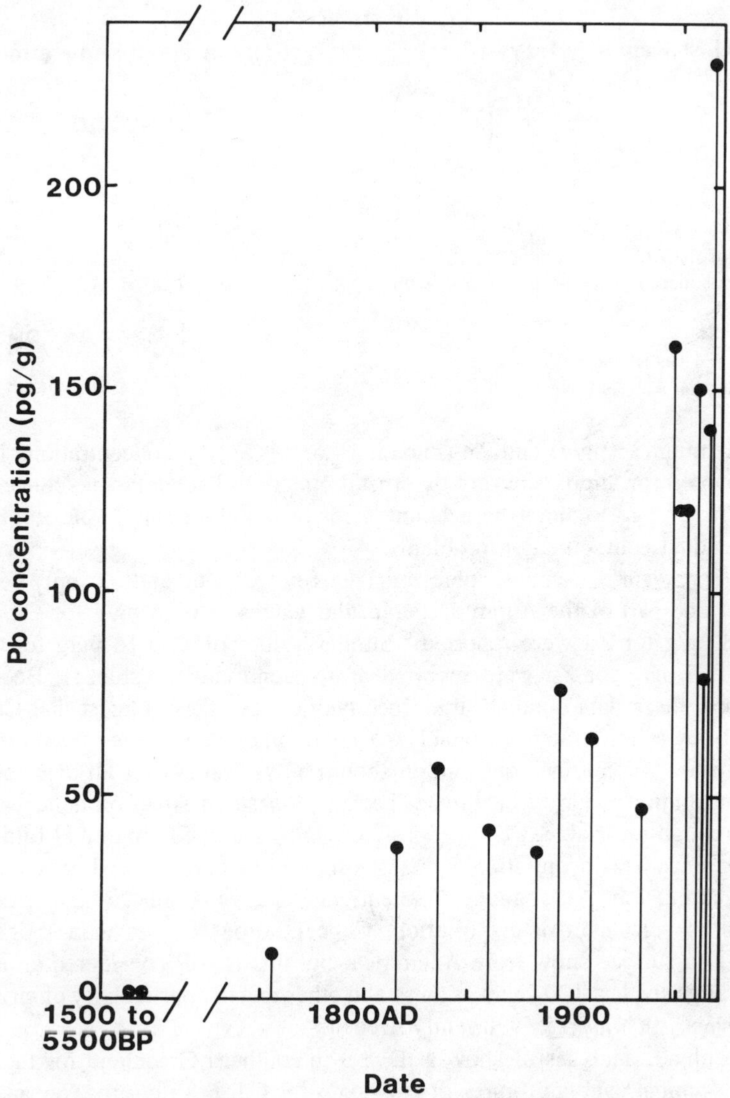

FIGURE 4. Variations in Pb concentrations in the region of Camp Century, Greenland.[14,20] (Reprinted by permission from Wolff, E. W. and Peel, D. A., *Nature*, 313, 535, 1985. Copyright (c) 1985 Macmillan Magazines Ltd.)

following legislation, particularly in the U.S. Such a decrease has also been seen in records of Pb in corals[28] in the north Atlantic. It is perhaps surprising if such a large decrease is seen by 1984 unless the North America source is dominant for south Greenland. This finding, therefore, requires confirmation by further analyses and indeed a new complete record for the last several hundred years is a priority for research.

In summary, an increase of at least 100-fold was observed for Pb in Greenland snow from ancient times to the 1960s. This may now have reduced in response to smaller emissions.

C. OTHER METALS

Data for other metals are even more sparse than for Pb. There are no data considered reliable for preindustrial snow and ice except for some analyses of ice from Dome C, Antarctica.[29] These showed raised concentrations of metals during the last glacial maximum.

TABLE 1
Most Reliable Heavy Metal Concentrations in Polar Snow and Ice

	Concentrations (pg/g)			
	Cd	Cu	Pb	Zn
Antarctica				
Recent snow (part years only)[9,11,25]	0.2	2	2-7	4
Preindustrial ice[22]			0.5	
Highest concentration reported from Wisconsin glaciation[21,29]	1.0	33	29	62
Greenland				
1983—84[23]	0.74	6.2	28	27
1960s and 1970s[14,15,26,27]			200	
Preindustrial ice[14,20]			1	

For Cu (maximum 33 pg/g) and Zn (maximum 62 pg/g), the concentrations in this period can be accounted for almost entirely by crustal material. The Holocene concentrations (Cu 2.7 pg/g, Zn 5.7 pg/g) cannot be accounted for by known natural sources, and could be reflecting residual contamination problems.

For Antarctic surface snow, there are data for Cd, Cu, and Zn only for part-years' snowfall. At one part of the Antarctic Peninsula, values of 0.26 pg/g for Cd, 1.8 pg/g for Cu, and 3.3 pg/g for Zn were reported.[9] Similar values of $<$ 0.16 pg/g for Cd, 2.3 pg/g for Cu and 5.5 pg/g for Zn were reported at a second site[25] (Table 1). Because of their limited nature, these data contain large uncertainties, but they suggest that Cu and Zn are enriched only by a small factor (probably $<$6) over amounts expected from crustal material alone (Figure 5). The marine contribution should be very small for all these elements. The volcanic contribution is hard to estimate, because measured emission data for metals from volcanoes are highly variable. Cd appears significantly enriched (up to 200-fold) over crustal material, but since its concentration is the lowest of any metal reported, this may still reflect residual contamination. Alternatively, the enrichment may be due to high-temperature processes, such as volcanism or to pollution. Concentrations of less than 1 pg/g have been found for Hg in surface snow from Antarctica, but this is still considered an upper limit.[30] A record through the last 100 years of these and other metals, particularly of environmentally interesting ones, such as Hg, is urgently needed.

For Greenland, analyses of snow at Dye 3 in southern Greenland for the period 1983 to 1984 gave annual average figures of 0.74 pg/g for Cd, 6.2 pg/g for Cu, and 27 pg/g for Zn[23] (Table 1). Cu and, for some samples, Zn have concentrations very close to those expected from crustal material alone (based on Al concentrations). This suggests that their source is largely natural, and that any increase is confined to a factor of two to three at most. It may be expected that Zn concentrations have fallen from a maximum in the period 1940 to 1970, according to estimates of Zn production in the U.S.[31] Cd is again substantially enriched over levels expected from crustal material alone. It is impossible clearly to assign the cause of this at present. Some data for Hg[32] suggest values of 2 to 15 pg/g, with no trend between 1927 and 1971. However, it is considered[18,19] that these values may still include contamination. It is evident that in Greenland, there is also an urgent need to obtain reliable historical records of these and other metals through the industrial period.

D. LIMITATIONS OF THE DATA PRESENTED

The reader will by now be very disappointed at the paucity of the data presented. Some further caveats must also be added.

1. Total Metal

None of the data described have been produced by methods involving a digestion likely to have analyzed total metal. Most analyses were carried out at close to pH 2[11,13,21,22] or at

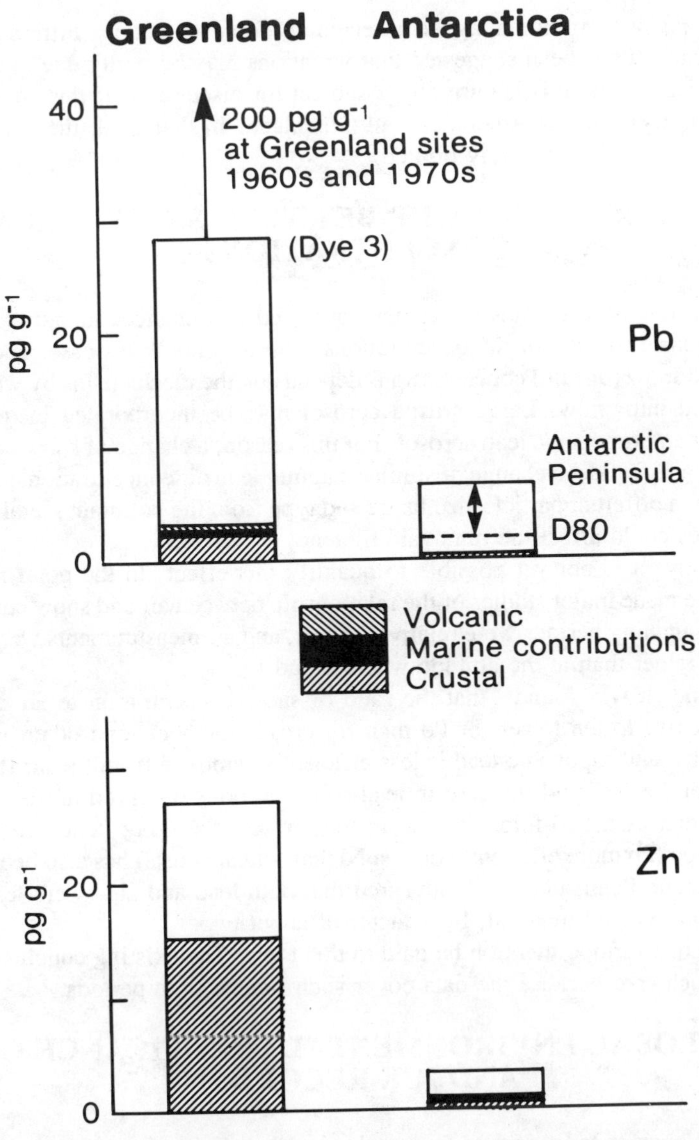

FIGURE 5. The concentrations of Pb and Zn in recent snow in Antarctica and Greenland.[9,11,23] The shaded areas are estimates of the natural inputs of metals to the snow. The unshaded area is the metal unaccounted for by known natural sources. Measured preindustrial Pb concentrations are about 0.5 pg/g for Antarctica, 1 to 2 pg/g for Greenland.

pH 3.[9,23,25] Evidence has been presented that these methods will analyse all but the most firmly bound aluminosilicate material.[9] Results through a glacial cycle, showing the close relationship between Pb and crustal material,[13,21,22] suggest that at pH 2, even crustal Pb is analyzed. However, comparative analyses using full digestion are required to show whether all the metal is being analyzed or not.

2. Representativeness of the Samples

Some of the data presented are for part years only, so that the data must be unrepresentative if there are seasonal variations in concentration. There may also be differences between different sites on the same ice sheet, as appears to be the case for Pb in Antarctic surface snow.[11] Concentration may be affected to some extent by differences in snow accumulation rate. It has been suggested that variations may be partly due to the emissions from local sources,[11] and this is currently a subject for discussion. If this were the case, it might apply to other sites in Antarctica and Greenland. In that case, the value of records from these areas would be severely limited.

V. THE RELATIONSHIP BETWEEN SNOW AND AIR CONCENTRATIONS

Throughout this paper, it has so far been assumed that changes in snow concentration reflect proportionate changes in air concentrations. This may not be the case. The relationship between air and snow (or rain) concentrations depends on the mechanisms by which particles are incorporated into snow. Large crustal aerosol may be incorporated in quite different ways to smaller anthropogenic lead aerosol. For this reason, a change of snow concentrations does not necessarily reflect a change of similar magnitude in air concentrations, as the natural metal may be in a different particle size range and type from the pollutant metal. In addition, climatic changes could affect the removal efficiency of aerosol.

Unfortunately, it is not yet possible to quantify this effect. In the polar regions, only two groups have made major studies of the relationship between air and snow concentrations. In both cases, sampling periods were relatively short, and air measurements were made close to the ground, rather than at the altitude where cloud forms.

In Greenland, it was found[33] that the ratio of snow concentration to air concentration was a factor of five to ten lower for Pb than for crustal elements and other heavy metals. This suggests that anthropogenic lead is less efficiently removed from the air than is crustal material, so that the 100-fold increase in lead concentrations observed in the snow[14] might represent as much as a 500-fold increase in the air. On the other hand, an effect in the opposite direction (Pb more efficiently deposited than silicate dusts) has also been reported.[20]

In the Antarctic Peninsula,[25,34] it appeared that both lead and zinc were less efficiently removed than was crustal material, by a factor of about two.

It is urgent that further attention be paid to this topic. The existing conclusions must be treated with much care because the data cover such limited time periods.

VI. THE GLOBAL ENVIRONMENTAL SIGNIFICANCE OF SNOW AND ICE RECORDS

It has often been stated that data from polar ice sheets give a record of global changes in pollution and of global background levels. These statements are clearly simplistic. The barrier to interhemispheric transport of material is strong, and the relative concentrations of metals found in Greenland and Antarctica (Figure 5) show a clear contrast. 90% of pollution is emitted in the northern hemisphere and this is shown by the Pb data in particular.

The records cannot even truly be said to give a hemispheric record. In Greenland, there is some suggestion that changes in Pb concentration may be reflecting primarily North American emissions,[23] whereas other parts of the Arctic seem particularly sensitive to Eurasian emissions.[35] There are parts of the Arctic which are at an altitude such that they appear not to have experienced increases (seen elsewhere in the Arctic) in other pollutants such as sulfate,[36] and this may turn out to be the case for Pb also.

There is also a seasonal variation in the input of metals to the Arctic, with a winter-spring maximum when the polar front is further south.[23,37,38] Thus the Greenland samples are probably sampling a different part of the Arctic troposphere at different times of year.

In the more isolated Antarctic, assuming that local emissions are not important, the concentrations may indeed represent more of a smoothed average from the southern hemisphere. Nevertheless, this should not be taken as meaning that a preindustrial concentration for Antarctica is identical to a preindustrial concentration for the whole southern hemisphere. The Antarctic circumpolar vortex provides an efficient barrier to the ingress of particles, be they anthropogenic or natural.

What can be said is that the snow records present a vivid picture of the input of pollutants to even the most pristine environment and, therefore, a marker as to how widespread any problem may be. The variation between different parts of the ice sheet provides very useful information on transport pathways. Most of all, where sources can be assigned to the preindustrial concentrations, as for Pb in Antarctica, it becomes possible to calculate in a similar fashion the preindustrial air concentration in other parts of the world. This may be particularly applicable to the oceanic atmosphere, and to the ocean waters below. The snow and ice records are of less relevance to more isolated shallow seas, where local terrestrial influences may confuse the picture, and where in any case the marine environment may be dominated by fluvial inputs.

VII. ANALYSES FROM GLACIERS IN TEMPERATE REGIONS

It should be possible to obtain records from glaciers outside the polar ice caps, provided these are high and cold enough that no melting takes place. In practice, the attempts that have been made have suffered from problems similar to those found in polar regions. A review[39] concluded that the available data were sparse and generally questionable, and they will not be considered further here. Such records do, however, have great potential importance. They will of course reflect regional rather than global or hemispheric changes. A collection of such records from various regions would however be just as valuable as a single 'global' record.

VIII. CONCLUSIONS AND EXPECTED FUTURE ADVANCES

From the large amount of work carried out on snow and ice samples, it has been possible to extract only very limited information on historical changes in global heavy metal pollution. In the absence of temporal records for most elements other than Pb, most inferences are based on a comparison of estimated natural inputs with measured present-day concentrations.

It has been possible to conclude that, for Pb, there has been a small (5- to 12-fold) increase in concentrations in Antarctic snow, and a larger (100-fold) increase in concentrations in Greenland. This latter figure may be decreasing in response to legislation reducing usage of leaded gasoline (petrol). There is slight evidence that these increase factors might translate to larger figures still for increases in the atmosphere. However, there remain questions about the data and about the extent to which they are affected by local sources. Nonetheless, the increases seen appear to demonstrate the global nature of lead pollution. Values in ancient ice have allowed an assignation of the natural sources of lead, which in turn allows in principle a calculation of the natural atmospheric input to individual regions of the world. Finally for Pb, we are able to see that there have been periods in the past (during the ice age maximum) when Pb concentrations in the Antarctic were far higher than today (up to 30 pg/g). Similar calculations based on measured Al values in the glacial maximum in Greenland[40] suggest that we may expect even higher Pb concentrations in such samples for Greenland than for Antarctica. These values may help to set modern levels in context.

For other metals, only present-day concentrations can be estimated. These suggest that, in both Greenland and Antarctica, increases in Cu and Zn can only have been modest. In the last glacial maximum, crustal material was the main source of both Cu and Zn. Sources of Cd are still unclear. For one important metal, Hg, there are still no reliable data.

In the next few years, we can expect to see detailed records for several metals over the last one to two centuries. Suitable samples have already been collected in Antarctica, and collections are planned for Greenland. Further study of samples through a complete glacial cycle may also be expected from Antarctica for metals other than Pb. This should help to elucidate the natural sources of these metals. Lead isotope analysis should also help in determining source areas.[41,42] If the above analyses are successful, it should be possible to extend the work to provide regionally significant records. Finally, there will need to be much more understanding of the relationship between snow records and air records. This will require detailed programs of simultaneous air and snow sampling.

The field of heavy metal analysis in snow is emerging from a dark era of contaminated samples and confusing data. There is some hope that a similar review in 3 years time would be much stronger on data and lighter on excuses!

REFERENCES

1. **Lorius, C., Jouzel, J., Ritz, C., Merlivat, L., Barkov, N. I., Korotkevich, Y. S., and Kotlyakov, V. M.,** A 150,000-year climatic record from Antarctic ice, *Nature,* 316, 591, 1985.
2. **Barnola, J. M., Raynaud, D., Korotkevich, Y. S., and Lorius, C.,** Vostok ice core provides 160,000 year record of atmospheric CO_2, *Nature,* 329, 408, 1987.
3. **Neftel, A., Moor, E., Oeschger, H., and Stauffer, B.,** Evidence from polar ice cores for the increase in atmospheric CO_2 in the past two centuries, *Nature,* 315, 45, 1985.
4. **Legrand, M.,** Chemistry of Antarctic snow and ice, *J. Physique,* 48, C1-77, 1987.
5. **Mayewski, P. A., Lyons, W. B., Spencer, M. J., Twickler, M., Dansgaard, W., Koci, B., Davidson, C. I., and Honrath, R. E.,** Sulfate and nitrate concentrations from a south Greenland ice core, *Science,* 232, 975, 1986.
6. **Mulvaney, R. and Peel, D. A.,** Anions and cations in ice cores from Dolleman Island and the Palmer Land plateau, Antarctic Peninsula, *Ann. Glaciol.* 10, 121, 1988.
7. **Boutron, C. F. and Patterson, C. C.,** The occurrence of lead in Antarctic recent snow, firn deposited over the last two centuries and prehistoric ice, *Geochim. Cosmochim. Acta.,* 47, 1355, 1983.
8. **Wolff, E. W., Landy, M. P., and Peel, D. A.,** Preconcentration of cadmium, copper, lead, and zinc in water at the 10^{-12} g/g level by absorption onto tungsten wire followed by flameless atomic absorption spectrometry, *Anal. Chem.,* 53, 1566, 1981.
9. **Wolff, E. W. and Peel, D. A.,** Closer to a true value for heavy metal concentrations in recent Antarctic snow by improved contamination control, *Ann. Glaciol.* 7, 61, 1985.
10. **Mart, L.,** Seasonal variations of Cd, Pb, Cu and Ni levels in snow from the eastern Arctic Ocean, *Tellus,* 35B, 131, 1983.
11. **Boutron, C. F. and Patterson, C. C.,** Relative levels of natural and anthropogenic lead in recent Antarctic snow, *J. Geophys. Res.,* 92, 8454, 1987.
12. **Boutron, C. F. and Batifol, F. M.,** Assessing laboratory procedures for the decontamination of polar snow or ice samples for the analysis of toxic metals and metalloids, *Ann. Glaciol.,* 7, 7, 1985.
13. **Boutron, C. F., Patterson, C. C., Petrov, V. N., and Barkov, N. I.,** Preliminary data on changes of lead concentrations in Antarctic ice from 155,000 to 26,000 years BP, *Atmos. Environ.,* 21, 1197, 1987.
14. **Murozumi, M., Chow, T. J., and Patterson, C.,** Chemical concentrations of pollutant lead aerosols, terrestrial dusts and sea salts in Greenland and Antarctic snow strata, *Geochim. Cosmochim. Acta.* 33, 1247, 1969.
15. **Herron, M. M., Langway, C. C., Jr., Weiss, H. V., and Cragin, J. H.,** Atmospheric trace metals and sulfate in the Greenland ice sheet, *Geochim. Cosmochim. Acta.* 41, 915, 1977.
16. **Dickson, E. M.,** Mercury and lead in the Greenland ice sheet: a reexamination of the data, *Science,* 177, 536, 1972.
17. **Robinson, J.,** Lead in Greenland snow, *Ecotoxicol. Environ. Saf.,* 5, 24, 1981.
18. **Wolff, E. W. and Peel, D. A.,** The record of global pollution in polar snow and ice, *Nature,* 313, 535, 1985.
19. **Boutron, C. F.,** Atmospheric toxic metals and metalloids in the snow and ice layers deposited in Greenland and Antarctica from prehistoric times to present, in *Toxic Metals in the Atmosphere,* Nriagu, J. O. and Davidson, C. I., Eds., John Wiley & Sons, New York, 1986, 467.

20. **Ng, A. and Patterson, C. C.,** Natural concentrations of lead in ancient Arctic and Antarctic ice, *Geochim. Cosmochim. Acta.* 45, 2109, 1981.
21. **Boutron, C. F. and Patterson, C. C.,** Lead concentration changes in Antarctic ice during the Wisconsin/Holocene transition, *Nature,* 323, 222, 1986.
22. **Boutron, C. F., Patterson, C. C., Lorius, C., Petrov, V. N. and Barkov, N. I.,** Atmospheric lead in Antarctic ice during the last climatic cycle, *Ann. Glaciol.* 10, 5, 1988.
23. **Wolff, E. W. and Peel, D. A.,** Concentrations of cadmium, copper, lead and zinc in snow from near Dye 3 in south Greenland, *Ann. Glaciol.* 10, 193, 1988.
24. **Boutron, C. F. and Patterson, C. C.,** Corringendum: Lead concentration changes in Antarctic ice during the Wisconsin/Holocene transition, *Nature,* 326, 626, 1987.
25. **Dick, A. L.,** Trace elements in Antarctic Snow and Air, Ph.D. thesis, Council for National Academic Awards, London, 1987.
26. **Boutron, C.,** Trace element content of Greenland snows along an east-west transect, *Geochim. Cosmochim. Acta.* 43, 1253, 1979.
27. **Davidson, C. I., Chu, L., Grimm, T. C., Nasta, M. A., and Qamoos, M. P.,** Wet and dry deposition of trace elements onto the Greenland ice sheet, *Atmos. Environ.,* 15, 1429, 1981.
28. **Shen, G. T. and Boyle, E. A.,** Lead in corals: reconstruction of historical industrial fluxes to the surface ocean, *Earth Plan. Sci. Lett.,* 82, 289, 1987.
29. **Batifol, F., Boutron, C., and de Angelis, M.,** Changes in copper, zinc and cadmium concentration in Antarctic ice during the past 40,000 years, *Nature,* 337, 544, 1989.
30. **Dick, A. L., Sheppard, D. S., and Patterson, J. E.,** Mercury content of Antarctic surface snow: initial results, *Atmos. Environ.,* in press.
31. **Shen, G. T., Boyle, E. A., and Lea, D. W.,** Cadmium in corals as a tracer of historical upwelling and industrial fallout, *Nature,* 328, 794, 1987.
32. **Appelquist, H., Jensen, K. O., Sevel, T., and Hammer, C.,** Mercury in the Greenland ice sheet, *Nature,* 273, 657, 1978.
33. **Davidson, C. I., Santhanam, S., Fortmann, R. C., and Olson, M. P.,** Atmospheric transport and deposition of trace elements onto the Greenland ice sheet, *Atmos. Environ.,* 19, 2065, 1985.
34. **Dick, A. L.,** Trace elements in simultaneously sampled aerosol and snow from the Antarctic Peninsula, *Ann. Glaciol.,* 10, 201, 1988.
35. **Barrie, L. A.,** Arctic air chemistry: an overview, in *Arctic Air Pollution,* Stonehouse, B., Ed., Cambridge University Press, Cambridge, 1986, 5.
36. **Holdsworth, G. and Peake, E.,** Acid content of snow from a mid-troposphere sampling site on Mount Logan, Yukon Territory, Canada, *Ann. Glaciol.,* 7, 153, 1985.
37. **Barrie, L. A. and Hoff, R. M.,** Five years of air chemistry observations in the Canadian Arctic, *Atmos. Environ.,* 19, 1995, 1985.
38. **Heidam, N. Z.,** Crustal enrichments in the Arctic aerosol, *Atmos. Environ.,* 19, 2083, 1985.
39. **Batifol, F. M. and Boutron, C. F.,** Toxic metals and metalloids in high altitude alpine glaciers snow and ice, in *Proc. 3rd Eur. Symp. Physico-Chemical Behaviour of Atmospheric Pollutants,* Versino, B. and Angeletti, G., Eds., D. Reidel, Dordrecht, 1984, 471.
40. **Cragin, J. H., Herron, M. M., Langway, C. C., Jr., and Klouda, G. A.,** Interhemispheric comparison of changes in the composition of atmospheric precipitation during the late Cenozoic era, in *Polar Oceans,* Dunbar, M. J., Ed., Arctic Institute of North America, Calgary, 1977, 617.
41. **Sturges, W. T. and Barrie, L. A.,** Lead 206/207 isotope ratios in the atmosphere of North America as tracers of US and Canadian emissions, *Nature,* 329, 144, 1987.
42. **Maring, H., Settle, D. M., Buat-Ménard, P., Dulac, F., and Patterson, C. C.,** Stable lead istope tracers of air mass trajectories in the Mediterranean region, *Nature,* 330, 154, 1987.

Chapter 12

EVIDENCE FROM BIOLOGICAL SAMPLES FOR HISTORICAL CHANGES IN GLOBAL METAL POLLUTION

Robert W. Furness, David R. Thompson, and Paul M. Walsh

TABLE OF CONTENTS

I. INTRODUCTION

It has been argued that global, as distinct from local, pollution by heavy metals is most likely to arise with lead, mercury, and cadmium.[1-4] Relative to natural emissions, anthropogenic inputs of these elements are high[3,4] and atmospheric transport may be followed by deposition far from industrial sites. Attempts to quantify changes in heavy metal contamination of remote areas using ice cores or snow samples have been hindered by the extremely low concentrations of these elements in ice and snow and consequent technical difficulties of avoiding contamination of samples and of measuring such low levels with accuracy (as discussed by Wolff in this volume).

Since plants and animals may accumulate these metals to concentrations five to nine orders of magnitude higher than found in snow and ice samples, it is easy to measure levels in biological samples with little likelihood of sample contamination due to handling by the analyst. Where samples of biological material collected at dated times in the past can be obtained, it should be possible to compare metal levels in contemporary samples with levels found in the past. Such studies should not be considered as an alternative to analysis of ice cores, but as a valuable source of independent evidence to support data from ice and snow samples. However, a number of difficulties arise.[5] In particular, there are rather few suitable biological samples available for analysis. It is essential that samples should not have been treated (for example by use of preservatives) in such a way that metal levels will have been altered. The collection site for samples must be known so that geographical variations may be taken into account if these are pronounced, and the date of collection must not be open to doubt. The tissues must be suitable for the measurement of the metal of interest; for example, bird feathers or mammalian hair may provide excellent data on mercury levels, but cadmium is generally not transferred into feathers and hair in detectable amounts[6-8] and in studies where levels of cadmium and lead in feathers or hair have been measured, the levels found show little or no correlation with cadmium and lead levels in other organs.[8-10]

Since the nonessential heavy metals cadmium, lead, and mercury usually show high coefficients of variation in populations,[10-12] any attempt to compare population levels between past and present day samples requires a large sample size of both contemporary and historical tissues before any statistically meaningful comparison may be made. Where metal levels may vary, perhaps by an order of magnitude between individuals within any one population, attempts to test whether or not metal levels have changed as a result of human activities should take particular care not to draw conclusions from small numbers of tissue samples. Other factors, such as changes in the dietary composition or migration patterns of animals may also alter their metal levels and so care must be taken to avoid erroneous conclusions due to a lack of consideration of such possible alternative explanations.

In this chapter, we firstly review papers that have used historical biological material to assess time trends in metal levels in the environment, either to detect instances of local pollution or to seek evidence for global trends, and then we assess the likely value of this approach in current and future studies.

II. REVIEW OF PUBLISHED WORK

Rather few studies have been made in which metal levels in historical biological samples have been analyzed, and all have considered only mercury, since the tissue distribution of cadmium and lead makes those elements more difficult to measure in historical samples.

Three papers[13-15] describe mercury levels in museum specimens of fish collected 60 to 100 years ago compared with levels in present-day specimens. Miller et al.[13] analyzed total mercury levels in seven tuna (five skipjack *Katsuwonus pelamis*, one albacore *Thunnus alalunga*, and one bluefin *Kishinoella tonggol*) collected between 1878 and 1909 from a wide range of Atlantic and Pacific sites, two tuna collected in 1970 to 1971 from the Pacific, three samples of tinned albacore tuna, a (physically deformed) swordfish *Xiphias gladius* head preserved in isopropanol since 1946 and six fresh swordfish from California. They found an order of magnitude difference in mercury levels between the six 1970 to 1971 swordfish, with the 1946 sample falling within this range. Similarly, the museum and contemporary tuna showed variations between 0.13 and 0.64 μg/g wet weight, with no significant difference between groups. Barber et al.[14] measured total mercury levels in recent and 90-year-old benthopelagic fish. The museum samples consisted of one *Antimora rostrata* collected in 1883 and one *Aldrovandia macrochir* collected in 1886. Both fish had mercury levels similar to those of nine and five contemporary individuals of those species, and these authors also concluded that the high levels in these fish species "are natural and may not be related to anthropogenic inputs of mercury to the oceans." Based on such meager sample sizes such a conclusion appears rash. Even if a difference did exist, given the high degree of individual variation in metal levels it would be virtually impossible to demonstrate a statistically significant difference between museum and contemporary fish with such a small sample size. Indeed, Rowland[15] cautioned that the results did not show the absence of oceanic mercury pollution but suggested that any anthropogenic increase in this deep ocean habitat was unlikely to be of a factor of as much as five. Nevertheless, these papers seem to have been taken by the scientific community to indicate that no global increase in mercury levels in marine ecosystems has occurred, despite the fact that the small sample sizes analyzed would prevent an increase from being readily detectable.

Analysis of polar bear *Ursus maritimus* hair for mercury has shown that mean levels within populations vary from 2.5 μg/g for 41 animals on the southern shore of Hudson's Bay to 18.5 μg/g for 5 animals at Amundsen Gulf[16] with no overlap in individual values sampled at these 2 localities. Intermediate values were obtained at other sampling sites in between and further northeast. The authors presented data for mercury levels in hair of 18 animals collected 1910 to 1927 but of these, 8 were not adults. Since young animals have lower mercury levels than adults,[10] these 8 cannot be considered further. Eight of the adults were collected from the vicinity of Amundsen Gulf and so were comparable with the contemporary sample of five from that area. However, separate measurements of total organic and inorganic mercury levels in the hair of the historical samples showed clear evidence of contamination during preservation or storage with inorganic mercury, such that one young bear had hair levels of 53.3 μg/g of inorganic mercury but only 0.3 μg/g of organic mercury. Although the authors washed the museum hair samples thoroughly, in our experience, if stored samples become at all damp, much inorganic mercury contamination will become bound to hair too strongly to be removed by any washing procedure. Normally polar bear hair contains very little inorganic mercury, the organic fraction representing over 80% of the total[16] as found in human hair. Unfortunately, Eaton and Farant did not analyze inorganic and organic mercury levels separately in the contemporary samples and so a direct comparison between organic mercury levels in the museum and recent samples is not possible. If we assume that 85% of the total mercury in the contemporary samples was organic as the authors suggest is normal, then the comparison (Table 1) indicates a statistically significant difference, with the modern samples having four times as much mercury as the museum samples, and yet Eaton and Farant concluded "observed high mercury levels in the Canadian Arctic reflect geological sources and are not connected with anthropogenic release." Geographical variation in mercury levels in polar bears around the Canadian Arctic seems to be at least as great as this historical increase, and may be attributed to differences in diets since different seal species have quite different mercury levels.[10,16,17]

Hansen, (see Chapter 13) describes a study of levels of metals in hair samples from 15th century mummified Eskimos compared with today's Eskimos. That study shows evidence of increases in mercury and lead levels that were attributed to increased global

TABLE 1
Mercury Levels (μg/g) in Hair Samples from Adult Polar Bears Collected in the Amundsen Gulf Area of the Canadian Arctic

Period	Sample size	Mean	S.D.	Range
1910—27	8	4.6	1.8	1.8—7.5
1977—80	5	18.5	14.5	9.3—44.3

Note: It is assumed that organic mercury levels represented 85% of total mercury for museum samples.

pollution, while some changes in selenium, iron, and bromine levels appeared to be related to a reduction in the relative importance of marine foods in the diet, a trend which would also have tended to reduce, rather than increase, mercury and lead exposure.[18]

Mercury and other heavy metals are taken up by bog plants to a particularly high degree and are strongly retained by the humic substances of peat and so peat cores may be used to examine historical changes in metal fluxes.[19] Madsen showed that the deposition rate of mercury on two peatbogs in Denmark increased in parallel with the known increase in consumption of mercury in European Community countries between 1740 and 1980, and suggested that some brief periodically elevated deposition rates could be attributed to major volcanic events occurring in Iceland.

Several papers have considered the extent to which changes in mercury levels in the environment may be assessed by analysis of time series of feather samples from bird populations.[20-25] Use of alkyl-mercury seed dressings in Scandinavia from the 1940s to the 1960s led to clearly elevated mercury levels in feathers of terrestrial and freshwater birds compared to levels in pre-1940 museum specimens.[20] This study did not distinguish between organic and inorganic mercury levels in feathers and so encountered problems due to contamination of some museum specimens with inorganic mercury. The authors arbitrarily discounted anomalously high values of total mercury from museum samples as due to contamination, though this procedure requires a detailed knowledge of the results to be expected in order that anomalous data can be omitted.

Study of total mercury levels in feathers of black guillemots *Cepphus grylle*, common murres *Uria aalge*, and thick-billed murres *U. lomvia* from the Baltic Sea, Kattegat, and Atlantic[21] showed that mercury levels had increased in the species in the Baltic Sea, from levels of around 1 to 2 μg/g in common guillemot primaries from before 1900 to 2 to 8 μg/g around 1960 to 1974 and from 2 to 5 μg/g in black guillemot primaries before 1900 to 9 to 25 μg/g around 1960 to 1974.[21] This work used neutron activation analysis to measure total mercury levels, and yet the authors made no mention of problems with post-mortem contamination during museum preparation or storage, which we have found to be a problem in some historical collections (see below). Elevated mercury levels in Baltic seabirds can be attributed to use of mercury seed dressings in Scandinavia and other local sources of pollution, and there is some evidence to suggest that levels decreased slightly after mercury seed dressing became less common.[21] Examination of changes in mercury levels in the birds around the Faeroes and Greenland suggest that an increase in mercury levels may have occurred in those areas too, but to a much smaller extent than in the Baltic; from around 1 μg/g in black guillemots from Greenland before 1900 to around 2 μ/g in 1973, and from 0 to 1 μg/g in primaries of common murres from the Faeroes before 1920 to 0.5 to 1.5 μg/g in 1973.[21]

After the outbreak of Minamata disease in the 1960s in Japan, caused by mercury discharge into the Shiranai Sea at Minamata Bay,[22] an analysis was made of mercury levels in feathers of around 100 birds from this area collected between 1955 and 1980 and these values were compared with analyses of feathers from similar birds from *uncontaminated* sites in China and Korea. The authors concluded that feathers provided a reliable measure of geographical differences in mercury levels in the ecosystem, and that analysis of time series from museum collections should provide evidence for trends in mercury levels over periods of decades.[22]

III. FUTURE PROSPECTS AND PROBLEMS

A. CADMIUM

Since cadmium tends to accumulate to high concentrations in kidney tissue and is generally not transferred to the more stable tissues,[9] such as feathers and hair, the usefulness of preserved specimens in determining any historical changes in cadmium levels is likely to be limited.

B. MERCURY

As a result of being both chemically and physically stable, together with their relative abundance in preserved collections, feathers and fur should make suitable study tissues for the assessment of historical changes in mercury levels on a global scale. By contrast, soft, internal organs are prone to physical decay and chemical change and as such are likely to be less reliable for this purpose.

The elucidation of historical trends in mercury levels in the environment by use of time series of feather samples from appropriate bird populations[20-25] has been undertaken largely without consideration of the form of mercury incorporated into the feather keratin molecule, or of the consequences of mercury contamination via preservation processes. Additional, inorganic mercury used in the treatment of preserved specimens will result in erroneous total mercury values.

An initial investigation into the relationship between organic and inorganic feather mercury[26] has revealed that organic mercury represents about 100% of the total mercury level. This result is based on a comparison of total and organic feather mercury levels in a sample of 95 seabirds of various species. The variation around this figure is relatively large due to the destructive nature of the analytical technique, making direct comparisons within a given sample difficult. However, by performing an initial fractionation of the feather mercury to remove the intrinsic organic mercury only, a more meaningful assessment of the historical mercury concentration can be made. This technique should effectively overcome any contamination problems associated with study skin preservation processes, and the subjective discarding of high mercury values. Studies of the effects of storage, washing and exposure of hair samples or feathers to UV-B and other factors indicate that mercury is firmly bound in feathers and levels do not change with a variety of harsh treatments.[27,28]

Recent work on the dynamics of mercury in birds and of the relationships between molt patterns and mercury levels in feathers[29-32] gives a good basis for studies of time series of mercury levels in seabird feathers, and progress in this area can be expected.

C. LEAD

Lead has been shown to accumulate to relatively high concentrations in bone and teeth of marine vertebrates.[34-36] Other similarly hard and relatively stable tissues, such as horns and antlers, are likely to show a similar pattern. By taking into account both the possibility of lead contamination subsequent to sample collection, and the distribution of lead between these and other internal tissues, historical changes in global lead levels in a range of appropriate biological samples could be investigated. Although lead levels in hair have been shown to be quantifiable,[37] the relationships between lead concentrations in hair (and feathers) and internal tissues are poorly documented.[9] Without a detailed knowledge of such distribution patterns and relationships, a meaningful interpretation of any hair/feather lead data will be difficult. Given the high industrial usage of lead in recent decades, it would be valuable to develop the possibility of using selected hard tissues of preserved biological samples to determine changes in lead concentrations over time.

REFERENCES

1. **Gill, G. A.,** On the Marine Biogeochemistry of Mercury, Ph.D. thesis, University of Connecticut, Storrs, CT, 1986.
2. **Nriagu, J. O.,** Global inventory of natural and anthropogenic emissions of trace metals to the atmosphere, *Nature London,* 279, 409, 1979.
3. **Lantzy,R. J. and Mackenzie, F. T.,** Atmospheric trace metals: global cycles and assessment of man's impact, *Geochem. Cosmochim. Acta,* 43, 511, 1979.
4. **Bryan, G. W.,** Pollution due to heavy metals and their compounds, in *Marine Ecology,* Kinne, O., Ed., John Wiley & Sons, New York, Vol 5, 1984, 1289.
5. **Walsh, P. M.,** The use of seabirds as monitors of heavy metal levels in the marine environment, this volume, chap. 10.
6. **Goede, A. A. and de Bruin, M.,** The use of bird feather parts as a monitor for metal pollution, *Environ. Pollut.,* 8, 281, 1984.
7. **Honda, K., Yamamoto, Y., Hidaka, H., and Tatsukawa, R.,** Heavy metal accumulations in adelie penguin, *Pygoscelis adeliae,* and their variations with the reproductive processes, *Mem. Natl. Inst. Polar Res. Spec. Issue,* 40, 443, 1986.
8. **Hansen, J. C.,** A survey of human exposure to mercury, cadmium and lead in Greenland, *Medd. Groenl.,* 3, 1, 1981.
9. **Goede, A. A. and de Voogt, P.,** Lead and cadmium in waders from the Dutch Wadden Sea, *Environ. Pollut.,* 37, 311, 1985.
10. **Thompson, D. R.,** Heavy metal levels in marine vertebrates, this volume, chap. 5.
11. **Muirhead, S. J. and Furness, R. W.,** Heavy metal concentrations in the tissues of seabirds from Gough Island, South Atlantic Ocean, *Mar. Pollut. Bull.,* 19, 278, 1988.
12. **Schneider, R., Steinhagen-Schneider, G., and Drescher, H. E.,** Organochlorines and heavey metals in seals and birds from the Weddell Sea, in *Antarctic Nutrient Cycles and Food Webs,* Siegfried, W. R., Condy, P. R., and Laws, R. M. Eds., Springer-Verlag, Berlin, 1985.
13. **Miller, G. E., Grant, P. M., Kishore, R., Steinkruger, F. J., Rowland, F. S., and Guinn, V. P.,** Mercury concentrations in museum specimens of tuna and swordfish, *Science,* 175, 1121, 1972.
14. **Barber, R. T., Vijayakumar, A., and Cross, F. A.,** Mercury concentrations in recent and ninety-year-old benthopelagic fish, *Science,* 178, 636, 1972.
15. **Rowland, F. S.,** Mercury levels in swordfish and tuna, *Biol. Conserv.,* 5, 52, 1973.
16. **Eaton, R. D. P. and Farant, J. P.,** The polar bear as a biological indicator of the environmental mercury burden, *Arctic,* 35, 422, 1982.
17. **Norstrom, R. J., Schweinsberg, R. E., and Collins, B. T.,** Heavy metals and essential elements in livers of the polar bear *(Ursus maritimus* in the Canadian Arctic, *Sci. Total Environ.,* 48, 195, 1986.
18. **Hansen, J. C.,** Human exposure to metals through consumption of marine foods: a case study of exceptionally high intake among Greenlanders, this volume, chap. 13.
19. **Madsen, P. F.,** Peat bog records of atmospheric mercury deposition, *Nature,* 193, 127, 1981.
20. **Berg, W., Johnels, A., Sjostrand, B., and Westermark, T.,** Mercury content in feathers of Swedish birds from the past 100 years, *Oikos,* 17, 71, 1966.
21. **Appelquist, H., Drabaek, I., and Asbirk, S.,** Variation in mercury content of guillemot feathers over 150 years, *Mar. Pollut. Bull.,* 16, 244, 1985.
22. **Doi, R., Ohno, H., and Harada, M.,** Mercury in feathers of wild birds from the mercury-polluted area along the shore of the Shiranui Sea, Japan. *Sci. Total Environ.,* 40, 155, 1984.
23. **Somer, E. and Appelquist, H.,** Changes and differences in mercury level in the Baltic and Kattegat compared to the North Atlantic using *Uria* sp. (guillemot sp.) and *Cepphus grylle* (black guillemot) as indicators. Paper presented at the 9th Conf. Baltic Oceanographers, Kiel, West Germany, 1974.
24. **Johnels, A. G., Edelstam, C., Olsson, M., and Westermark, T.,** Mercury as an environmental poison in Sweden, *Fauna Flora,* 63, 172, 1968.
25. **Edelstam, C., Johnels, A. G., Olsson, M., and Westermark, T.,** Ecological aspects of the mercury problem, *Nord. Hyg. Tidskr.,* 50, 14, 1969.

26. **Thompson, D. R. and Furness, R. W.,** unpublished data.
27. **Appelquist, H., Asbirk, S., and Drabaek, I.,** Mercury monitoring: mercury stability in bird feathers, *Mar. Pollut. Bull.*, 15, 22, 1984.
28. **Suzuki, T. and Tamamoto, R.,** Organic mercury levels in human hair with and without storage for eleven years, *Bull. Environ. Contam. Toxicol.*, 28, 186, 1982.
29. **Honda, K., Nasu, T., and Tatsukawa, R.,** b Seasonal changes in mercury accumulation in the Black-eared Kite, *Milvus migrans lineatus, Environ. Pollut. Ser.* A, 42, 325, 1986.
30. **Braune, B. M. and Gaskin, D. E.,** A mercury budget for the Bonaparte's Gull during autumn moult, *Ornis Scand.*, 18, 244, 1987.
31. **Braune, B. M.,** Comparison of total mercury levels in relation to diet and molt for nine species of marine birds, *Arch. Environ. Contam. Toxicol.*, 16, 217, 1987.
32. **Furness, R. W., Muirhead, S. J., and Woodburn, M.,** Using bird feathers to measure mercury in the environment: relationships between mercury content and moult, *Mar. Pollut. Bull.*, 17, 27, 1986.
33. **Furness, R. W.,** The use of allometric predictions of plumage mass in estimating heavy metal budgets for seabirds, unpublished data.
34. **Reid, M. and Hacker, C. S.,** Spatial and temporal variation in lead and cadmium in the Laughing Gull, *Larus atricilla, Mar. Pollut. Bull.*, 13, 387, 1982.
35. **Braham, H. W.,** Lead in the California sea lion *(Zalophus californianus), Environ. Pollut.*, 5, 253, 1973.
36. **Hutton, M.,** Accumulation of heavy metals and selenium in three seabird species from the United Kingdom, *Environ. Pollut.*, 26, 129, 1981.
37. **Mason, C. F. and Reynolds, P.** Organochlorine residues and metals in otters from the Orkney Islands, *Mar. Pollut. Bull.*, 19, 80, 1988.

Chapter 13

HUMAN EXPOSURE TO METALS THROUGH CONSUMPTION OF MARINE FOODS: A CASE STUDY OF EXCEPTIONALLY HIGH INTAKE AMONG GREENLANDERS

Jens C. Hansen

TABLE OF CONTENTS

I. INTRODUCTION

The basis for existence of Eskimos in Greenland has traditionally for centuries been hunting of smaller cetaceans as well as large baleen whales. This hunting pattern, however, also included exploitation of other marine organisms, such as seals, walrus *Odobenus rosmarus*, and some terrestrial animals like caribou *Rangifer tarandus* and polar bear *Ursus maritimus*. The relative importance of the various prey species varied with time and season and to some degree from locality to locality.

During the first decades of this century, seal hunting decreased drastically especially in the southwest region, at the same time cod *Gadus morhua* appeared and a transition to fishing took place and has rapidly developed since 1950. No similar developments took place in North, Northwest, and East Greenland; these regions are still hunting districts. At present about 20% of the population (approximately 50,000) lives in these regions.[1]

During this century, an increasing amount of imported food has become available and today every hamlet has a grocer's shop. This means that even in the most traditional hunting districts imported food now plays an important part in the human diet. A recent survey carried out among hunting families in North Greenland showed that only around 25% of the total energy intake originated from local marine food.[2] However, meat and organs from marine mammals are still important food items in these districts.

The Eskimo diet, in which marine mammals are dominant as meat suppliers, places man at the end of a long food chain with the risk of higher exposure to cumulative toxic elements, such as methyl mercury and cadmium than is the case in industrialized countries, where the intake of marine food in general is relatively low. Consequently a high exposure level of mercury and cadmium could be expected. In the late sixties, Dalgaard Mikkelsen[3] analyzed blood and hair samples for mercury from nine Greenlanders living in a village. Blood concentrations ranged from 35 to 85 μg/l, with a mean of 56 μg/l, hair concentrations from 13.4 to 38.8 μg/g, with a mean of 26.6 μg/g. Later, Hansen and Nygaard[4] reported blood mercury concentrations in 25 Greenlanders living in Upernavik and Sukkertoppen. Mean concentrations in the inhabitants of these two towns were given as 105 and 92 μg/l, respectively. Both these preliminary reports clearly indicated a high exposure level among Greenlanders.

Because of concern that pollution with heavy metals from the lead-zinc mine, Marmorilik in West Greenland, could indicate an impact on human health, a number of muscle and liver samples from seals and whales were analyzed by the Greenlandic Fisheries Investigations, Copenhagen.[5] A total of 89 seals and 7 whales caught on the west coast and a control sample of 7 seals from the east coast, were analyzed for total mercury, methyl mercury, cadmium, copper, lead, and zinc. Based on these data, an estimation of human dietary intake of mercury, cadmium, and lead was made by the Danish National Food Institute.[6] On the basis of these calculations, it was suspected that the WHO provisional tolerable weekly intakes (PTWI) of mercury and cadmium were exceeded among Greenlandic hunters. In the most heavily exposed individuals, mercury on average exceeded PTWI by a factor of 14, cadmium by a factor of 12, whereas lead intake did not exceed the PTWI. This suggested that not only mercury but also cadmium exerted a potential environmental hazard to the Inuit population of Greenland. For further evaluations of the human exposure situation, the Ministry of Greenland in Copenhagen in 1978 took the initiative to start a survey. This was started in 1979 under the auspices of the Commission for Scientific Research in Greenland. The individual studies are summarized in Table 1. The sampling places are indicated in Figure 1.

In addition to the studies given in the table, analyses have been carried out on eight hair samples originating from well-preserved mummies from the Umanak district, dated to the year 1470 ± 50 by the ^{14}C method.[16] Preliminary results from these samples have been

TABLE 1
Subprograms Carried Out in the Period 1979 to 1985

Sampling region	Year of sampling	Number of samples	Sex	Analyses: Blood Hg	Blood Cd	Blood Pb	Blood Se	Blood SCE	Hair	Ref.
West Greenland	1979	111	M	*	*				*	7
East Greenland	1979	26	M						*	8—12
	1981	178	M and F	*	*	*	*	*	*	8—12
West and East Greenland	1982	2 × 98	F	*		*				14
Denmark	1979	7	M	*	*	*			*	11
	1981	29	M and F	*	*	*	*			11
North Greenland	1983	97	M and F	*	*	*	*		*	13
North, East, and West Greenland	1983/85	2 × 367	F	*		*	*			15

Note: Samples from Denmark were of Eskimos living in Denmark. SCE = sister chromatid exchange. Samples References 14 and 15 were from mothers and their young babies.

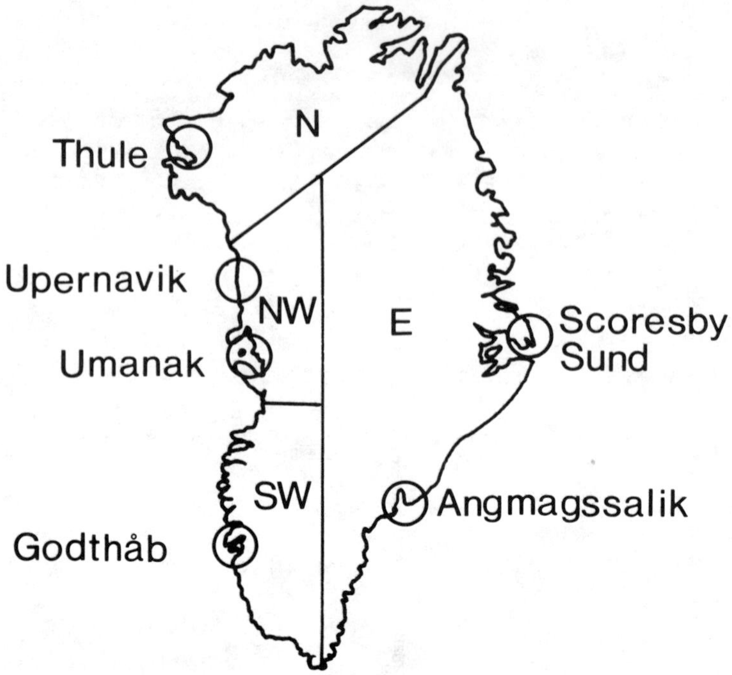

FIGURE 1. Map of Greenland with indications of sampling places and geographical regions.

published.[17] A more detailed description is under preparation. Four hair samples from the late 18th century taken in Angmagssalik during the Gustav Holm Expedition have also been analyzed.

II. MERCURY EXPOSURE

A. EXPOSURE LEVELS IN ADULTS

In all regions studied, the determining factor for mercury exposure was found to be the daily intake of meat from marine mammals. It has been shown by Johansen[5] that in meat from Greenlandic seals and whales, approximately 70% of the total mercury is in the methylated state, which means that the main exposure is in the form of methyl mercury. On a regional basis as well as within the regions, a close relationship between meat consumption and blood and hair concentrations of mercury has been found. As the extent to which meat from marine mammals is part of the diet is geographically determined according to differences in lifestyle, a northward trend of increasing blood concentrations was found. Figure 2 gives the accumulated frequencies of whole-blood concentrations of mercury found in the four regions.

The high mercury concentrations found in North Greenland could be suspected to be a result of marine pollution. There is, however, nothing to support this. On average, the meat consumption will supposedly parallel the number of animals caught, on which statistics do exist. From these statistics, a mean figure of meat available per person per day can be calculated. When this figure is plotted against the median mercury concentration in blood, as seen in Figure 3, a straight line is observed, indicating that the actual mercury concentrations in animal species do not vary significantly from one region to another. The blood mercury concentrations are thus obviously determined primarily by the consumption of meat of marine origin, and are lowest in the southwest part of the country, where a median of

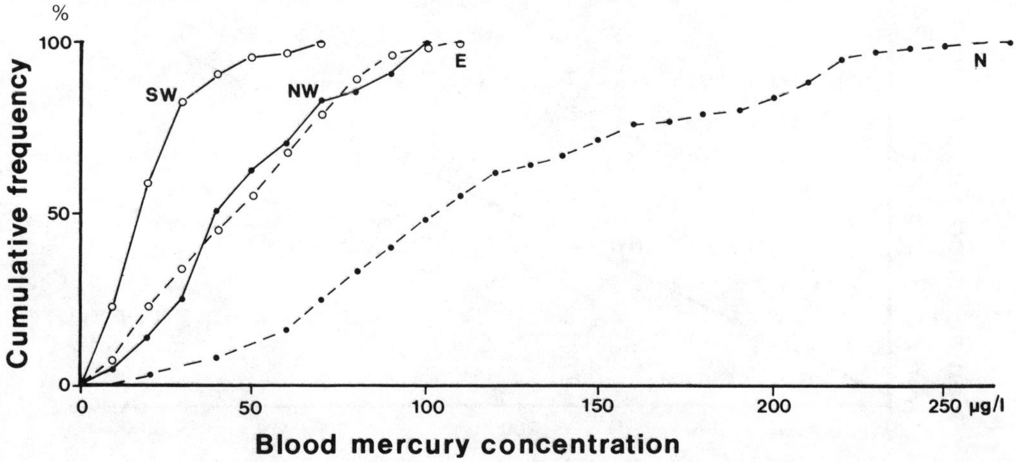

FIGURE 2. Accumulated frequencies of blood mercury concentrations in adults in the four regions; N: North, E: East, NW: Northwest, SW: Southwest.

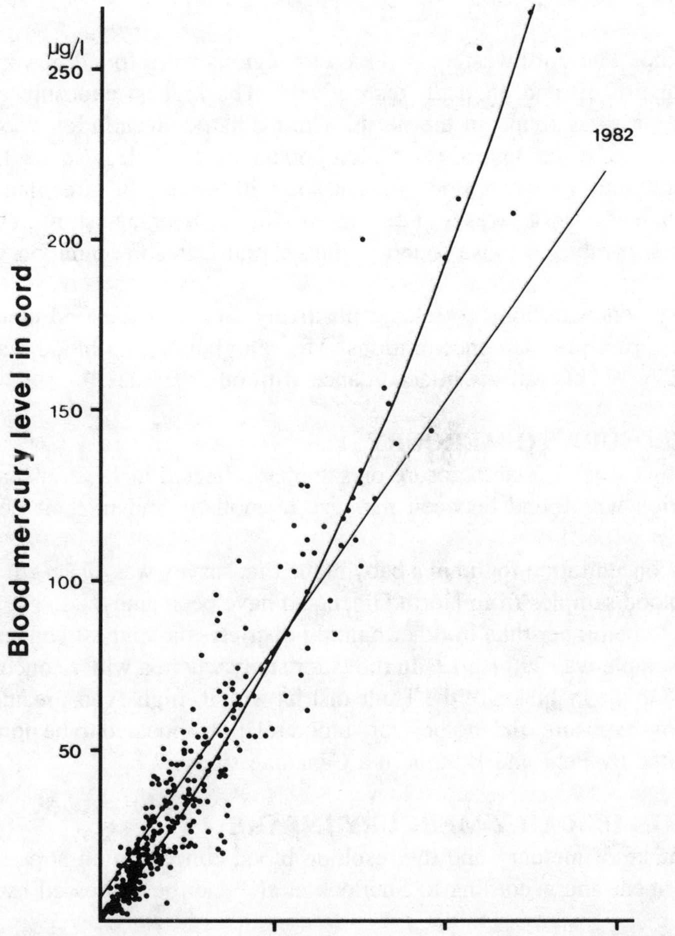

FIGURE 3. Relationships between blood mercury concentrations in mothers and in cord blood samples. The straight line is the regression line found in the first study.[15]

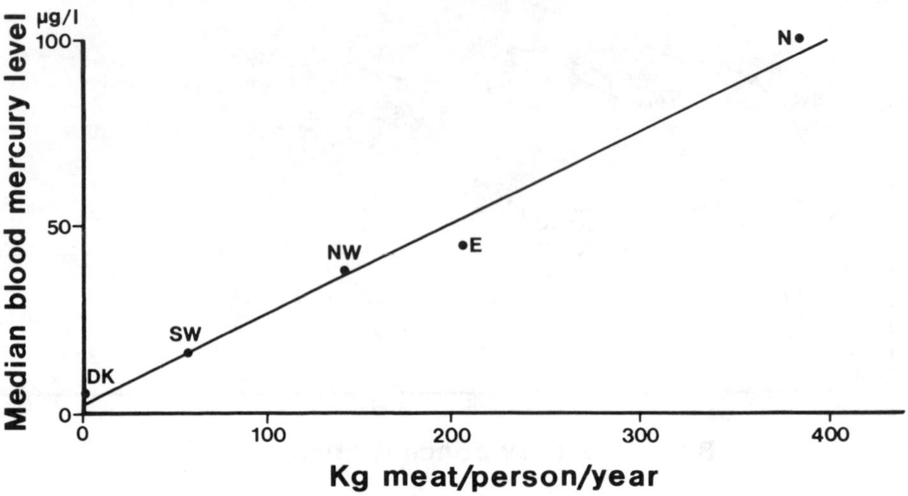

FIGURE 4. Relationships between regional median blood concentration and estimated amount of available meat from marine mammals.

17 μg/l was found. The northwestern and eastern regions were found to vary in a similar way with medians of 39 and 45 μg/l, respectively. The highest exposure as reflected in blood concentrations was found in the north. The median concentration was 102 μg/l. In North Greenland, 16% of the investigated adult population exceeded 200 μg/l, which WHO regards as the minimum toxic blood concentration in adults. In Greenlanders living in Denmark, a median of 5.0 μg/l was found. Danes living in Greenland showed blood mercury concentrations comparable to those found in the general Danish population with a median of 2 μg/l.[18]

Hair mercury concentrations correlated positively with whole-blood concentrations as well as with the serum-plasma concentrations. The ratio hair/whole blood was found to be approximately 250.[7,8] This ratio is in accordance with other studies.[19]

B. FETAL EXPOSURE TO MERCURY

In the investigations of fetal exposure on samples collected in 1982, a significant linear positive correlation was found between mercury in mothers and in their newborn babies ($r = +0.8639$, $p < 0.001$).

The highest concentration found in a baby in the first survey was 177 μg/l. In an ongoing study, 21 cord blood samples from North Greenland have been analyzed. The concentration level was found to be higher than in other hunting districts, the highest concentration found in a cord blood sample was 296 μg/l.[15] In the latter study where a wider concentration range was obtained due to the inclusion of the Thule district with its high exposure and the Godthab district with its low exposure, the mother/cord blood relation appeared to be nonlinear (Figure 4) as also indicated by Fehr and Dennis in a Canadian study.[20]

C. ESTIMATION OF DAILY MERCURY INTAKE

The daily intake of mercury and the resulting blood concentration at equilibrium seem rather well described, and according to Sherlock et al.[21] can be expressed by the equation:

$$\text{Hg} - \text{B}\ (\mu\text{g/l}) = 0.8 \times \text{daily intake}\ (\mu\text{g}) \qquad (1)$$

By applying this equation to the median blood mercury concentrations found in the different Greenlandic regions, the daily intake can be estimated, as the exposure through the diet can

TABLE 2
Estimated Daily Intake of Mercury in Four Greenlandic Regions

	Estimated mean daily intake of Hg (μg)	Estimated daily intake of Hg in (μg) high consumers	Intake/PTWI Mean	Intake/PTWI High
North	128	238	4.5	8.3
Northwest	49	78	1.7	2.7
Southwest	25	34	0.9	1.2
East	56	68	2.0	2.4

Note: Mercury intake given as a mean of total study groups and a mean of the group with the highest exposure, i.e., six meals of local food per week.

TABLE 3
Estimated Meat Consumption per Capita

Region	Available per capita per year (kg)	Available per capita per day (g)	Estimated daily consumption (g)	% Eaten of total available
North	384	1070	341	32
Northwest	143	390	130	33
Southwest	57	160	65	41
East	204	560	150	27

be regarded as being rather constant and the probands consequently were in a steady state. In Table 2, estimated daily mean intake is indicated in the total study group in various regions and in subgroups with the highest consumption of marine mammals, i.e., six meals per week or more. When comparing the estimated intake with the PTWI suggested by WHO,[22] it is seen that nearly all the groups exceed the PTWI: 0.3 mg per person per week total mercury with a maximum of 0.2 mg per person per week as methyl mercury.

As the median mercury levels correlate with the amount of meat from marine mammals available in the various regions (Figure 4), it seems justified to suppose that the concentrations of mercury in food do not vary considerably from one region to another. Preliminary results from analyses of meat and organs from North Greenland have shown a median mercury concentration in seal meat of 0.38 mg/kg. Supposing this value to be valid in general, it can be used to calculate the amount of meat eaten daily based upon the estimates of daily intake of mercury. The calculated figures are shown in Table 3. The estimated amount eaten is seen to correspond to approximately one third of the amount available. An exception is seen in Southwest Greenland, where a higher (41%) proportion is eaten. This difference can probably be explained by the fact that there are no sledgedogs in this region. In the northern regions, sledgedogs will consume a substantial part of the available meat. The proportion of one third eaten of the amount available is in good agreement with the results of a food survey, carried out by Helms in East Greenland[23] indicating a daily consumption of 180 g of sealmeat, which is very close to one third of the available meat of 204 kg/year = 560 g/d (cf. Figure 3.)

In the calculations of mercury intake, only meat is considered in spite of the fact that liver contains much more mercury (7.5 mg/kg in seal liver but 0.38 mg/kg in meat). In liver, approximately 70% of the mercury is present in inorganic form.[5] Furthermore, mercury in seal liver is reported to be associated with selenium in a molar ratio close to one.[24,25] Possibly mercury in liver is less easily absorbed than is mercury in muscle as the blood

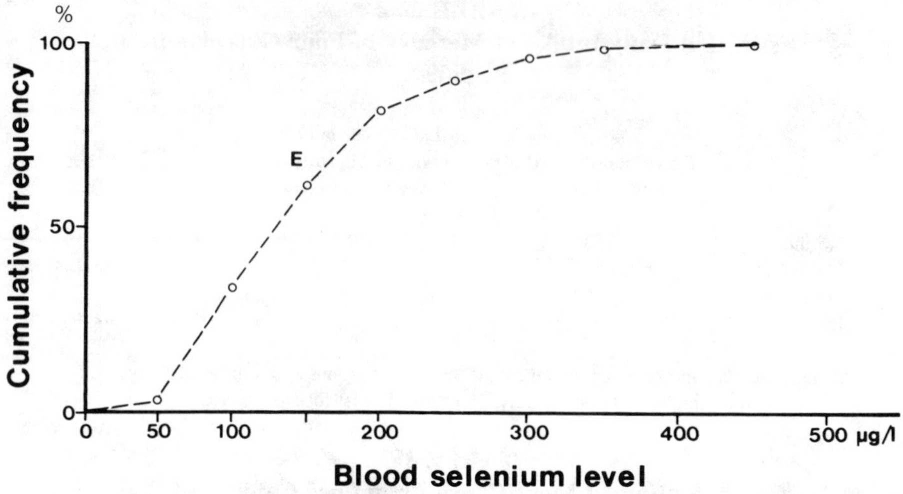

FIGURE 5. Accumulated frequencies of blood selenium concentrations in Angmagssalik, East Greenland.

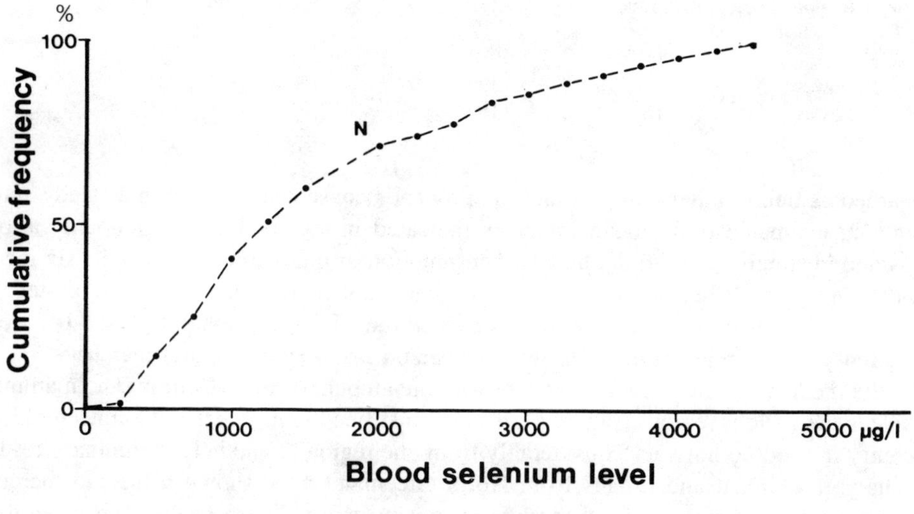

FIGURE 6. Accumulated frequencies of blood selenium concentrations in Thule, North Greenland.

mercury concentrations can be explained through meat consumption alone. For this reason, liver is excluded from the calculations of mercury intake.

III. SELENIUM

A. BLOOD SELENIUM LEVELS IN ADULTS

Due to the fact that selenium analyses were included in the programs at a later stage, only adults in North and East Greenland have been investigated. Like mercury, blood selenium concentrations were found to be closely related to the amount of marine mammal meat eaten. Figures 5 and 6 show the accumulated frequencies of concentrations among adults in the districts of Angmagssalik and Thule. The concentrations in the two districts are very different with median concentrations of 130 and 1225 µg/l, respectively. At the moment, the difference in amount of an order of magnitude cannot be explained, but is probably a result of interspecies variations in selenium concentrations among food animals combined with a different eating pattern in the two districts.

The molar ratio of selenium to mercury in whole blood was found to vary on the East coast. The lowest ratios were found in the Angmagssalik district with mean ratios from 6.9 to 16.6 depending on the concentration levels.[10] In North Greenland a mean Se/Hg ratio of 34 was found in the town (Quaanaaq) and a mean ratio of 94 in samples from the villages. In babies, the Se/Hg molar ratio was in all cases 1.

With regard to selenium in hair, it is concluded, based on experiences both in Angmagssalik and Thule, that hair selenium does not reflect daily intake as clearly as does blood selenium and is of limited use as an indicator of exposure. Contrary to what is found in blood, the Se/Hg molar ratio in hair was in all cases below one which indicates that mercury is more readily taken up in hair than is selenium.

B. FETAL EXPOSURE TO SELENIUM

Selenium, like mercury, passes the placental barrier. The mean cord blood selenium concentration was, however, found to be lower than that of the mothers indicating that selenium does not pass the placental barrier as freely as does mercury. The baby/mother ratio was independent of differences in concentration levels between districts, being constant with a mean of 0.88 (0.86 to 0.93).

C. SELENIUM EXPOSURE

The blood selenium concentrations found in adults from North Greenland were higher than can be explained through meat ingestion. A median blood concentration of 1225 μg/l corresponds according to Schrauzer and White[26] to a calculated daily intake of 1284 μg. The difference might be due to ingestion of especially selenium-rich food items in the North, such as the skin of narwhals *Monodon monoceros*, mutuk. A few samples of mutuk have been analyzed and a content as high as 15 mg/kg has been found.[27]

The blood selenium concentrations found in Eskimos are high as compared to values from several other countries. They are, in the district of Angmagssalik, about the same level as those found in the U.S.[28] In general, whole-blood concentration values vary a great deal according to locality, due to varying geochemical conditions. McKenzie et al.[29] found selenium mean blood values of 42 to 83 μg/l in New Zealand, which is known as an area low in selenium, while a mean selenium value of 320 μg/l has been reported from South Dakota, a selenium-rich area.[30] In Denmark, a mean of 89 μg/l has been found in 20 samples.[31]

The high selenium concentrations found in the Thule district, the highest being 3000 μg/l, are close to what in the Peoples Republic of China have been found in patients with chronic selenosis.[32] The mean whole-blood concentrations found among residents in a high selenium area with chronic selenosis was 3200 μg/l with a range of 1300 to 7500 μg/l. There is no guarantee of identical analytical quality between the two studies, but the figures from Thule are surprising and need consideration. Yang et al.[32] suggest a possible adaptive response in humans exposed to a high selenium intake, and this could easily apply to Greenlanders, who have been dietarily exposed through millenia. So far no signs or symptoms of chronic selenosis have been recorded among the Greenlanders in Thule.

D. SELENIUM AS A PROTECTIVE AGENT AGAINST TOXIC EFFECTS OF MERCURY

Selenium is now recognized to be an essential nutrient. As part of the active groups in the enzyme glutathione peroxidase, it is, together with other antioxidants, such as vitamin E, part of the defence mechanism against cellular oxidative damage. Beside this function, it may play a protective role with regard to the toxic effects of heavy metals, mercury, cadmium, and probably also lead.

No human data exist on the protective role of selenium against mercury toxicity, but it

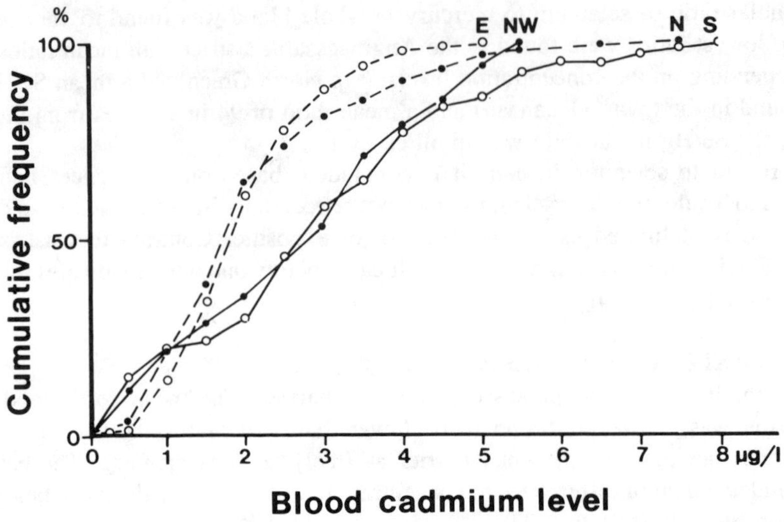

FIGURE 7. Accumulated frequencies of blood cadmium concentrations in the four regions.

is tempting to assume that a rich selenium supply will protect against possible adverse effects of dietary mercury exposure. Recent studies, however, seem to have challenged this theory. Magos et al.[33] studied the Se-Hg interaction using inorganic selenite and organic selenium and found that organic selenium affected the organ distribution of mercury differently than selenite. Furthermore, the speciation of mercury was affected by the form of selenium. A mercury-selenium compound, presumably Hg-Se, accounted for a greater proportion of total mercury in tissues after selenite administration than after a dose of organic selenium. These results suggest that laboratory experiments with selenite overestimate the protective effect of dietary selenium against inorganic mercury and probably also against methyl mercury. Makito et al.[34] have shown that inorganic mercury interacts selectively with simultaneously administered selenium to form a high molecular weight complex, but does not react with the selenium already bound to proteins. A direct warning not to underestimate heavy metal exposure in the Arctic should be taken from the information provided by Wulf et al.[12] that sister chromatid exchange frequency was related to blood concentrations of mercury and cadmium, whereas no influence of selenium could be revealed by multiple linear regression analyses in spite of the fact that selenium is present in blood in surplus to mercury[19] and in spite of the protective effect demonstrated *in vitro*.

IV. CADMIUM

A. EXPOSURE LEVELS IN ADULTS

Blood cadmium concentrations have only been determined in studies of adult populations. In all regions investigated, smoking was found to be the major determinant of whole-blood cadmium concentration.

The accumulated frequency distribution of blood cadmium concentrations found in the four Greenlandic regions is shown in Figure 7. The figure clearly demonstrated differences in concentration levels between on the one hand the eastern and northern region with median concentrations of 1.8 and 1.9 µg/l, respectively, and on the other hand the northwest and southwest regions with median concentrations of 2.8 and 2.7 µg/l, respectively.

The influence of diet on blood cadmium concentrations was not as pronounced as that of smoking in spite of the reported high dietary intake.[5] When arithmetric mean cadmium

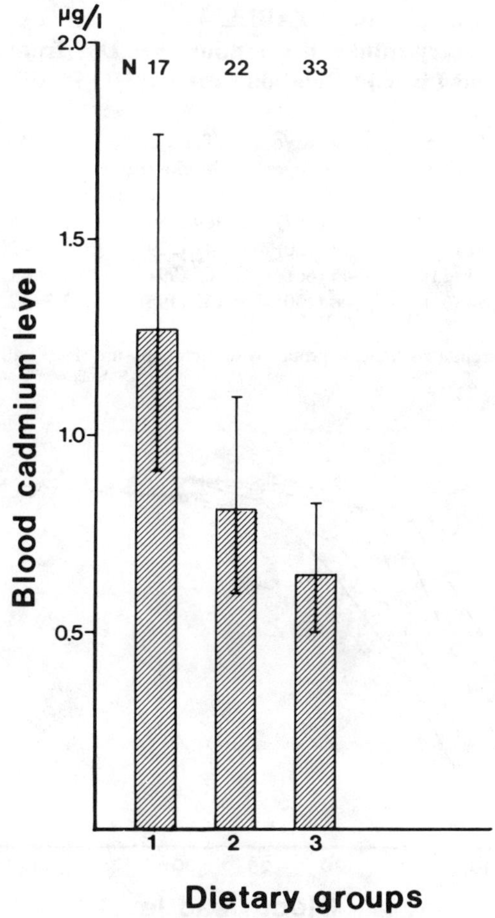

FIGURE 8. Blood cadmium concentrations according to eating habits in nonsmokers from all regions. Arithmetic mean and 2 × SEM indicated. (1) Persons eating more than six meals of marine food per week; (2) two to six meals of marine food per week; (3) one meal or less of marine food per week.

blood concentrations from nonsmokers in all districts were calculated for each eating group as shown in Figure 8, it is seen that the diet does influence the concentration, and the mean concentration in group 1 (eating more than six meals of local food per week) is significantly higher than the group 3 mean (eating one or less meals of local food per week). Analyses of cadmium in hair showed concentrations not related to blood concentrations. Hair cadmium is, therefore, considered useless as an index of exposure.

B. DAILY INTAKE OF CADMIUM

Analyses of food items have shown a mean cadmium concentration of 0.36 mg/kg in meat and of 20.0 mg/kg in liver.[41] Probably cadmium from both meat and liver is absorbed to the same degree, and in an estimation of daily dietary cadmium intake, both must be taken into consideration. The approximate meat intake in different regions has been calculated form the mercury data (cf. Table 3). The evaluation carried out by the Danish National Food Institute[6] based on interviews indicates a liver consumption of 3.6% of meat intake. By using this figure, the daily intake for the total study groups, as well as for those with the highest consumption of local food, has been calculated. The data are given in Table 4, with indication of the intake/PTWI ratios (PTWI for cadmium: 400 to 500 µg per person per week).

TABLE 4
Estimated Dietary Intake of Cadmium per Day from Seal Meat and Liver in Relation to the WHO PTWI

	Cd μg/d from meat	Cd μg/d from liver	Total daily intake (μg)	Weekly intake PTWI
North	123 (229)	246 (457)	369 (686)	5.2—6.5 (9.7—12.0)
Northwest	47 (79)	94 (150)	141 (229)	2.0—2.5 (3.2—4.0)
Southwest	23 (23)	46 (66)	69 (99)	1.0—1.2 (1.4—1.7)
East	54 (65)	108 (130)	162 (195)	2.3—2.8 (2.7—3.4)

Note: Data in parentheses indicate groups with high exposure (PTWI 400 to 500 μg per person per week).

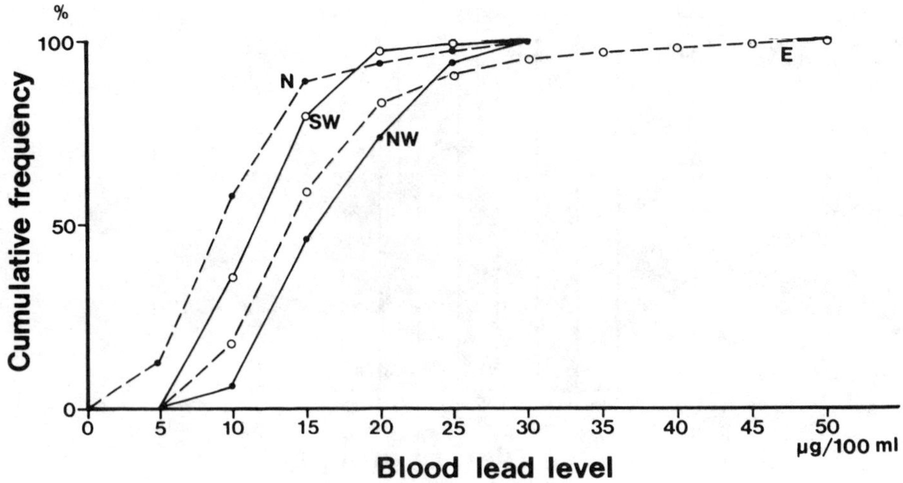

FIGURE 9. Accumulated frequencies of blood lead concentrations in adults in the four regions.

In spite of the high dietary intake, only moderately elevated blood cadmium concentrations have been recorded (Figure 8). At the highest exposure level in Thule (North Greenland), an effect of dietary cadmium was observed.[13] However, the main exposure factor is smoking, a conclusion also drawn in a study in New Zealand[36] where large quantities of oysters with a high cadmium content are consumed.

V. LEAD

A. ADULT AND FETAL EXPOSURE LEVEL

Lead concentrations in blood could not be related to eating habits in any district, indicating that dietary lead intake from marine mammals is not the main source of exposure. Other food items might, however, be of importance. Neither could smoking habits be related to lead concentrations except in pregnant women, where a moderate relationship with smoking was demonstrated.[14] Age dependency was demonstrated in Angmagssalik.

The accumulated blood lead concentration frequencies found in the various Greenlandic regions are shown in Figure 9. The highest individual blood lead concentrations were found in Angmagssalik, East Greenland, where five samples showed concentrations above 35μg/100 ml, which has been fixed as the upper acceptable concentration in nonoccupationally exposed adults in Europe (EEC Directive 29/3/77).

TABLE 5
Trace Element Concentrations in 15th and 20th Century Human and Animal Hair Samples

	Human samples				Animal samples			
	15th cent.		20th cent.		15th cent.		20th cent.	
Element	No.	Mean	No.	Mean	No.	Mean	No.	Mean
Hg	6	3.1 (2.3—4.2)	22	9.8 (6.8—13.5)	9	0.6 (0.5—0.7)	10	2.6 (1.8—3.7)
Cd	6	0.5 (0.3—0.7)	22	0.4 (0.3—0.5)	9	0.3 (0.3—0.5)	10	0.8 (0.6—1.0)
Pb	6	0.7 (0.3—2.1)	22	6.0 (3.5—10.3)	9	0.4 (0.3—0.4)	10	1.9 (1.2—1.8)
Se	5	2.8 (2.3—3.5)	10	0.8 (0.7—0.9)	9	2.3 (1.9—2.9)	8	2.4 (2.0—3.0)

Note: Means and 95% confidence intervals (in brackets) are indicated; units μg/g.

The relatively high blood lead level found among Greenlandic Eskimos is remarkable, as no car traffic or polluting industries exist in the country. The reason must probably be sought in long distance atmospheric transport of lead-containing particles in connection with dietary factors facilitating lead absorption.[14] In cord blood, the concentrations of lead were about the same level as in mothers.[14] A correlation between concentrations of lead in hair and blood has been reported[37,38] but was not confirmed in the Greenlandic studies.[9,13]

Cadmium and lead concentrations in hair co-vary significantly. This has been demonstrated both in West Greenland and in North Greenland. In East Greenland, only a few hair samples have been analyzed. In the Thule data, the coefficient of correlation between the two elements is $r = +0.6944$, $p < 0.001$. The mean concentrations of 1.4 and 0.08 μg/g for lead and cadmium, respectively, are rather low compared to concentrations reported in Western societies[39] and also compared to the results from West Greenland where mean values of 3.7 μg/g of lead and 0.3 μg/g of cadmium were found.[7] No attempt to estimate daily lead intake has been made as lead exposure in Greenland is apparently not of dietary origin.

VI. THE HISTORICAL DEVELOPMENT OF METAL EXPOSURE

Through analyses of hair samples from eight well-preserved Inuit mummies found at Umanak in Northwest Greenland and dated with the ^{14}C method to 1470 ± 50 A.D.,[14] it has been possible to compare trace element levels between 15th and 20th century Greenlanders. Comparison analyses have also been performed on animal hair samples originating from the fur clothes of the mummies and from present-day animal hair samples from the same species. The value of trace element concentrations in hair as an indicator of individual exposure is questionable for some elements, especially lead and cadmium[7,13] while mercury levels in fur can be regarded as a good indicator of exposure. With these limitation, differences can, however, be taken as an indication of changes of diet and/or environment.

The samples were analyzed by atomic absorption spectrometry at the Institute of Hygiene, University of Aarhus, for mercury, lead, copper, and selenium and by energy dispersive X-ray spectrometry at the Division of Toxicology, University of Rochester, for mercury, lead, iron, zinc, bromine, copper, manganese, and calcium. The two methods gave comparable results for the elements analyzed by both methods. The results for mercury, cadmium, lead, and selenium are given in Table 5.

The most striking feature for the comparison between old and present-day samples is that the concentration of mercury and lead has increased considerably with time, while the

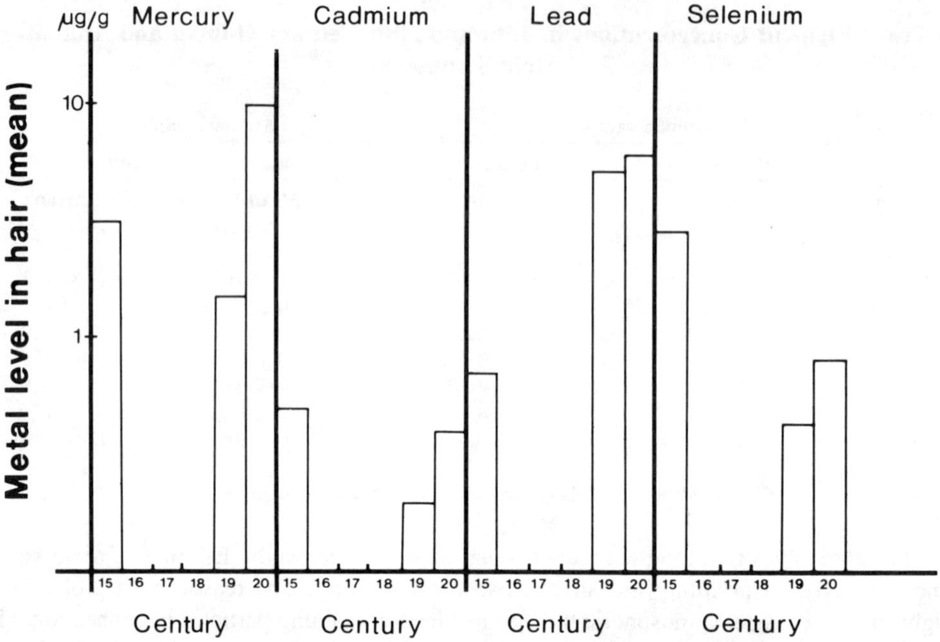

FIGURE 10. Mean metal concentrations in human hair samples from the 15th century (Qilakitsoq N:6), the 19th century (Angmagssalik N:4), and from present-day samples (Umanak N:22).

two essential elements selenium and iron, and also bromine, have decreased. These differences can be explained in part by changes in diets. Fifteenth century Eskimos lived almost exclusively on marine products (as demonstrated by a $^{13}C/^{12}C$ ratio technique[16] whereas today's Eskimos live on a mixed diet of local and imported food. Increases in mercury and lead levels seem likely to be due to increased global atmospheric pollution, perhaps mediated through food chain bioaccumulation. Analyses of four 19th century hair samples collected in East Greenland in 1883 by the Gustav Holm Expedition have also been carried out. With respect to all elements analyzed, except for lead, these four samples showed lower values than those found in the 15th century Eskimos.[40] The reason for this discrepancy seems to be reasonably explained by the fact that the East Greenland Eskimos were suffering from a severe famine at the time of arrival of the expedition. Comparison between hair element concentrations in samples from the 15th, 19th, and 20th centuries are illustrated in Figure 10.

The finding that the lead levels in the 19th century samples are nearly the same as those found in present-day Eskimos is noteworthy and supports the idea that the actual lead exposure is not of dietary origin although the absorption and bodyburden are probably influenced by dietary factors. During those 500 years, the mercury concentrations in human hair have increased, but still the 15th century level of 3μg/g is high compared to the present day Danish hair mercury level of 0.6 μg/g.[18] This shows that in a population depending on Arctic marine food, a relatively high mercury exposure has probably always been present. Assuming the same hair/blood ratio as found today, mean blood mercury in the 15th century Eskimos can be estimated to be about 10 μg/l

VII. DISCUSSION

In general, the traditional hunting trade has decreased in Greenland during this century resulting in a decreasing amount of local food available per capita. However, analyzing the statistics on catches in relation to population size in detail for the last 20 years, it appears that the available amount of meat per capita has increased significantly in all regions during this period except in Southwest Greenland. The most pronounced increase has taken place in North Greenland, largely due to an increased catch of whales.

The present increase in hunting can be interpreted as a result of the activities of the Inuit movement, which has worked hard, and partly successfully, to preserve the Inuit culture by diminishing the influence from Europe, in particular, Denmark. That influence had tended to turn young people from their old customs towards European lifestyles. The current return to the traditional hunting trade means an increasing exposure to toxic metals and probably also increasing health hazards. Special attention should be shown to fetal exposure to mercury. The decrease in worldwide mercury discharge observed since the mid-seventies will not be reflected in marine food chains for several decades, and consequently the human exposure level can be expected to be undiminished in years to come.

With regard to exposure of Greenlanders to toxic heavy metals, the most serious problems are associated with mercury as an environmental pollutant. Selenium attracts interest as a possible antagonist to heavy metal toxicity. Cadmium and lead seem not to be of any immediate toxicological importance. It is noteworthy, however, that in Greenlanders, the lead concentration is the same as in European industrialized countries. These findings show the need for further research on transport and pathways, especially with regard to long distance transport and bioaccumulation. The high dietary cadmium intake is not, or is only to a minor degree, reflected in blood and hair concentrations, also a fact calling for further elucidation.

The mercury exposure level necessitates medical as well as legal attention. Mercury exposure is found to be related to the intake of meat from marine mammals, of which especially the predatory whale species, are of importance. The concentration found in whale meat is not alarmingly high. Preliminary data show a median value of 1.1 mg/kg, which is not much in excess of what has been legally fixed in several countries as the maximum tolerable concentration in foodstuffs. The problem in Greenland arises from the high per capita meat intake in North Greenlanders. The actual blood concentrations found here are hardly of any immediate risk to the adult population, whereas the fetal exposure might be a real hazard with blood concentrations as high as 296 μg/l found in cord blood samples. Selenium as an effective antagonist has been seriously questioned by the finding of a positive correlation between SCE (sister chromatid exchange) and blood mercury concentrations, but without relation to the selenium content.[12] The exposure level thus indicates that a neurophysiological follow up study of children in the hunting districts is needed.

The most straightforward solution to the mercury exposure problem would seem to be a legal regulation of maximally permitted concentrations in meat and organs for human consumption. This would, however, encounter serious practical problems. Primarily, laboratory investigations are not possible in the country, as no laboratory equipped for this kind of analysis exists in Greenland at the moment. Secondly, the hunters are probably not motivated to wait for laboratory tests before consuming their catch. Besides this, it would raise practical problems to send samples to a laboratory and to store the meat while waiting for results.

Comparison with mercury concentrations among Inuits from earlier centuries seems to indicate that an increase in exposure has occurred which could be due to global increases in mercury pollution. The present exposure level for Eskimos is undoubtedly at a level where action must be taken. The findings in the present surveys indicate that pollution with heavy metals is of global importance, and point at the need for an internationally coordinated effort, firstly to control global pollution, and secondly to define realistic maximally tolerable exposure levels for humans, taking into account specific local conditions; in this respect, studies in Greenland have proven to be of great importance.

REFERENCES

1. **Kapel, F. O. and Petersen, R.,** Subsistence Hunting — the Greenland Case, *Rep. Int. Whal. Commn,* Special Issue 4, 7, 51, 1982.
2. **Helms, P.,** personal communication, 1987.
3. **Carstensen, J. and Poulsen, E.,** Public Health Aspects of Environmental Pollution with Mercury and Cadmium in Scandinavia, in *Problems of the Contamination of Man and his Environment by Mercury and Cadmium,* Commission of the European Communities, Luxembourg, July 1973.
4. **Hansen, J. C. and Nygaard, S.-P.,** Humane kviksølvkoncentrationer pa Grønland, *Nord. Hyg. Tidskr.,* 55, 149, 1974.
5. **Johansen, P.,** Heavy metals in marine mammals and heavy metal intake in humans in Greenland, in *Proc. 5th Int. Symp. Circumpolar Health,* Harvald, B. and Hart Hansen, J. P., Eds. Nordic Council for Medical Research, Rep. Ser. 33, 1981, 540.
6. Miljøministeriet, Statens Levnedsmiddelinstitut: Overslagsberegning af den vestgrønlandske fangerbefolknings belastning med cadmium (Cd), bly (Pg) og kviksølv (Hg) og en toksikologisk vurdering af denne, (in Danish), Report from the National Danish Food Institute, 1978.
7. **Hansen, J. C.,** A survey of human exposure to mercury, cadmium and lead in Greenland, *Medd. Grøland, Man Soc.,* 3, 1, 1981.
8. **Hansen, J. C., Wulf, H. C., Kromann, N., and Albøge, K.,** Human exposure to heavy metals in East Greenland, I. Mercury, *Sci. Total Environ.,* 26, 233, 1983.
9. **Hansen, J. C., Kromann, N., Wulf, H. C., and Albøge, K.,** Human exposure to heavy metals in East Greenland, II. Lead, *Sci. Total Environ.,* 26, 245, 1983.
10. **Hansen, J. C., Kromann, N., Wulf, H. C., and Albøge, K.,** Selenium and its interrelation with mercury in wholeblood and hair in an East Greenlandic population, *Sci. Total Environ.,* 38, 33, 1984.
11. **Hansen, J. C., Wulf, H. C., Kromann, N., and Albøge, K.,** Cadmium concentrations in blood samples from an East Greenlandic population, *Dan. Med. Bull.,* 32, 277, 1985.
12. **Wulf, H. C., Kromann, N., Kousgaard, N., Hansen, J. C., Niebuhr, E., and Albøge, K.,** Sister chromatid exchange (SCE) in Greenlandic Eskimios. Dose-relationship between SCE and seal diet, smoking, and blood cadmium and mercury concentrations, *Sci. Total Environ.,* 48, 81, 1986.
13. **Hansen, J. C. and Sloth Pedersen, H.,** Environmental exposure to heavy metals in North Greenland, *Arct. Med. Res.,* 4, 12, 1986.
14. **Hansen, J. C., Christensen, R. B., Allermand, H., Albøge, K., and Rasmussen, R.,** Concentrations of mercury, selenium and lead in blood samples from mothers and their newborn babies in four Greenlandic hunting districts, *Medd. Grøland, Man Soc.,* 6, 1, 1984.
15. **Hansen, J. C., Fenger, H. J., Christensen, R. B., Nielsen, F., Baunsgaard, P., and Albøge, K.,** Mercury in Greenlandic Inuit mothers and infants, in press.
16. **Hart Hansen, J. P., Meldgaard, J., and Nordqvist, J., Eds.,** *Qilakitsoq. De grønlandske mumier fra 1400-tallet,* Chr. Ejlers Forlag, Copenhagen 1985.
17. **Hansen, J. C.,** Trace metal concentration in hair from ancient and present-day Greenlanders, in *Proc. 5th Int. Symp. Circumpolar Health,* Harvald, B. and Hart Hansen, J. P., Eds., Nordic Council for Medical Research, Rep. Ser. 33, 1981, 543.
18. **Bach, E.,** Voksnes belanstning med kviksølv. Delrapport om en befolkningsundersøgelse i Miljøstyrelsens undersøgelsesprogram for tunge metaller, Dansk Institut for Klinisk Epidemiologi, 1981.
19. **Clarkson, T. W., Hamada, R., and Amin-Zaki, L.,** Mercury, in *Changing Metal Cycles and Human Health,* J. O. Nriagu, Ed., Dahlen Konferenzen, Springer-Verlag, Berlin 1984, 285.
20. **Fehr, F. and Dennis, C. A. R.,** Investigations into the use of health hazard of mercury in Saskatchewan, PIEH No. 4, the Praries Institute of Environmental Health, Saskatchewan, Canada, 1975.
21. **Sherlock J., Hislop, J., Newton, D., Topping, S., and Whittle, K.,** Elevation of mercury in human blood from controlled chronic ingestion of methylmercury in fish, *Human Toxicol.,* 3, 117, 1984.
22. WHO, Food Addition Series No. 4., World Health Organization, Geneva, 1972
23. **Helms, P.,** Kostundersøgelse i Angmagssalik, *Forsk. Grønl.* 1, 10, 1981.
24. **Koeman, J. H., Peeters, W. H. M., Koudstall-Hol, C. H. M., Tjioe, P. S., and de Goeij, J. J. M.,** Mercury-selenium correlation in marine mammals, *Nature,* 245, 385, 1973.
25. **Koeman, J. H., van den Ven, W. S. M., de Goeij, J. J. M., Tjioe, P. S., and van Haaften, J. L.,** Mercury and selenium in marine mammals and birds, *Sci. Total Environ.,* 3, 279, 1975.
26. **Schrauzer, G. N. and White, D. A.,** Selenium in human nutrition: dietary intakes and effects of supplementation, *Bioinorg. Chem.,* 8, 303, 1978.
27. **Hansen, J. C.,** unpublished data, 1988.
28. **Allaway, W. H., Kubota, J., Losee, F., and Roth, M.,** Selenium, molybdenum, and vanadium in human blood, *Arch. Eviron. Health,* 16, 342, 1968.

29. **McKenzie, R. L., Rea, H. M., Thomson, D. C., and Robinson, M. F.,** Selenium concentration and glutathione peroxidase activity in blood of New Zealand infants and children, *Am. J. Clin. Nutr.*, 32, 1413, 1978.
30. Howe, Sister Marmion, Selenium in the blood of South Dakota Indians, *Arch. Environ. Health,* 34,444, 1979.
31. **Jensen, Egeskov, G., Gissel-Nielsen, G., and Clausen, J.,** Leucocyte glutathione peroxidase activity and selenium level in multiple sclerosis, *J. Neurol. Sci.*, 48, 61, 1980.
32. **Yang, G., Wang, S., Zhou, R., and Sun, S.,** Endemic selenium intoxication of humans in China, *J. A. Clin. Nutr.*, 37, 872, 1983.
33. **Magos, L., Clarkson, T. W., and Hudson, A. R.,** Differences in the effects of selenite and biological selenium on the chemical form and distribution of mercury after the simultaneous administration of $HgCl_2$ and selenium in rats, *J. Pharmacol. Exp. Ther.*, 228, 478, 1984.
34. **Makito, M., Akira, N., and Nobumasa, J.,** Studies on possibility of interaction between endogenous selenium and inorganic mercury in rats (Japanese), *Eisei Kagaku,* 30, 229, 1984.
35. **Morimoto, K., Tijima, S., and Koizumi, A.,** Selenite prevents the induction of Sister-Chromatid exchanges by methyl mercury and mercuric chloride in human whole-blood cultures. *Mutat. Res.*, 102, 183, 1982.
36. **Sharma, S., and Singh, R.,** Selenium in soil, plant and animal systems, *CRC Crit. Rev. Environ. Control,* 13(1), 23, 1983.
37. **Klevay, L. M.,** Hair as a biopsy material. III. Assessment of environmental lead exposure, *Arch. Environ. Health,* 26, 169, 1973.
38. **Chattopadhyay, A., Roberts, T. M., and Jervis, R. E.,** Scalp hair as a monitor of community exposure to lead, *Arch. Environ. Health,* 32, 226, 1977.
39. **Iyengar, G. V., Kollmer, W. E., and Bowen, H. J. M.,** *The Elemental Composition of Human Tissues and Body Fluids,* Verlag Chemie, Weinheim, New York, 1978.
40. **Hansen, J. C.,** Heavy metals exposure in Greenland — A health hazard to the population?, in *Geomedical Research in Relation to Geochemical Registration,* J. Lag, Ed., Universitetsforlaget, Oslo, 137, 1984.
41. **Hansen, J. C.,** unpublished data.

INDEX

D

N

Q

R

S

T